THE PARADOX OF
THE ORGANISM

THE PARADOX OF THE ORGANISM

ADAPTATION AND INTERNAL CONFLICT

EDITED BY

J. ARVID ÅGREN
MANUS M. PATTEN

HARVARD UNIVERSITY PRESS

CAMBRIDGE, MASSACHUSETTS

LONDON, ENGLAND

2025

EU GPSR Authorised Representative
LOGOS EUROPE, 9 rue Nicolas Poussin, 17000, LA ROCHELLE, France
E-mail: Contact@logoseurope.eu

Library of Congress Cataloging-in-Publication Data
Names: Ågren, J. Arvid, editor. | Patten, Manus M., 1980– editor.
Title: The paradox of the organism : adaptation and internal conflict /
edited by J. Arvid Ågren, Manus M. Patten.
Description: Cambridge, Massachusetts ; London, England :
Harvard University Press, 2025. | Includes bibliographical references and index.
Identifiers: LCCN 2025011151 | ISBN 9780674296381 (hardcover) |
ISBN 9780674302044 (ebook) | ISBN 9780674302068 (pdf)
Subjects: LCSH: Adaptation (Biology). | Organisms. | Evolution (Biology). |
Biology—Philosophy.
Classification: LCC QH546 .P23 2025 | DDC 578.4—dc23/eng/20250710
LC record available at https://lccn.loc.gov/2025011151

For David Haig

CONTENTS

Introduction: Organismality and Its Discontents   I
J. Arvid Ågren and Manus M. Patten

I. CONSIDERATIONS OF CONFLICT

1. Resolving the Paradox of the Organism with Inclusive Fitness   9
Thomas W. Scott and Stuart A. West

2. In What Sense Can There Be Conflict Between the Levels of Selection?   32
Pierrick Bourrat

3. Internal Conflicts That Organisms Can Live With   53
Manus M. Patten

II. IMPLICATIONS OF CONFLICT

4. A Metaphysical Paradox of Organismal Identity   75
Ellen Clarke and Will Morgan

5. The Conflict Behind the Conflicts: Organisms as Agents and as Ecosystems   96
Philippe Huneman

6. Cooperation and Conflict in Metazoan Development: Toward a Neo-Darwinian Embryology   125
David Haig

7. Selfish Genetic Elements and the Evolution of Sex Determination and Inheritance Systems   200
Martijn A. Schenkel, Laura Ross, and Nina Wedell

III. APPLICATIONS OF CONFLICT

8. **Clinical Applications of the Paradox of the Organism**   235
   Amy M. Boddy and J. Arvid Ågren

9. **Internal Conflict in Genomes and in Persons**   255
   Samir Okasha

10. **Host–Symbiont Conflict in Reef–Building Corals and the Risk of Coral Bleaching**   273
    Dakota E. McCoy, Brendan Cornwell, Sönke Johnsen, and Jennifer A. Dionne

   Conclusion: The Eternal Disquiet Within   328
   Manus M. Patten and J. Arvid Ågren

CONTRIBUTORS   339

INDEX   341

THE PARADOX OF
THE ORGANISM

Introduction

Organismality and Its Discontents

J. ARVID ÅGREN AND MANUS M. PATTEN

Multicellular bodies are an intimate integration of their constituent parts. From photosynthesis to flight, it is only through the cooperation and coordination of genes, genomes, and cells that we get the intricate adaptations that characterize plants, animals, and fungi. The idea of organisms as unified bodies goes back a long way. Thinkers from Aristotle to William Paley believed that the function of body parts could only be understood in light of the whole organism. Similarly, much progress has been made in biology, as well as medicine, by treating parts of organism like parts of machines and asking what their functions are. By now, the research of most biologists rests on the idea that organisms are made up of parts adapted to serve the interests of the individual organism.

The unified appearance, however, is illusory—bodies are far from harmonious. Like any collective, multicellular organisms are vulnerable to exploitation from within by selfish genetic elements and selfish cell lineages, which either enhance their own proliferation at the expense of other genes and cells or distort the development of traits away from organismal optima. It was this observation that led Richard Dawkins to coin the term "the paradox of the organism" to highlight how the existence of organisms as seemingly coherent, functional units is at odds with the possibility for selfish parts to pursue agendas that undermine the integrity of the organism. As he put it:

> The paradox of the organism is that it is not torn apart by its conflicting replicators but stays together and works as a purposeful entity, apparently on behalf of all of them. Not only is it not torn apart; it functions as such a convincingly unified whole that biologists in general have not seen that there is a paradox at all! (Dawkins 1990, S64)

In many ways, within-organism conflicts are exactly what we ought to expect. As Leo Buss made clear in *The Evolution of Individuality*, "Individuality is a derived character" (Buss 1987, 4). What he meant was that at some point in the past, the complex multicellular organisms that we observe all around us had to become individuals made of parts. The evolution of new levels of individuality is what gave life its hierarchical structure: genes in genomes; multiple genomes in eukaryotic cells; cells in multicellular organisms; and organisms in social groups. Each new level in the hierarchy was the product of what are now known as major transitions in individuality, where entities that could previously survive and reproduce on their own could now only do so as part of a new higher-level collective individual.

Since Dawkins and Buss, the evolutionary forces governing transitions in individuality have been probed in a series of books, including *The Major Transitions in Evolution* (Maynard Smith and Szathmáry 1995), *Darwinian Dynamics* (Michod 1999), *Principles of Social Evolution* (Bourke 2011), and *Domains and Major Transitions of Social Evolution* (Boomsma 2022). In recent years there has also been an increased interest in the antecedents of these ideas, with several modern authors reaching back to much earlier work that can be seen to have anticipated the major-transitions paradigm. For example, we have seen the reissue of Julian Huxley's 1912 book *The Individual in the Animal Kingdom* with a new introduction in 2022, and Wilhelm Roux's 1881 *Der Kampf der Theile im Organismus* was translated and reissued in English for the first time as *The Struggle of Parts* in 2024. The authors of the chapters of this book pick up on the deep challenges laid out in both the modern and earlier work.

The title of our introduction is a play on the title of Sigmund Freud's *Civilization and Its Discontents* (originally *Das Unbehagen in der Kultur*, 1930), which explores the tension between human desires for individuality and their commitments to civilized society. In this book we do something similar by examining those elements of bodies—the selfish elements, specifically—that can be seen as unhappy with the constraints of organismal living. Much as Freud thought a focus on this tension would reveal something about human nature, we think there are lessons about organisms that can come from studying the friction between them and their various selfish elements. However, these internal conflicts have traditionally been researched piecemeal and so we lack a common framework for their study. For example, what do different within-organism conflicts have in common, and, most importantly, how are they managed so that organisms retain their primacy as the ultimate beneficiaries of adaptation?

There is another spirit in which our title may be taken, though. The second sense of "discontent" that runs through this volume captures how some scholars from both biology and philosophy feel about the concept of organismality. Within-organism conflicts raise questions of who or what is in control of the process of adaptation. While it is widely accepted that organisms have agency and that they pursue goals of fitness maximization, there is only a nascent, perhaps reluctant appreciation that genes and cells may do so as well. Moreover, accepting that organisms are compromises between multiple competing agents reveals the frailty of several core concepts in evolutionary biology. In particular, it forces a reconsideration of what exactly an organism is. If it is not a tightly integrated, unified machine, then what is it, and should it retain its special role in adaptive explanations?

The central argument of the book is that within-organism conflict has significant implications both for the way that organisms work and for the way we understand evolutionary theory. Selfish elements, and the internal conflicts they foment, are not two-headed snakes: cute and curious, but of limited relevance to the grand story of biology. Instead, we must recognize that they are core features of both organisms and the organism concept writ large.

This book consists of ten contributed chapters organized into three parts. Part I, "Considerations of Conflict," discusses the concept of conflict and introduces the current thinking about what allows organisms to be the central unit of adaptations, despite the threats from conflicts within. In Chapter 1, Scott and West use inclusive fitness theory to argue that the so-called parliament of genes offers a resolution to the paradox of the organism by aligning the fitness interests of genes in the same ways that members of a parliament may cooperate to achieve selfish goals. In Chapter 2, Bourrat analyzes what we mean when we talk of conflict between levels. Bourrat distinguishes between metaphorical and genuine conflict in these cases, with the latter requiring that we consider the selfish parts as though they are separable from the organism and its interests. In Chapter 3, Patten teases apart two kinds of internal conflict—one that is merely harmful to organismal fitness and another that completely threatens organisms' unity of purpose—and notes that the two kinds of conflict bear on the major evolutionary transitions in individuality in different ways.

Part II, "Implications of Conflict," deals with the various ways that internal conflicts drive the evolution of biological systems and shape our notions of organismality. Both Chapter 4 (Clarke and Morgan) and Chapter 5 (Huneman) draw on internal conflicts to explain what an organism is exactly. Clarke and Morgan wrestle with how to define an organism in the face of the changes a body inevitably undergoes with the passage of time. Taking insight from Dawkins's paradox of the organism, they find a solution in the distinction between ordinary, law-abiding genes and selfish genetic elements. Huneman contends that organisms exist on a spectrum from unified agents to ecosystems (i.e., loose associations of unlike parts) depending on the burden placed on them by the selfish elements within. Chapter 6 (Haig) and Chapter 7 (Schenkel, Ross, and Wedell) look at cells and genomes from a paradox-of-the-organism perspective. Haig shows how the separation of germ and somatic cells create both an arena for internal conflicts and a way to resolve them. In particular, he reverses the common narrative by arguing that the separation is not to protect the germline from the mutational pressures of the soma but to protect the bodily functions of the soma from the genetic conflicts of the germline. Schenkel, Ross, and Wedell then address the issue of reproduction and inheritance. By considering empirical examples of how those fundamental processes operate outside the traditional lab model systems favored by experimental biologists, both chapters develop an account of bodily functions that puts internal conflict at the heart.

Finally, Part III, "Applications of Conflict," looks at how we can take the foundational concepts in the first part of the book and apply them to practical matters, ranging from cancer to the human mind to conservation. Chapter 8 (Boddy and Ågren) and Chapter 9 (Okasha) apply the paradox of the organism to human biology. Boddy and Ågren discuss a variety of maladies that stem from internal conflicts and show that whereas the traditional body-as-machine approach in medicine runs into difficulty explaining such phenomena, a paradox-of-the-organism perspective offers a way forward. Okasha then probes the link between internal genetic conflicts and internal psychological conflicts, such as cognitive dissonance. By arguing that a link exists at a conceptual rather than mechanistic level, a paradox-of-the-organism analysis provides a way to unify physical and mental conflicts. Finally, Chapter 10 (McCoy et al.) focuses on coral and their symbionts to link internal conflicts to macroevolutionary patterns and processes. McCoy et al. demonstrate that the implications of internal conflicts extend far beyond the bodies that form the battleground, and accounting for them is crucial for efforts to conserve endangered species.

As reflected in the author list, this book is based on the belief that the most exciting bits of contemporary research happen at the intersection between biology and philosophy. An underlying goal is therefore to encourage more of that kind of work. As editors, we wanted to stimulate discussion and interaction between biologists and philosophers and encourage a new approach to the study of conflict and organismality. The intended audience of the book is anyone who is drawn to the same zone, especially early career researchers—including undergraduates and graduate students—in evolutionary biology, philosophy of biology, and beyond.

Our work on this book was supported by the John Templeton Foundation (#62220) as part of the Agency, Directionality, and Function cohort program. We are grateful to all staff associated with the program, especially Alan Love for his leadership. The opinions expressed in this book are those of the authors and not those of the John Templeton Foundation. We would also like to thank our editor at Harvard University Press, Rachel Field, who believed in this project from the beginning, and Ian Sherman, who shared advice on how best to approach edited volumes.

The book is the result of a working group formed in 2022 that was devoted to the paradox of the organism. The group comprised biologists and philosophers who met virtually as well as in person at an interdisciplinary conference in Washington, DC, in the fall of 2022. A key part of the working group was a novel seminar course on the topic designed by the editors for biology and philosophy students; it was taught by Patten at Georgetown University in the fall of 2022 and 2023. Throughout the course, members of the working group participated in virtual discussions centering on both their prior work and on the draft chapters of this volume. We are grateful to all participants of the conference and the courses for their engaging comments and stimulating discussions on the issue of this book.

References

Boomsma, J. (2022). *Domains and Major Transitions of Social Evolution*. Oxford University Press.

Bourke, A. F. G. (2011). *Principles of Social Evolution*. Oxford University Press.

Buss, L. (1987). *The Evolution of Individuality*. Princeton University Press.

Dawkins, R. (1990). Parasites, desiderata lists and the paradox of the organism. *Parasitology, 100,* 63–73.

Freud, S. (1930/2010). *Civilization and Its Discontents*. W. W. Norton.

Huxley, J. (1912/2022). *The Individual in the Animal Kingdom*. MIT Press.

Maynard Smith, J., and Szathmáry, E. (1995). *The Major Transitions in Evolution*. Oxford University Press.

Michod, R. (1999). *Darwinian Dynamics*. Princeton University Press.

Roux, W. (1881/1924). *The Struggle of Parts*. Harvard University Press.

PART I

CONSIDERATIONS OF CONFLICT

Resolving the Paradox of the Organism with Inclusive Fitness

THOMAS W. SCOTT AND STUART A. WEST

ABSTRACT   Our modern explanation for organism design—that natural selection leads organisms to maximize their inclusive fitness—relies on the
assumption that we can effectively ignore the consequences of genetic
conflict within individuals. This assumption seems to be contradicted by
widespread evidence for selfish genetic elements, which may compromise
inclusive fitness, leading to the "paradox of the organism." We argue that
the paradox of the organism is resolved by four factors: (1) many selfish genetic elements do not distort traits; (2) there is little internal conflict over most
traits; (3) conflict over helping behaviors tends not to appreciably compromise the organism's inclusive fitness; and (4) residual conflict is suppressed
by the parliament of genes.

Introduction

A striking feature of living organisms is that they appear designed or
adapted for their environments (Paley 1829). This was once taken as evidence for the existence of God (a "designer"), but Charles Darwin provided a naturalistic explanation with his theory of natural selection
(Darwin 1859). Darwin argued that heritable characters that are associated with greater reproductive success will tend to accumulate in biological populations, leading organisms to appear as if they were designed to
maximize their reproductive success (Figure 1.1). Ronald Fisher, the
biologist and mathematician, formalized Darwin's theory of natural
selection, showing how natural selection would increase and therefore
ultimately maximize reproductive success (Fisher 1930).

Darwin's explanation for biological design has remained largely intact, aside from one revision. The biologist William Hamilton showed
that the quantity that is improved by natural selection, and apparently
maximized by organisms, is not reproductive success per se, but inclusive fitness (Hamilton 1964). An organism's inclusive fitness is the sum
of its direct and indirect fitness (Figure 1.2). Direct fitness is the number

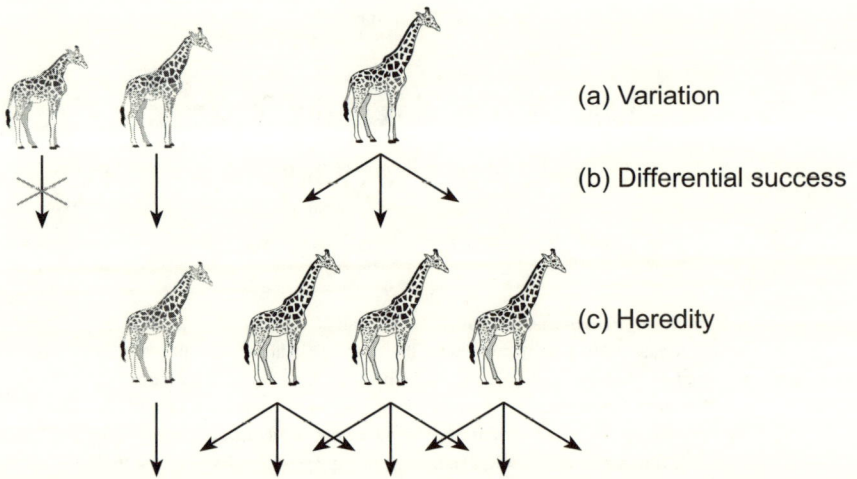

Figure 1.1. Natural selection. Natural selection operates if three conditions are satisfied: variation, differential success linked to variation, and heredity. Here, we illustrate with an example: the evolution of long necks in giraffes. (a) Initially, there are natural variations in giraffes' neck lengths. (b) Longer-necked giraffes have access to more food, high up in the trees, and so live longer to have more offspring. (c) Giraffes' offspring resemble their parents. As a result of (a), (b), and (c), the population gradually shifts to be dominated by long-necked giraffes (adaptation). Reformatted from S. R. Levin, T. W. Scott, H. S. Cooper, and S. A. West, "Darwin's Aliens," *International Journal of Astrobiology, 18* (2019), 109, figure 1.

of adult offspring that the organism would have in the absence of social interactions. Indirect fitness sums the offspring gained or lost by other individuals as a consequence of the organism's social behavior, where these gains or losses are weighted by the relatedness (genetic similarity) between the organism and affected individual.

Inclusive fitness captures how individuals are able to influence the transmission of their genes to future generations by either influencing their own reproductive success (direct fitness) or the reproductive success of other individuals with which they share genes (indirect fitness) (Hamilton 1970). The assumption that organisms appear to maximize their inclusive fitness is the starting point for the vast majority of adaptation research in behavioral and evolutionary ecology (Alcock 2009; Westneat and Fox 2010; Davies et al. 2012). Consequently, the assumption of inclusive fitness maximization underlies our modern evolutionary explanations for many traits, including foraging behavior, resource competition, sexual selection, parental care, sex allocation, signaling, and cooperation (Stephens and Krebs 1986; Stearns 1992; West 2009;

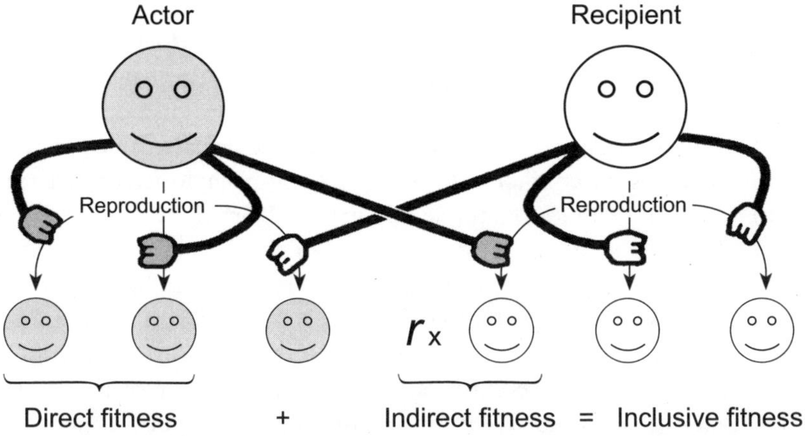

Figure 1.2. Inclusive fitness. Inclusive fitness is the sum of direct and indirect fitness. Social behaviors affect the reproductive success of self and others. The impact of the actor's behavior (shaded hands) on its reproductive success (shaded offspring) is the direct fitness effect. The impact of the actor's behavior (shaded hands) on the reproductive success of social partners (blank offspring), weighted by the relatedness (r) of the actor to the recipient, is the indirect fitness effect. Inclusive fitness does not include all of the reproductive success of relatives (blank offspring), only that which is due to the behavior of the actor (shaded hands). Also, inclusive fitness does not include all of the reproductive success of the actor (shaded offspring), only that which is due to its own behavior (shaded hands). Reformatted from S. A. West, A. S. Griffin, and A. Gardner, "Evolutionary Explanations for Cooperation," *Current Biology, 17* (2007), R661–R672, figure 3, with permission from Elsevier.

Bourke 2011). The success of work in such areas supports the assumption of inclusive fitness maximization, in spite of occasional technical challenges to the inclusive fitness concept (Birch 2017; Levin and Grafen 2019).

However, a problem with the idea that organisms appear to maximize inclusive fitness is that it assumes that we can effectively ignore the consequences of genetic conflict within individuals (Grafen 2006). Specifically, it assumes that all genes in an organism work together to improve an organism's inclusive fitness. This assumption seems to be contradicted by selfish genetic elements, which are genes that propagate themselves selfishly at the expense of the organism's inclusive fitness. Selfish genetic elements appear to be common (Burt and Trivers 2006). Does this undermine our modern explanation for organism design, as well as most research in behavioral and evolutionary ecology? The problem posed by selfish genetic elements for organism design is known as the paradox of the organism (Dawkins 1990; Patten et al. 2023).

In this chapter, we argue that the paradox of the organism is resolved by four factors. First, many selfish genetic elements do not gain their selfish advantage by shifting organism traits away from values where inclusive fitness is maximized, and so do not compromise organism design. Second, for many traits, selfish strategies are not available for genes, meaning there is no conflict. Third, for helping behaviors, conflict tends not to cause trait values to appreciably deviate from the values where organismal inclusive fitness is maximized. This is because the extent to which relatedness varies across the genome tends either to be low, or symmetrical in a way that approximately cancels out, recovering organismal inclusive fitness maximization. Fourth, when there are other forms of genetic conflict, this can be suppressed by the parliament of genes. This means that organism design can prevail in spite of internal conflict.

Many Selfish Genetic Elements Do Not Distort Traits

Many selfish genetic elements, such as transposons and meiotic drivers, do not need to manipulate organism traits in order to give themselves a selfish propagation advantage (Eshel 1985). These selfish genetic elements do not compromise organism design, because (1) they do not shift traits away from inclusive fitness maxima and (2) the cost of such drivers makes them disfavored across the entire genome, leading to selection to attenuate that cost (Scott and West 2019). For example, the *Segregation Distorter* (SD) meiotic driver in *Drosophila melanogaster* gains its advantage in heterozygous males by disrupting the proper development of rival sperm, and not by trait distortion (Larracuente and Presgraves 2012). Any organism-level fitness costs associated with SD would be opposed by SD as well as across the rest of the genome (Zanders and Unckless 2019).

Organism design is therefore only at risk of being compromised by a subset of selfish genetic elements, which gain their selfish benefit by manipulating a trait in a specific direction to increase their own transmission. Examples include sex ratio distorters and certain helping genes, considered below. Selfish genetic elements that gain their selfish benefit in this way, which have been called "trait distorters" by Scott and West (2019) and Patten et al. (2023), can have substantial influences on the traits of organisms, even when at fixation. However, trait distortion is only favored at the minority of genes in the genome that can gain a selfish advantage from manipulating a trait ("cabals of the few"), whereas non-trait-distorting selfish genetic elements may arise anywhere in the genome. Consequently, trait distorters may be relatively rare among selfish genetic elements (Leigh 1971; Cosmides and Tooby 1981; Scott and West 2019).

There Is No Conflict over Most Traits

For most traits, such as foraging or predator avoidance, the only way that a gene can increase its transmission to the next generation is by increasing the fitness of the organism carrying it, so there will be no conflict within organisms (West and Gardner 2013). Consequently, we would expect conflict to be limited to traits where transmission can be altered. In recent years, many have claimed that helping (cooperation) will be subject to appreciable conflict, shifting the evolved level of helping away from the value wherein organismal inclusive fitness is maximized, in ways we will explain (Jansen and van Baalen 2006; Nogueira et al. 2009; Gardner and Úbeda 2017). We will argue, in contrast to this, that conflict will not lead to appreciable deviation from organismal inclusive fitness maximization in the evolutionary long term.

Helping

Helping behaviors are when an organism pays a fitness cost (i.e., it has fewer offspring) so that another individual can gain a fitness benefit (i.e., it gains adult offspring). Providing help serves to change the inclusive fitness of the organism (actor) by $rb - c$, where c is the cost to the organism (actor), b is the benefit to the recipient, and r is the relatedness (genetic similarity) between the actor and recipient. The helping behavior is favored if it increases the actor's inclusive fitness (Taylor 1989; Grafen 2006; Levin and Grafen 2019, 2021; Lehmann and Rousset 2020; Scott and Wild 2023). This occurs when the following condition, known as Hamilton's rule, is satisfied: $rb > c$ (Hamilton 1963, 1964).

One potential cause of conflict within organisms is that genetic relatedness may vary across the genome (Grafen 1985). For instance, in principle, an organism and its social partner may have the same alleles at a first variable locus, leading to maximal relatedness ($r_1 = 1$), but different alleles at a second variable locus, leading to submaximal relatedness ($r_2 < 1$). In this scenario, helping will be favored at the first locus when $r_1 b > c$, and favored at the second locus when $r_2 b > c$. Because $r_1 > r_2$, there are some scenarios where helping is favored at the first locus but disfavoured at the second locus, leading to conflict. Technically, this will occur when $r_1 > c/b > r_2$.

A variety of mechanisms have been proposed for generating relatedness variation across the genome, including (1) greenbeards, (2) horizontal gene transfer, and (3) genomic imprinting. We will examine the extent to which these mechanisms generate conflict within organisms over helping behaviors.

Greenbeards

Greenbeard genes are those that recognize copies of themselves in other individuals and help those individuals (Gardner and West 2010). Empirical examples include the *traA* greenbeard gene in cellular slime molds, which causes cells to aggregate with and help other cells that share the greenbeard variant, and the *Gp-9* greenbeard gene in fire ants, which causes worker ants to recognize and kill queens that lack the greenbeard, thereby indirectly helping queens that have the greenbeard (Madgwick et al. 2019).

For greenbeard loci, relatedness between the organism (actor) and the recipient of help will be maximal (= 1). Provided that the actor's social partners are not clones of itself, relatedness will be submaximal at other loci across the genome, which can generate conflict with the greenbeard locus over helping (Biernaskie et al. 2011). Importantly, if greenbeards turn out to have a large influence over organismal helping behaviors, then this could shift helping behaviors away from the values where organismal inclusive fitness is maximized.

However, there are two reasons why greenbeard genes might not actually cause helping behaviors to deviate appreciably from values where organismal inclusive fitness is maximized. The first reason is that their phenotype (effect) is likely to derive only from a single gene, limiting the sophistication of greenbeard-encoded helping behaviors. Importantly, this means that greenbeards may be unlikely to facilitate the kind of complex adaptations characteristic of adapted organisms, which require many cooperating genes to build (West and Gardner 2013). A caveat to this is that, if the greenbeard resides on a region of the chromosome where recombination is suppressed (e.g., an inversion), the greenbeard phenotype may incorporate the effects of more than just a few genes, allowing it to build more complex adaptations. This is the case for the *Gp-9* greenbeard "supergene" in fire ants, which resides on an inversion comprising a variety of genes influencing colony founding strategy, mating behavior, dispersal, queen morphology, and colony size (Keller and Ross 1998; Wang et al. 2013; Kay et al. 2022).

The second reason why greenbeard genes might not appreciably distort helping behaviors is that relatedness might not often be very different between greenbeard loci and non-greenbeard loci. Therefore, there may not actually be much conflict over helping. The reason for this is that when a greenbeard identifies a copy of itself in another individual, this other individual is likely to be a genuine pedigree relative (family member), like a sibling or clonemate, meaning relatedness is high across all loci, not just the greenbeard! For instance, in bacteria, where most examples of greenbeards have been found, evolutionarily stable social interactions are generally between clones, meaning relatedness approaches 1 across *all* loci, not just the greenbeard locus (West et al. 2006; Dewar et al. 2021, 2024; Belcher et al. 2022). Relatedness will only be starkly different between greenbeard and non-greenbeard loci when the greenbeard is able to identify a copy of itself in a pedigree nonrelative (i.e., someone who isn't a family member). But this will be unlikely in general, because populations are viscous, meaning organisms are relatively unlikely to encounter pedigree nonrelatives, even if a greenbeard is being used to pick out social partners (Scott et al. 2022, 2023a).

This is supported by theoretical and empirical work. Theoretically, we recently showed that, when conditional helping based on a recognition locus evolves, relatedness at the recognition locus tends to converge on pedigree relatedness (e.g., $1/2$ for siblings, $1/8$ for cousins). Importantly, we found that this is true even when conditional helping and recognition is underpinned by a single "greenbeard" gene (Scott et al. 2022; Scott 2024). This implies that greenbeard genes will not appreciably shift helping behaviors away from values where organismal inclusive fitness is maximized, except in some restrictive scenarios (Jansen and van Baalen 2006; Rousset and Roze 2007). The conflict between greenbeards and the rest of the genome is often likely to be negligible.

Even though greenbeard genes have been shown to underpin a range of social traits, from social organization in fire ants, to cooperative slug formation in slime molds, to bacteriocin (weapon) production in bacteria, it has not been demonstrated in any of these cases that the greenbeards are what caused these traits to evolve (Madgwick et al. 2019). In the absence of such evidence, we cannot rule out possibility that the social traits would evolve anyway (for instance by kin selection to help pedigree relatives), with the greenbeards evolving secondarily or as a by-product. In other words, there is currently no empirical evidence that greenbeards shift helping behaviors away from values where organismal inclusive fitness is maximized (Madgwick et al. 2019).

Horizontal Gene Transfer

The growth and success of many bacterial populations depends on the evolution of helping behaviors, particularly the production of cooperative "public goods" (West et al. 2006). Public goods are molecules whose secretion provides a benefit to the local group of cells. Examples include iron-scavenging siderophores, exotoxins that disintegrate host cell membranes and elastases that break down connective tissues (Hale 1991; Jones et al. 1993; Griffin et al. 2004).

In bacteria, traits are either encoded by the chromosome or by rings of DNA called plasmids. Plasmids and chromosomes are transmitted vertically to daughter cells, but in addition to this, plasmids can be transmitted horizontally to neighboring cells. Horizontal gene transfer can cause groups of socially interacting bacterial cells to each acquire the same plasmid, causing relatedness at the plasmid loci to increase relative to the chromosome loci (Figure 1.3) (Smith 2001; Nogueira et al. 2009; Mc Ginty et al. 2011, 2013; Rankin et al. 2011a, 2011b; Birch 2014, 2017; Dimitriu et al. 2014; Bakkeren et al. 2022; Lee et al. 2022, 2023).

Horizontal gene transfer can generate conflict between the plasmid and chromosome over helping behaviors like the production of public goods, with the plasmid preferring a greater investment in public goods (Mc Ginty and Rankin 2012). This could result in bacterial cooperation being shifted away from organismal inclusive fitness maximization. This idea gained traction on the back of theoretical and empirical support. Theoretically, it was shown that when a plasmid invades a population, plasmid relatedness is initially higher than chromosomal relatedness, resulting in an increased short-term investment in public goods (Nogueira et al. 2009; Mc Ginty et al. 2013). Empirically, public goods genes were shown to be overrepresented on plasmids relative to the chromosome in *E. coli* (Nogueira et al. 2009).

However, we recently showed that the extent to which plasmids or other mobile elements distort helping behaviors is likely to be limited. Theoretically, we showed that horizontal gene transfer only appreciably elevates plasmid relatedness when plasmids are rare, where there are many plasmid-free cells available to infect. In contrast, when plasmids are common, there

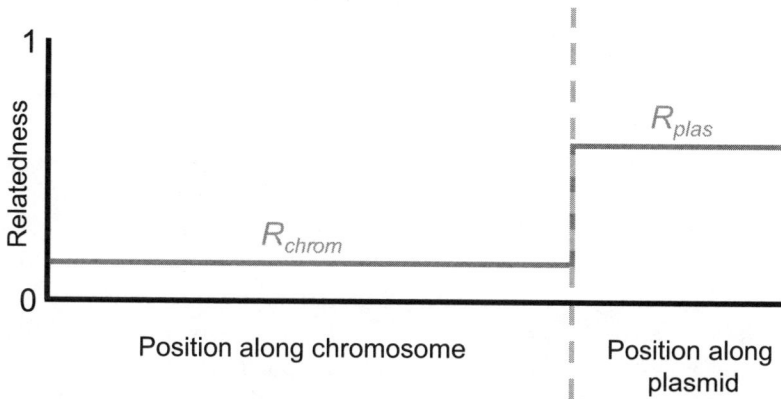

Figure 1.3. Plasmid relative to chromosome relatedness. Horizontal gene transfer may allow relatedness at plasmid loci (R_{plas}) to increase above relatedness at chromosomal loci (R_{chrom}). T. W. Scott, S. A. West, A. E. Dewer, and G. Wild, "Is Cooperation Favored by Horizontal Gene Transfer?," *Evolution Letters*, 7, no. 3 (2023) 113–121, figure 1(a). CC BY 4.0.

are few opportunities for horizontal gene transfer, meaning plasmid relatedness is no longer appreciably elevated above chromosome relatedness. Consequently, while horizontal gene transfer can allow helping behaviors to invade, plasmids will ultimately evolve to either be rare and cooperative, or common and noncooperative, meaning the long-term level of plasmid-mediated cooperation is consistently low. Consequently, horizontal gene transfer has relatively little influence over whether helping behaviors are maintained in the long run (Dewar et al. 2021; Scott et al. 2023b).

Empirically we showed, using comparative genomic analyses across fifty-one bacterial species, that extracellular proteins are (1) not overrepresented on plasmids compared to chromosomes; and (2) not more likely to be carried by plasmids that transfer at higher rates (Dewar et al. 2021). These analyses have been backed up by a more recent study, looking at a wider range of cooperative genes, in 146 species (Dewar et al. 2024). Therefore, in contrast to previous claims, horizontal gene transfer does not appear to significantly shift bacterial helping behaviors away from organismal inclusive fitness maximization, and so bacterial design is not appreciably compromised.

Genomic Imprinting

In sexually reproducing organisms, relatedness may be different depending on whether it is measured at a paternally or maternally inherited gene

(Haig 2002). For instance, for an organism that is deciding whether to help its uncle, relatedness will be 0 for a maternally inherited gene. This is because there is no chance that a maternally inherited gene will be identical by descent to a gene in the father or, therefore, the father's brother. Conversely, relatedness will be 1/2 for a paternally inherited gene. This is because there is a 100% chance that a paternally inherited gene will be identical by descent to a gene in the father, and therefore a 50% chance that it will be identical by descent to a gene in the father's brother. The familiar coefficient of pedigree relatedness between an organism and its uncle is recovered as the average over relatedness coefficients for paternally and maternally inherited genes: 1/4.

In this example, paternally inherited genes will favor greater investment in uncle-helping than maternally inherited genes. Technically, because maternally inherited genes have a relatedness of zero, they will never favor costly uncle-helping. More generally, conflict over helping between maternally and paternally inherited genes may arise in sexually reproducing organisms whenever there are asymmetric kin interactions—that is, when social partners are more related through the maternal or paternal side of the pedigree (e.g., mothers, fathers, aunties, uncles, cousins), as opposed to being equally related through both maternal and paternal channels (e.g., offspring, grand-offspring) (Haig and Graham 1991).

Conflict between paternally and maternally inherited genes plays out via genomic imprinting (Haig and Westoby 1989; Haig 1997, 2002, 2014; Gardner 2014; Patten et al. 2014; Gardner and Úbeda 2017). Genomic imprinting occurs when an allele has different epigenetic marks, and corresponding expression levels, when maternally and paternally inherited. There are many examples where genomic imprinting has apparently arisen in response to conflict, with paternally inherited genes upregulating the amount of help given to paternal relatives and downregulating the amount of help given to maternal relatives, and vice versa for maternally inherited genes (Haig 2002). For instance, in mammalian fetuses, several maternally inherited imprinted genes inhibit fetal growth to increase the fitness of mothers, who are related to such genes by 1 (e.g., *Igf2r, CDKN1C*, and *Grb10*). Conversely, several paternally inherited imprinted genes enhance fetal growth at the expense of the fitness of mothers, who are related to such genes by 0 (e.g., *IGF2, KCNQ1OT1, Air*) (Haig 2004).

Despite this potential for conflict between paternally and maternally inherited genes, and its actualization in instances of genomic imprinting, it is unlikely to shift helping behaviors away from the values where organismal inclusive fitness is maximized. This is because the conflict between paternally and maternally inherited genes is symmetrical—no side wields more power than the other. The conflict is symmetrical because meiosis means that 50% of all genes in an organism at any given time, and across evolutionary time, will be paternally inherited, and 50% will be maternally inherited. Selection on each of the two sets of genes to shift organism trait values will be equal and opposite and will exactly cancel out in general. Helping behaviors should therefore evolve in accordance with the average relatedness across paternally and maternally inherited genes, which is equal to the pedigree (family) relatedness.

Common Ancestry

We have argued that the extent to which greenbeard genes and horizontal gene transfer generate relatedness variation across the genome will be low. We have additionally argued that relatedness variation between paternally and maternally inherited genes will cancel out (symmetrical conflict), with little overall effect on long-term adaptation. In general, mechanisms such as these that increase relatedness at a subset of genes within the genome (genetic assortment mechanisms) are likely to be far less important for adaptation than common ancestry (interacting with pedigree relatives), which increases relatedness at all genes in the genome. There are two main reasons for this.

First, common ancestry, unlike genetic assortment mechanisms, leads to more or less equal genetic relatedness across the majority of the genome, such that most loci will be pulling in the same direction when constructing adaptations (Figure 1.4) (West and Gardner 2013). This "agreement" between genes is important for generating the types of multigene, complex adaptations that are characteristic of organism design. Relatedness may vary stochastically across the genome, but given that the genes underpinning multigene traits are generally scattered across the genome, the average relatedness associated with the multigene traits will be that associated with common ancestry (pedigree relatedness). Consequently, multigene trait evolution will typically proceed in accordance with pedigree relatedness (Grafen 1985, 1990; West and Gardner 2013).

Second, genetic assortment mechanisms, unlike common ancestry, rely on there being a fixed genetic architecture. For instance, they require that helping behaviors are influenced by genes with specific additional qualities,

Kinship

Greenbeard

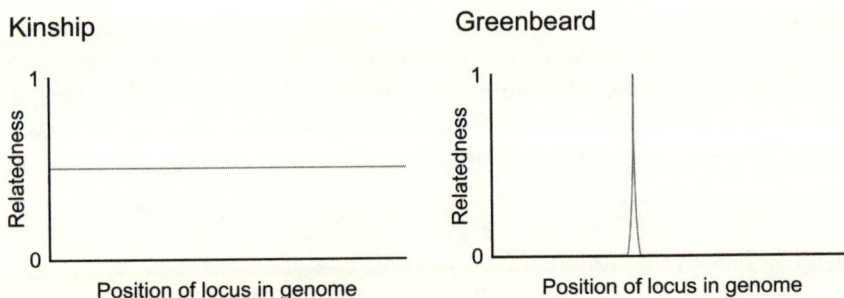

Figure 1.4. Relatedness through common ancestry and genetic assortment. *Left*: Common ancestry: the genetic relatedness between outbred diploid siblings is the same at all autosomal loci. *Right*: If genetic similarity is caused only by a single greenbeard gene, then relatedness is expected to drop off on either side of the greenbeard. The rate of drop-off will depend on recombination rates, population structuring, etc. Reprinted from S. A. West and A. Gardner, "Adaptation and Inclusive Fitness," *Current Biology*, 23 (2013), R577–R584, figure 4, with permission from Elsevier.

like being physically linked to recognition loci (as is the case for greenbeard genes), or being mobile (as is the case for plasmid-mediated cooperation). This is a potential problem because the genetic architecture may evolve (Hammerstein 1996; Rubenstein et al. 2019). The genetic details (linkage, mobility, etc.) might evolve away, meaning the outcome of evolution is the same as if the genetic details were not present in the first place. Alternatively, the genetic details may have evolved in response to the phenotype being selected, rather than having a causal link in the evolution of the phenotype—in other words, causality could be in the opposite direction. In contrast, common ancestry provides a more generally applicable way to generate relatedness, which doesn't require extra genetic factors.

The empirical data on social traits support the hypothesized special role of common ancestry, with studies showing that pedigree relatedness matters for a wide range of traits, including sex allocation, policing, conflict resolution, cooperation, altruism, spite, parasite virulence, cannibalism, dispersal, alarm calls, and eusociality (West 2009; Gardner and West 2010; Westneat and Fox 2010; Bourke 2011; Davies et al. 2012).

Conflict Arises Over Traits Like the Offspring Sex Ratio

We have argued that, despite recent claims to the contrary, helping behaviors will not be under appreciable conflict, and so helping behaviors will not be appreciably shifted away from inclusive fitness maximization. Organism design will therefore be largely preserved with respect to helping behaviors. However, for some other traits like the offspring sex ratio, selfish strategies are available to some genes, resulting in selfish trait distorters, generating conflict (Burt and Trivers 2006; Gardner and Úbeda 2017; Ågren and Patten 2022).

In large, outbred populations, where males and females are equally costly to produce, autosomes (i.e., genes not on sex chromosomes) will tend to prefer an equal ratio of female to male offspring (Fisher 1930; West 2009; Gardner 2023). The reason for this is that, if individuals happen to produce more male than female offspring, males will compete more intensely for mates than females, meaning females have greater fitness, leading to selection for a female-biased offspring sex ratio. Selection for a female-biased sex ratio will continue until there is no longer any male bias in the population. Conversely, if individuals happen to produce more female than male offspring, females will compete more intensely for mates than males, meaning males have greater fitness, leading to selection for a male-biased offspring sex ratio, which will persist until the female bias dissipates. In this way, any deviations from an equal offspring sex ratio are reversed by selection on autosomes.

However, sex chromosomes are inherited disproportionally through one sex. For instance, X chromosomes are inherited primarily thorough females (XX) rather than males (XY). Conversely, Y chromosomes are inherited only through males (XY), not females (XX). This mode of inheritance means that there is a selfish strategy available to genes on sex chromosomes that can allow them to increase their own propagation to the next generation at the expense of the inclusive fitness of the organism (Hamilton 1967). Specifically, genes on X chromosomes can bias the sex ratio toward a greater production of females (XX), and genes on Y chromosomes can bias the sex ratio toward a greater production of males (XY).

The paradox of the organism is particularly stark in light of the offspring sex ratio. Theoretical models based on organismal inclusive fitness maximization have explained a wide range of natural variation in sex ratio, and yet there have been many reported cases of selfish sex ratio distorters, on both X and Y chromosomes, particularly in fruit flies (*Drosophila*) (Charnov 1982; Leigh et al. 1985; Jaenike 2001; West 2009; Helleu et al. 2015). The occurrence of organismal design of the sex ratio on the one hand, and selfish sex ratio distorters on the other, seems paradoxical. We argue that even when we observe conflict over organism traits like the sex ratio, the consequences of this for organism design are likely to be minor. We give three reasons for this.

Selfish Genetic Elements Are Less Capable of Building Complex Adaptations

Opportunities for selfish transmission via trait distortion will only be available to a small number of genes within the genome ("cabals of the few") (Leigh 1971; Cosmides and Tooby 1981; Scott and West 2019). For the vast majority of genes, the only way to increase transmission to the next generation is to increase the fitness of the organism carrying it. For instance, only genes residing on sex chromosomes can gain a selfish transmission advantage from biasing the offspring sex ratio. Autosomes, which are far more numerous, maximize their own transmission when the offspring sex ratio is equal. The complex adaptations that are characteristic of organism design are underpinned by multiple loci, and so we would expect cabals of selfish genetic elements to be less capable of producing them, owing to insufficient genetic material (Gardner and Welch 2011; West and Gardner 2013).

The Parliament of Genes

When a gene increases its own transmission to the detriment of most of the other genes in the genome, these other genes will have a united interest in suppressing the selfish genetic element. Furthermore, because these other genes are far more numerous, they will be likely to win the conflict. Consequently, we would expect conflict within individuals to ultimately resolve itself in favor of the majority of genes in the genome, analogously to how a parliament implements policies according to a majority consensus of the members (Leigh 1971). We would expect the "parliament of genes" to pull traits toward the values where the majority of genes are propagated most efficiently, corresponding to organismal inclusive fitness maximization (Grafen 2006).

The parliament of genes idea was originally proposed by the biologist Egbert Leigh (1971), and it has gained theoretical and empirical support. We obtained the following three theoretical results (Scott and West 2019). First, larger selfish trait distortions are more likely to be suppressed than smaller ones. Consequently, trait distorters either lead to small trait distortions, with minor inclusive fitness consequences, or are suppressed. Second, selection on trait distorters favors the evolution of higher levels of trait distortion, which favors their suppression. Consequently, trait distorters evolve to bring about their own demise (Figure 1.5). Third, if selfish trait distortion is favored at only a small proportion of the genome (small cabal), as tends to be the case empirically, the average level of trait distortion over evolutionary time is low.

Empirical support for the parliament of genes idea comes from the relatively large literature on female-biased sex ratio distorters (X drivers)

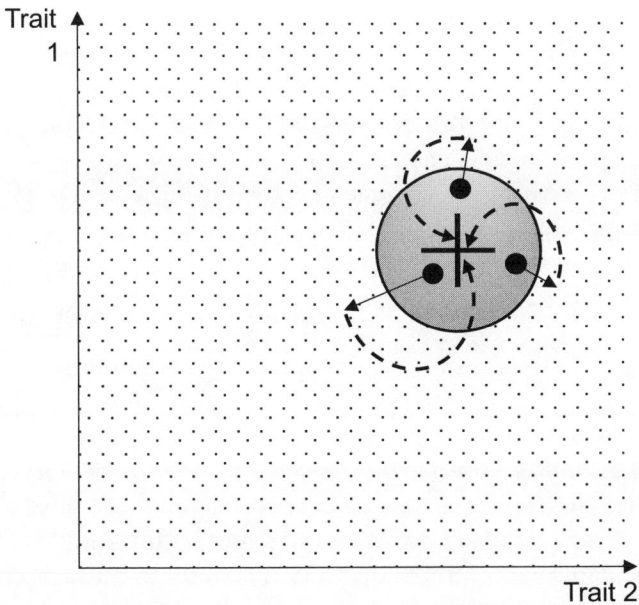

Figure 1.5. Selfish genetic elements evolve to be suppressed by the parliament of genes. The cross represents the position in phenotype space, here defined with respect to traits 1 and 2, which maximizes the inclusive fitness of an organism. The circle surrounding the cross represents the phenotype space where suppression of selfish genetic elements, which have distorted traits 1 or 2, would not be selected for. The surrounding area represents the phenotype space in which the parliament of genes is selected to suppress selfish genetic elements. The three dots represent three possible individuals, each with different weak selfish genetic elements that incur a small fitness cost. Because these deviations from individual fitness maximization are only slight, costly suppression of the weak selfish genetic elements does not evolve. However, the selfish genetic elements will evolve to become more distorting (solid arrows), bringing individuals into the area of phenotype space where they will be suppressed and organism inclusive fitness maximization (the black cross) is regained (dashed arrows). Reformatted from T. W. Scott and S. A. West, "Adaptation Is Maintained by the Parliament of Genes," *Nature Communications, 10* (2019), figure 6. CC BY 4.0.

in fruit flies (*Drosophila*). The scope of the parliament of genes to act against X drivers is shown by the following: (1) In most species in which an X driver is present, suppressors have been found on both the autosomes and the Y chromosome (Helleu et al. 2015). (2) Across natural populations of *D. simulans,* there is a positive correlation between the extent of sex ratio distortion and the extent of suppression (Atlan et al. 1997). (3) In both *D. mediopunctata* and *D. simulans* the presence of an X driver led to the experimental evolution of suppression (Carvalho et al. 1998; Capillon and Atlan 1999). (4) In natural populations of *D. simulans,* the prevalence of an X driver has been shown to sometimes decrease under complete suppression (Bastide et al. 2011). (5) Crossing different species of *Drosophila* has been shown to lead to appreciable sex ratio deviation by unlinking trait distorters from their suppressors, and hence revealing previously hidden trait distorters (Blows et al. 1999). Work on other sex ratio distorters has also shown that suppressors can spread extremely quickly from rarity, reaching fixation in as little as approximately five generations (Hornett et al. 2006).

Conflict Often Implies That Organisms Are Well Adapted

The very fact that we can observe conflict over traits like the offspring sex ratio often suggests that the organism has attained near perfection in its maximization of inclusive fitness (adaptation) (Figure 1.6) (West and Gardner 2013). If organisms were poorly fitted to their environments, it is likely that all their genes would be pulling in much the same direction, toward a distant set of closely coinciding optima. It is only when closing in on the organismal optimum that minor quantitative disagreements between genes would come into play. Empirically, one of the best ways to discover genetic conflicts is to detect departures from inclusive fitness maximization at the organism level. This implies that the conflict we observe in nature is often minor.

Resolving the Paradox of the Organism

In summary, our modern explanation for organism design—that natural selection leads organisms to maximize their inclusive fitness—relies on the assumption that we can effectively ignore the consequences of genetic conflict within individuals. This assumption seems to be contradicted by

Poorly adapted individual   Well-adapted individual

No conflict   Conflict

Phenotype   Optima

Phenotypic value   Phenotypic value

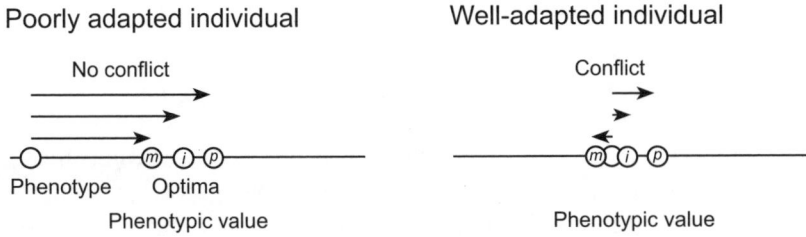

Figure 1.6. Conflict and inclusive fitness maximization. The figure shows a scenario where there is a conflict of interest over the optimum phenotype between genes with maternal origin (*m*), paternal origin (*p*), or which have no information about their origin (*i*). The latter also represents the perspective of a gene trying to maximize the inclusive fitness of the organism. *Left*: If organisms were poorly fitted to their environments, all their genes would be pulling in much the same direction, toward a distant set of closely coinciding optima. *Right*: It is only upon closing in upon the individual optimum that minor quantitative disagreements between genes come into play. Reformatted from S. A. West and A. Gardner, "Adaptation and Inclusive Fitness," *Current Biology*, 23 (2013), R577–R584, figure 5, with permission from Elsevier.

widespread evidence for selfish genetic elements, which may compromise inclusive fitness, leading to the paradox of the organism. We have argued that the paradox of the organism is resolved by four factors:

1. Many selfish genetic elements do not derive their selfish advantage from trait distortion, and so do not compromise organism design. In other words, organism design is not threatened by all selfish genetic elements, only the subset of selfish genetic elements that distort traits, which have been called "trait distorters" (Scott and West 2019; Patten et al. 2023). This resolves the paradox of the organism for selfish genetic elements that do not distort traits.

2. For most traits, like foraging and predator avoidance, all genes in the genome agree that the trait should evolve to maximize the inclusive fitness of the organism (West and Gardner 2013). Distortion of these traits is therefore not favorable, and so trait distorters do not arise. Organism design is only at risk of being compromised with respect to a minority of traits, where trait distorters may arise. For instance, there is no paradox around why genes in a bird genome cooperate to produce exquisitely designed flight. This resolves the paradox of the organism for most aspects of organism design (most traits).

3. It has been claimed that helping behaviors are vulnerable to trait distortion, compromising organism design with respect to helping

behaviors (Jansen and van Baalen 2006; Nogueira et al. 2009; Gardner and Úbeda 2017). The idea behind these claims is that the evolution of helping behaviors is often strongly influenced by the relatedness (genetic similarity) between actors and recipients, and relatedness can vary across the genome due to things like horizontal gene transfer and greenbeard recognition, leading to internal conflict. However, we have argued in opposition to these claims that over evolutionary time relatedness across genes in a genome will tend to equalize, owing to things like their common descent from a shared ancestor, and to the evolution of genetic architecture. Consequently, things like greenbeards and cooperative plasmids will generally not distort helping behaviors in the long run, and so will not compromise organism design. This resolves the paradox of the organism for helping behaviors.

4. Trait distortion can arise for a minority of traits, such as the offspring sex ratio. This residual conflict could in principle compromise organism design, but we have argued that it generally doesn't, because for such traits, trait distortion is favored at a small subset of loci in the genome ("cabals") and disfavoured at a large subset of loci ("the commonwealth") (Leigh 1971; Scott and West 2019). The sheer numerical advantage shifts the balance of power in the "parliament of genes" to favor the commonwealth. The cabal's motion to compromise organism design is rejected; trait distorters are suppressed. This resolves the paradox of the organism for the remainder of organism traits.

Overall, appreciable conflict over traits either does not arise, or is eliminated by the parliament of genes. Consequently, organism design prevails in spite of the possibility for internal conflict, resolving the paradox of the organism.

References

Ågren, J. A., and Patten, M. M. (2022). Genetic conflicts and the case for licensed anthropomorphizing. *Behavioral Ecology and Sociobiology, 76, 166.*

Alcock, J. (2009). *Animal Behavior.* Sinauer Associates.

Atlan, A., Merçot, H., Landre, C., and Montchamp-Moreau, C. (1997). The sex ratio trait in *Drosophila simulans:* Geographical distribution of distortion and resistance. *Evolution, 51*(6), 1886–1895.

Bakkeren, E., Gül, E., Huisman, J. S., et al. (2022). Impact of horizontal gene transfer on emergence and stability of cooperative virulence in *Salmonella typhimurium*. *Nature Communications, 13*, 1939.

Bastide, H., Cazemajor, M., Ogereau, D., Derome, N., Hospital, F., and Montchamp-Moreau, C. (2011). Rapid rise and fall of selfish *sex-ratio* X chromosomes in *Drosophila simulans*: Spatiotemporal analysis of phenotypic and molecular data. *Molecular Biology and Evolution, 28*(9), 2461–2470.

Belcher, L. J., Dewar, A. E., Ghoul, M., and West, S. A. (2022). Kin selection for cooperation in natural bacterial populations. *Proceedings of the National Academy of Sciences of the United States of America, 119*(9), e2119070119.

Biernaskie, J. M., West, S. A., and Gardner, A. (2011). Are greenbeards intragenomic outlaws? *Evolution, 65*, 2729–2742.

Birch, J. (2014). Gene mobility and the concept of relatedness. *Biology and Philosophy, 29*, 445–476.

Birch, J. (2017). *The Philosophy of Social Evolution*. Oxford University Press.

Blows, M. W., Berrigan, D., and Gilchrist, G. W. (1999). Rapid evolution towards equal sex ratios in a system with heterogamety. *Evolutionary Ecology Research, 1*, 277–283.

Bourke, A. F. G. (2011). *Principles of Social Evolution*. Oxford University Press.

Burt, A., and Trivers, R. (2006). *Genes in Conflict: The Biology of Selfish Genetic Elements*. Harvard University Press.

Capillon, C., and Atlan, A. (1999). Evolution of driving X chromosomes and resistance factors in experimental populations of *Drosophila simulans*. *Evolution, 53*, 506–517.

Carvalho, A. B., Sampaio, M. C., Varandas, F. R., and Klaczko, L. B. (1998). An experimental demonstration of Fisher's principle: Evolution of sexual proportion by natural selection. *Genetics, 148*, 719–731.

Charnov, E. L. (1982). *The Theory of Sex Allocation*. Princeton University Press.

Cosmides, L. M., and Tooby, J. (1981). Cytoplasmic inheritance and intragenomic conflict. *Journal of Theoretical Biology, 89*, 83–129.

Darwin, C. (1859). *On the Origin of Species by Means of Natural Selection, or, The Preservation of Favoured Races in the Struggle for Life*. J. Murray.

Davies, N., Krebs, J., and West, S. (2012). *An Introduction to Behavioural Ecology*. John Wiley and Sons.

Dawkins, R. (1990). Parasites, desiderata lists and the paradox of the organism. *Parasitology, 100*, S63–S73.

Dewar, A. E., Belcher, L. J., Scott, T. W., and West, S. A. (2024). Genes for cooperation are not more likely to be carried by plasmids. *Proceedings of the Royal Society B, 291*.

Dewar, A. E., Thomas, J. L., Scott, T. W., et al. (2021). Plasmids do not consistently stabilize cooperation across bacteria but may promote broad pathogen host-range. *Nature Ecology and Evolution, 5*, 1624–1636.

Dimitriu, T., Lotton, C., Beńard-Capelle, J., et al. (2014). Genetic information transfer promotes cooperation in bacteria. *Proceedings of the National Academy of Sciences of the United States of America, 111*(130), 11103–11108.

Eshel, I. (1985). Evolutionary genetic stability of Mendelian segregation and the role of free recombination in the chromosomal system. *The American Naturalist, 125,* 412–420.

Fisher, R.A. (1930). *The Genetical Theory of Natural Selection.* Clarendon Press.

Gardner, A. (2014). Genomic imprinting and the units of adaptation. *Heredity, 113,* 104–111.

Gardner, A. (2023). The rarer-sex effect. *Philosophical Transactions of the Royal Society B, 378.*

Gardner, A., and Úbeda, F. (2017). The meaning of intragenomic conflict. *Nature Ecology and Evolution, 1,* 1807–1815.

Gardner, A., and Welch, J. J. (2011). A formal theory of the selfish gene. *Journal of Evolutionary Biology, 24,* 1801–1813.

Gardner, A., and West, S. A. (2010). Greenbeards. *Evolution, 64,* 25–38.

Grafen, A. (1985). A geometric view of relatedness. *Oxford Surveys in Evolutionary Biology, 262,* 391–397.

Grafen, A. (1990). Do animals really recognize kin? *Animal Behaviour, 39,* 42–54.

Grafen, A. (2006). Optimization of inclusive fitness. *Journal of Theoretical Biology, 238,* 541–563.

Griffin, A. S., West, S. A., and Buckling, A. (2004). Cooperation and competition in pathogenic bacteria. *Nature, 430,* 1024–1027.

Haig, D. (1997). Parental antagonism, relatedness asymmetries, and genomic imprinting. *Proceedings of the Royal Society B, 264,* 1657–1662.

Haig, D. (2002). *Genomic Imprinting and Kinship.* Rutgers University Press.

Haig, D. (2004). Genomic imprinting and kinship: How good is the evidence? *Annual Review of Genetics, 38,* 553–585.

Haig, D. (2014). Coadaptation and conflict, misconception and muddle, in the evolution of genomic imprinting. *Heredity, 113,* 96–103.

Haig, D., and Graham, C. (1991). Genomic imprinting and the strange case of the insulin-like growth factor II receptor. *Cell, 64,* 1045–1046.

Haig, D., and Westoby, M. (1989). Parent-specific gene expression and the triploid endosperm. *The American Naturalist, 134,* 147–155.

Hale, T. L. (1991). Genetic basis of virulence in Shigella species. *Microbiological Reviews, 55,* 206–224.

Hamilton, W. D. (1963). The evolution of altruistic behavior. *The American Naturalist, 97,* 354–356.

Hamilton, W. D. (1964). The genetical evolution of social behaviour. I & II. *Journal of Theoretical Biology, 7,* 1–52.

Hamilton, W. D. (1967). Extraordinary sex ratios. *Science, 156,* 477–488.

Hamilton, W. D. (1970). Selfish and spiteful behaviour in an evolutionary model. *Nature, 228,* 1218–1220.

Hammerstein, P. (1996). Streetcar theory and long-term evolution. *Science, 273,* 1032.

Helleu, Q., Gérard, P. R., and Montchamp-Moreau, C. (2015). Sex chromosome drive. *Cold Spring Harbor Perspectives in Biology, 7,* a017616.

Hornett, E. A., Charlat, S., Duplouy, A. M. R., et al. (2006). Evolution of male-killer suppression in a natural population. *PLoS Biology, 4,* e283.

Jaenike, J. (2001). Sex chromosome meiotic drive. *Annual Review of Ecology and Systematics, 32,* 25–49.

Jansen, V. A. A., and van Baalen, M. (2006). Altruism through beard chromodynamics. *Nature, 440,* 663–666.

Jones, S., Yu, B., Bainton, N. J., et al. (1993). The lux autoinducer regulates the production of exoenzyme virulence determinants in *Erwinia carotovora* and *Pseudomonas aeruginosa. The EMBO Journal, 12,* 2477–2482.

Kay, T., Helleu, Q., and Keller, L. (2022). Iterative evolution of supergene-based social polymorphism in ants. *Philosophical Transactions of the Royal Society B, 377.*

Keller, L., and Ross, K. G. (1998). Selfish genes: A green beard in the red fire ant. *Nature, 394,* 573–575.

Larracuente, A. M., and Presgraves, D. C. (2012). The selfish segregation distorter gene complex of *Drosophila melanogaster. Genetics, 192,* 33–53.

Lee, I. P. A., Eldakar, O. T., Gogarten, J. P., and Andam, C. P. (2022). Bacterial cooperation through horizontal gene transfer. *Trends in Ecology and Evolution, 37,* 223–232.

Lee, I. P. A., Eldakar, O. T., Gogarten, J. P., and Andam, C. P. (2023). Recombination as an enforcement mechanism of prosocial behavior in cooperating bacteria. *iScience, 26,* 107344.

Lehmann, L., and Rousset, F. (2020). When do individuals maximize their inclusive fitness? *The American Naturalist, 195,* 717–732.

Leigh, E. G. (1971). *Adaptation and Diversity: Natural History and the Mathematics of Evolution.* Freeman, Cooper.

Leigh, E. G., Herre, E. A., and Fischer, E. A. (1985). Sex allocation in animals. *Experientia, 41,* 1265–1276.

Levin, S. R., and Grafen, A. (2019). Inclusive fitness is an indispensable approximation for understanding organismal design. *Evolution, 73,* 1066–1076.

Levin, S. R., and Grafen, A. (2021). Extending the range of additivity in using inclusive fitness. *Ecology and Evolution, 11,* 1970–1983.

Levin, S. R., Scott, T. W., Cooper, H. S., and West, S. A. (2019). Darwin's aliens. *International Journal of Astrobiology, 18,* 1–9.

Madgwick, P. G., Belcher, L. J., and Wolf, J. B. (2019). Greenbeard genes: Theory and reality. *Trends in Ecology and Evolution, 34,* 1092–1103.

Mc Ginty, S. É., Lehmann, L., Brown, S. P., and Rankin, D. J. (2013). The interplay between relatedness and horizontal gene transfer drives the evolution of plasmid-carried public goods. *Proceedings of the Royal Society B, 280,* 20130400.

Mc Ginty, S. É., and Rankin, D. J. (2012). The evolution of conflict resolution between plasmids and their bacterial hosts. *Evolution, 66,* 1662–1670.

Mc Ginty, S. E., Rankin, D. J., and Brown, S. P. (2011). Horizontal gene transfer and the evolution of bacterial cooperation. *Evolution, 65,* 21–32.

Nogueira, T., Rankin, D. J., Touchon, M., Taddei, F., Brown, S. P., and Rocha, E. P. C. (2009). Horizontal gene transfer of the secretome drives the evolution of bacterial cooperation and virulence. *Current Biology, 19,* 1683–1691.

Paley, W. (1829). *Natural Theology, or, Evidences of the Existence and Attributes of the Deity, Collected from the Appearances of Nature.* Lincoln and Edmands.

Patten, M. M., Ross, L., Curley, J. P., Queller, D. C., Bondduriansky, R., and Wolf, J. B. (2014). The evolution of genomic imprinting: Theories, predictions and empirical tests. *Heredity, 113,* 119–128.

Patten, M. M., Schenkel, M. A., and Ågren, J. A. (2023). Adaptation in the face of internal conflict: The paradox of the organism revisited. *Biological Reviews, 98,* 1796–1811.

Rankin, D. J., Mc Ginty, S. E., Nogueira, T., et al. (2011a). Bacterial cooperation controlled by mobile elements: Kin selection and infectivity are part of the same process. *Heredity, 107,* 279–281.

Rankin, D. J., Rocha, E. P. C., and Brown, S. P. (2011b). What traits are carried on mobile genetic elements, and why? *Heredity, 106,* 1–10.

Rousset, F., and Roze, D. (2007). Constraints on the origin and maintenance of genetic kin recognition. *Evolution, 61,* 2320–2330.

Rubenstein, D. R., Ågren, J. A., Carbone, L., et al. (2019). Coevolution of genome architecture and social behavior. *Trends in Ecology and Evolution, 34,* 844–855.

Scott, T. W. (2024). Crozier's paradox and kin recognition: Insights from simplified models. *Journal of Theoretical Biology, 581,* 111735.

Scott, T. W., Grafen, A., and West, S. A. (2022). Multiple social encounters can eliminate Crozier's paradox and stabilise genetic kin recognition. *Nature Communications, 13,* 3902.

Scott, T. W., Grafen, A., and West, S. A. (2023a). Host–parasite coevolution and the stability of genetic kin recognition. *Proceedings of the National Academy of Sciences of the United States of America, 120*(30), e2220761120.

Scott, T. W., and West, S. A. (2019). Adaptation is maintained by the parliament of genes. *Nature Communications, 10,* 5163.

Scott, T. W., West, S. A., Dewar, A. E., and Wild, G. (2023b). Is cooperation favored by horizontal gene transfer? *Evolution Letters, 7,* 113–120.

Scott, T. W., and Wild, G. (2023). How to make an inclusive-fitness model. *Proceedings of the Royal Society B, 290.*

Smith, J. (2001). The social evolution of bacterial pathogenesis. *Proceedings of the Royal Society B, 268,* 61–69.

Stearns, S. C. (1992). *The Evolution of Life Histories.* Oxford University Press.

Stephens, D. W., and Krebs, J. R. (1986). *Foraging Theory.* Princeton University Press.

Taylor, P. D. (1989). Evolutionary stability in one-parameter models under weak selection. *Theoretical Population Biology, 36,* 125–143.

Wang, J., Wurm, Y., Nipitwattanaphon, M., et al. (2013). A Y-like social chromosome causes alternative colony organization in fire ants. *Nature, 493,* 664–668.

West, S. (2009). *Sex Allocation.* Princeton University Press.

West, S. A., and Gardner, A. (2013). Adaptation and inclusive fitness. *Current Biology, 23,* R577–R584.

West, S. A., Griffin, A. S., and Gardner, A. (2007). Evolutionary explanations for cooperation. *Current Biology, 17,* R661–R672.

West, S. A., Griffin, A. S., Gardner, A., and Diggle, S. P. (2006). Social evolution theory for microorganisms. *Nature Reviews Microbiology, 4,* 597–607.

Westneat, D. F., and Fox, C. W. (2010). *Evolutionary Behavioral Ecology.* Oxford University Press.

Zanders, S. E., and Unckless, R. L. (2019). Fertility costs of meiotic drivers. *Current Biology, 29,* R512–R520.

CHAPTER TWO

In What Sense Can There Be Conflict Between the Levels of Selection?

PIERRICK BOURRAT

ABSTRACT   This chapter proposes an analysis of the idea of conflict between levels of selection. An evolutionary conflict occurs when two parties have opposing evolutionary interests. I argue that, in multilevel contexts, the idea of opposing interests is difficult to articulate. The root of the problem, I show, is that often when applying the idea of conflict between levels, higher-level entities are composed of lower-level ones, rendering the evolutionary fate of the former inevitably tied to the evolutionary fate of the latter. Nonetheless, I propose different interpretations of the ideas of conflict between levels corresponding to scenarios involving different compositional relationships between lower and higher levels. The upshot is that the term "conflict between levels" can either be understood in a metaphorical sense or be redescribed as a case where the conflict occurs between parties that do not exhibit a relationship of composition. Finally, I apply my analysis in the context of the evolution of cancer, where cancerous cells have been described as in conflict with the organism.

Introduction

The idea of evolutionary conflict is often invoked in evolutionary biology literature, particularly in situations where a phenotype is regarded as suboptimal from the perspective of its bearer, either because it is a straightforward maladaptation or because it is an extravagant adaptation—that is, it is costly to produce (Queller and Strassmann 2018). For instance, a case of evolutionary conflict occurs when an ant is infected by the lancet liver fluke (*Dicrocoelium dendriticum*), a parasite that manipulates the ant's behavior, causing it to climb grass blades and clamp its mandibles onto the top of the blade. This behavior increases the chances of the ant being eaten by a herbivore (Moore and Moore 2002, 55–56), thereby allowing the parasite to continue its life cycle. This behavior is a straightforward maladaptation from the perspective of the ant due to a conflict between the ant and the parasite over the ant's behavior. Adaptations such as the exaggerated begging behaviors of juveniles in some species of birds or the antlers of a moose are "extravagant" because they would not exist in the absence of opposing interests: between offspring and parents in the case of begging juveniles and between males in the case of the moose.

In the contexts where adaptation and appearance of design are regarded either as what evolution ought to explain or as a working hypothesis—what Godfrey-Smith (2001) calls explanatory and methodological adaptationism, respectively—the existence of suboptimal phenotypes represents a puzzle that must be solved. Invoking the existence of a conflict between two parties is one available solution to this puzzle. Other candidate explanations include the existence of some underlying genetic constraints for the bearer of the phenotype and that the phenotype evolved as a result of a chancy process (i.e., drift).

One context in which the idea of evolutionary conflict has been used is the multilevel selection theory literature and related topics, such as the transition from unicellular to multicellular organisms and the evolution of cancer from a multilevel selection perspective. In these works, levels of organization at which a process of selection can occur, such as the genetic level and the individual level, or the cellular level and the organismal level, are regarded as two parties that can be in conflict, in a way that is similar to how the ant and the lancet liver fluke are in conflict.

Instances of work that have used the idea of a conflict between levels of selection, often in passing, are numerous (e.g., Maynard Smith 1988; Maynard Smith and Szathmáry 1995; Tsuji 1995; Michod and Roze 2001; Taylor et al. 2002; Joseph and Kirkpatrick 2004; Okasha 2006, 2021; Folse and Roughgarden 2010; Rainey and Kerr 2010, 2011; Alizon et al. 2011; Wade 2016; Ratcliff et al. 2017). For instance, Charlesworth (2000), discussing Haldane's 1932 *The Causes of Evolution*, writes, "In addition to conflicts between different levels of selection, there is also the possibility of conflict between entities at the same level of organization, but which are subject to different rules of inheritance" (p. 493). In discussing the problem for the evolution of altruism, Wilson and Sober (1994) succinctly summarize their argument as follows: "Altruism involves a conflict between levels of selection. Groups of altruists beat groups of non-altruists, but non-altruists also beat altruists within groups" (p. 599).

While the case for evolutionary conflict between prey and predators, sexual conflicts, or hosts and pathogens—all of which occur between parties that are at a single level of organization (hereafter, "classical evolutionary conflicts")—is straightforward to conceptualize, situations where a conflict involves multiple levels are more perplexing. In this chapter, I offer an analysis of this type of situation and, more specifically, examine the sense in which there can be conflict between the levels of selection. I argue that if this phrase is understood as referring to a genuine evolutionary conflict in the same way that classical evolutionary conflicts (as defined above) are understood, it is a mistake to claim that a conflict between levels of selection can exist. However, I argue that the notion of conflict involved in the multilevel selection literature can be understood in a metaphorical sense as it involves counterfactual scenarios; that is, there *would* be a conflict under different conditions. This way of conceiving conflict between levels of selection yields a different type of explanation from that of classical evolutionary conflicts. However, I argue that it can, once clearly separated from the classical notion, provide insight into the evolutionary mechanisms that have been at play to allow for the emergence and maintenance of higher-level entities that result from evolutionary transitions in individuality, such as multicellular organisms and superorganisms.

From a purely philosophical stance, disambiguating the notion of evolutionary conflict between levels of selection is important. Mixing different notions of conflict between levels of selection can lead to misunderstandings and impede scientific progress. This type of misunderstanding has been identified as a problem for the more general idea of a "level of organization" (Brooks 2021). This chapter is intended to offer a conceptual clarification to facilitate discussion surrounding the idea of evolutionary conflict in multilevel settings, such as multilevel selection theory, evolutionary transitions in individuality, intragenomic conflict, and cancer, as seen through the lens of multilevel selection.

The chapter is divided into three sections. In the first section, I propose a definition of a classical evolutionary conflict and a simple "test" to detect whether, in a given setting, there is an evolutionary conflict between two types of entities. In the second section, I apply this definition to a multilevel setting. I show that it can be interpreted in at least two different ways. Under the first interpretation, different levels refer to different physical substrates, such as a cell and the *rest* of the body of an organism. Under the second interpretation, they refer to the same substrate, such as a cell and the *whole* body of an organism, including the focal cell. I argue that only the first interpretation aligns with the classical notion of evolutionary conflict and that under the second interpretation, conflict can only be understood in a metaphorical sense because no factual conflict occurs between the different levels. Finally, in the last section, I briefly illustrate how my analysis can be useful in the context of recent discussions of the role of multilevel selection in the evolution of cancer and the ways in which the metaphorical idea of conflict between levels of selection, once properly separated from the classical sense of evolutionary conflict, can nevertheless be useful.

Defining and Detecting Classical Evolutionary Conflict

In this section, I begin with a simple definition of an evolutionary conflict.

> *Classical Evolutionary Conflict.* A situation where two or more (biological) entities have opposing evolutionary interests with respect to a particular trait, where an evolutionary interest is measured in terms of fitness (long-term growth rate).

A few remarks are in order regarding this definition. First, the term "conflict" refers here to evolutionary "entities" at any level of organization and in any domain. This is so because I consider there to be no a priori reason why there could only be evolutionary conflicts at a particular level or in a particular domain. In a given particular setting, "entity" might, for instance, refer to a gene, an organism, a group, a cell, a chromosome, a cultural item, and so forth. Second, the idea that two entities have divergent interests over a particular trait implies that they actually interact, such as when the lancet liver fluke *manipulates* the ant's behavior.

Third, one might consider that the definition I provide refers to the idea of competition for resources rather than conflict. A famous case of competition for resources was described by Gause (1934): two species of *Paramecium, P. caudatum* and *P. aurelia,* compete for resources, and *P. aurelia* outcompetes *P. caudatum* because *P. aurelia* has an advantage in resource utilization. In response to this, it should be noted that situations of competition for resources represent instances of evolutionary conflict. In such cases, the trait over which there is a conflict is which entity any limiting resources should go to (see also Queller and Strassmann 2018). While considering resources as a trait is nonstandard, it is aligned with the view that phenotypes can extend beyond the physical boundaries of an organism (Dawkins 1982; Haig 2012). Finally, while I mainly focus here on the idea of conflict, part of what I argue can be straightforwardly applied to the ideas of evolutionary cooperation (or synergy) and independence, in which cases evolutionary interests are aligned and independent, respectively.

With these remarks in place, given the definition of classical evolutionary conflict provided in this chapter, I propose a "test" to detect the presence of such conflicts in a particular setting. As was mentioned earlier, if two or more entities have divergent evolutionary interests with respect to a particular trait, the value of this trait depends on some interactions between the entities, as each will "push" the trait value in a particular direction. Therefore, eliminating such putative interactions and comparing the resulting evolutionary success of the entities to a situation in which the entities have the opportunity to interact provides a means to assess whether any evolutionary conflict is occurring between them (see also McCoy et al., this volume, for examples of ways to quantify conflict between corals and microalgae using specific traits as proxies). More specifically, if at least one of the entities benefits from the elimination of interactions, this is evidence that the two entities are in evolutionary conflict with respect to that trait. If both do worse in the absence of interactions, they are in synergy with respect to that trait. Finally, if there is no change, there was neither conflict nor synergy over this trait. Depending on the entities studied, different ways to implement the test could be devised, each providing different degrees of evidence for evolutionary conflict.

One very crude but effective way to implement the test, inspired from Gause's (1934) famous experiment of competitive exclusion in *Paramecium* (Foster and Bell 2012), is illustrated in Figure 2.1. Suppose *A* and *B* are both microscopic organisms of the same species and that they reproduce asexually with perfect inheritance in the same environment. Suppose you know nothing about their biology except that they both need a particular nutrient that is available in limited quantities to grow. Besides this nutrient, everything they need to grow is always present in an optimal concentration at all times. You also know that, given their planktonic (i.e., free-living) lifestyle, any interaction between them will be mediated by their liquid environment. Crucially, you do not know whether they can synthesize this nutrient or whether there is some variation for this trait. You want to know whether these two entities are in evolutionary conflict over gathering this nutrient in the environment. To do so, you measure the long-term growth rate of the two entities in situations where they coexist in their natural environment (see Figure 2.1a)

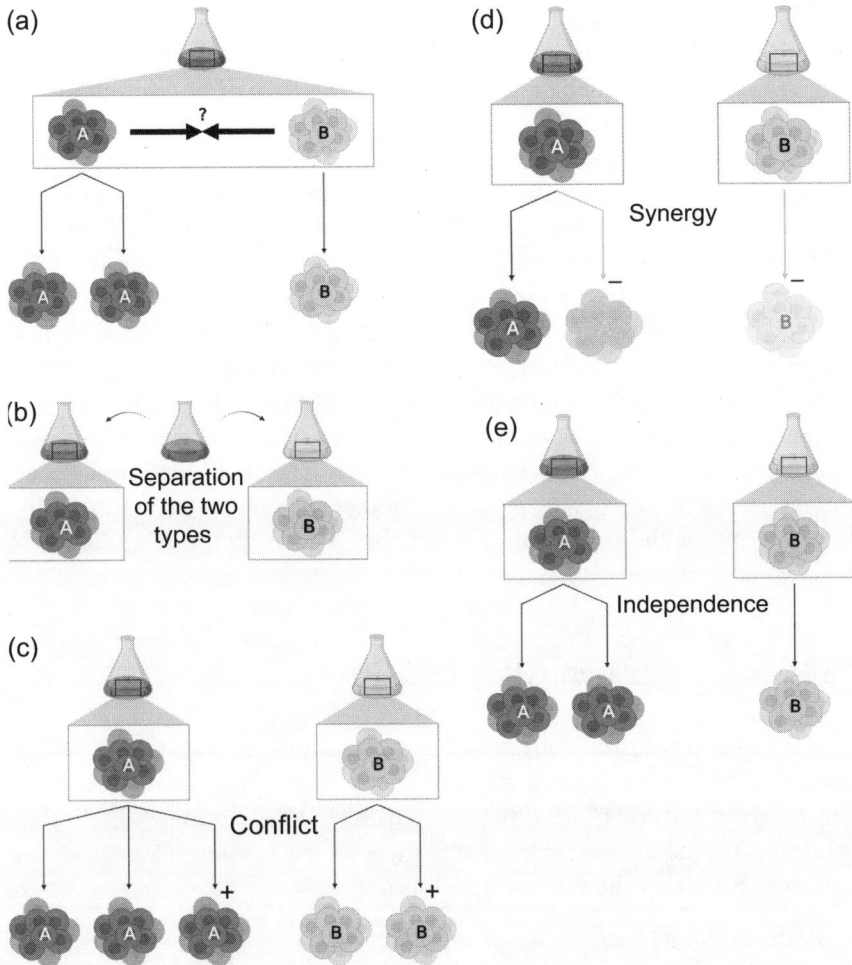

Figure 2.1. Possible implementation of a test for detecting an evolutionary conflict between two entities. (a) When two types of entities are in an environment where they have the opportunity to interact with respect to a trait, one way to assess whether there is an evolutionary conflict is to eliminate any opportunity for interactions between them by keeping them separate, as shown in (b). If the fitness (measured here as growth after some time) of at least one of the two entities is different when separated from the other entity, then there are interactions between the two entities. In particular, an evolutionary conflict occurs between two entities when at least one entity does better (in fitness terms) when separated from the other type, as in (c). Evolutionary cooperation or synergy occurs when both entities do worse in the absence of the other type, as in (d), or one does worse with no change for the other. Finally, if no change in fitness is observed as the result of the separation, as in (e), there is evolutionary independence of the two entities with respect to the trait.

and make the same measurement (in the same conditions) in situations where they are separated from one another (see Figure 2.1b). In this setting, if one or both of the two entities grows at a higher rate in the absence of the other, then there is strong evidence that an evolutionary conflict between the two entities occurs (see Figure 2.1c). This is because any interaction between two entities that prevents one or both from growing optimally is eliminated when the two types are separated. If both do worse, the two entities are cooperating or synergizing with respect to the trait (see Figure 2.1d). This is because any interaction between the two entities that allowed for the nutrient to be produced in a higher concentration by one entity is eliminated when they are separated. Finally, if there is no difference in growth rate, there is no conflict or synergy; in other words, there is evolutionary independence between the two entities with respect to the trait (see Figure 2.1e). The lack of changes between the two conditions indicates that there is no interaction between the two entities when they are in the presence of each other.[1]

Are Genuine Conflicts Between Levels of Selection Possible?

In the previous section, I proposed a definition of evolutionary conflict and described an associated test for determining whether there is conflict between two or more types of entities. In this section, I apply this definition to a situation where the entities refer to different levels of selection. This yields the following definition:

> *Evolutionary Conflict (levels of selection).* A situation where two or more (biological) entities at different levels of selection have divergent evolutionary interests with respect to a particular phenotype, where an evolutionary interest is measured in terms of fitness (long-term growth rate).

This definition can be understood in at least two different ways. In the first sense, the higher-level and lower-level entities (henceforth, "collectives" and "particles") are physically distinct entities (see Figure 2.2a),

---

[1] To be clear, the test presented here is intended to illustrate the point rather than present a surefire test. Nonetheless, given any setting, a test of this form could in principle be designed.

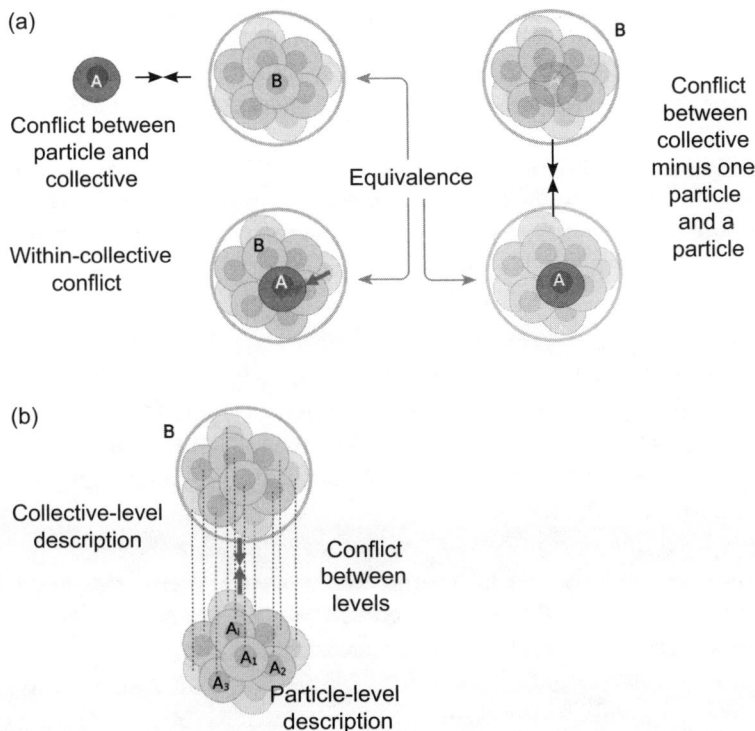

Figure 2.2. Two senses under which an evolutionary conflict between higher-level and lower-level entities can be understood. In both cases, A and B refer to the lower-level and higher-level entity, respectively. (a) Three instances in which A and B are physically distinct entities. This type of situation does not pose any conceptual problem, as it is on par with classical evolutionary conflict. (b) A and B are made of the same physical substrate so that B constitutively depends on multiple A. This poses a conceptual problem because any change in the long-term fitness of A will also lead to a change in fitness in the same direction of B.

such as a cell and an organism minus this cell. In the second sense, the collectives and particles are made of the same physical substrate (i.e., particles constitute a collective; see Figure 2.2b), such as an organism and the cells that compose this organism. A conflict between levels could also apply to more ambiguous cases where particles partially constitute a collective, such as when an organism is considered to be composed of more than its cells and includes, for example, an extracellular matrix. I will only discuss clear-cut cases here.

Discussing an evolutionary conflict between two or more levels (of selection) in the first sense is straightforward. One just needs to be aware

(1) that the entities at the "higher" and "lower" levels refer to levels of description (i.e., they are just ways to describe biological objects), and (2) that one could provide an equivalent description using a single level. This type of case is more accurately characterized as a case of conflict between entities *described* at different levels, rather than conflict between levels, where the conflict would be a phenomenon in the world rather than in the mind of the observer. In simple cases, a "collective" (B) can be equivalently redescribed in terms of "particles" (A) (see Figure 2.2a). For instance, assuming that a simple multicellular organism is only composed of cells and no other substrate, it can be redescribed in terms of cells. A multicellular organism (B) is a collection of cells (A) that have adhesive powers. This way of switching back and forth between the particle and collective levels of description is particularly interesting in the context of the first stages of an evolutionary transition in individuality (Godfrey-Smith and Kerr 2013; Bourrat 2023a). However, switching back to a particle-level description and retaining the full information of the system becomes more challenging in situations where the complexity of the collective increases. When this occurs, the only pragmatic description becomes a collective-level one (Bourrat 2023a). At any rate, whether an adequate lower-level description is available or not, one can always, *in principle,* apply the test presented in Figure 2.1c to cases where the two entities do not physically overlap, for A and B could potentially be separated. Note that whether this can be done in practice has no bearing on the conceptual distinction. Thus, these cases are very much like classical cases of evolutionary conflict at a single level.

I mentioned in the previous paragraph that the early phases of an evolutionary transition in individuality, such as cells exhibiting both a unicellular and colony mode of living, represent situations where this sense of evolutionary conflict between "levels" can be applied. As we shall see in "Subversion from Within and Conflicts Between the Levels of Selection," the evolution of cancerous cells within an organism can also be examined from this perspective.

Referring to evolutionary conflict in the second sense (i.e., where particles constitute a collective) is more problematic. As in the first sense, "higher" and "lower" levels also refer to levels of description, but because collectives are constituted by particles so that both descriptions refer to the same substrate (see Figure 2.2b), levels are not merely different ways to describe different objects—they are also compositional, so that the different description applies to one object (B) and a part of it (A). This distinction from the previous sense of conflict is important if one is committed to the standard scientific assumption that in a situation where there is a strict compositional relationship between different levels, the higher levels mereologically supervene—that is, constitutively depend—on the lower levels.

One implication of this assumption is the existence of the following constraint (more fully explored in Bourrat 2024): If one were to change some property of a single particle (A) that is part of a collective (B), there would necessarily also be a change in some property of B. If the particle-level property that is changed has an effect on the fitness of the particle, so that its long-term growth is affected, by virtue of being part of a collective, this implies that the long-term growth rate of the collective will also be altered in the same direction.[2] Concretely, altering the long-term reproductive success of a particle, when particles form collectives and collectives cannot grow indefinitely, necessarily translates into a higher number of collectives being produced in the long run (Bourrat 2021; Bourrat et al. 2022).

If the supervenience assumption were to be violated, nothing could prevent causal chains at the collective level from being created *ex nihilo*. That is, particles would not have to play any role in the existence of collective-level processes, the latter of which would be *strongly* emergent. However, the scientific consensus is that emergent phenomena are always *weak* so that any process at the collective level could always, in principle, be accounted for by processes at the particle level (see Bedau 1997 for more on the distinction between strong and weak emergence).

The constraint that a change in a particle's property affecting fitness would necessarily affect the fitness of the collective this particle belongs to in the same direction—and further, that in the long run, the particle and the collective would have the same evolutionary fate—poses a problem for the claim that an evolutionary conflict can exist between different levels of selection.[3] This is so because an evolutionary conflict presupposes not only that the fitnesses of the parties involved are distinguishable but also that they go in opposite directions. Both conditions

---

[2] This constraint is related to Kim's exclusion principle (Kim 1988); for a brief discussion, see Okasha (2006), 106–107.

[3] There are exceptions to this when a collective can grow indefinitely or have an infinite size, but both assumptions are idealizations and therefore do not undermine this reasoning (see Bourrat 2021 for details).

are precluded from the existence of a constitutive relationship between the particles and the collective. Therefore, the idea of conflict between levels cannot be understood literally in such a setting.

This problem can be appreciated more concretely by implementing the test presented in Figure 2.1c in the situation where a collective is strictly composed of particles. In attempting to perform the test, one would be confronted with the practical problem of being unable to physically separate A from B without changing the nature of B. It is not possible to separate a cell from its collective without changing the nature of the collective. From there, any increase in long-term reproductive output of the collective, once the particle has been removed, does not correspond to a comparison of the fitness of B in the presence and absence of A that could reveal an evolutionary conflict in the same way that there is a conflict between an ant and a lancet liver fluke. Instead, this corresponds to a comparison of the fitness of B and the fitness of another type of collective that does not contain A. Similarly, from the perspective of the particle, the comparison would be a comparison between a particle in the presence of a collective minus a particle and the absence of it (leading us back to the setting presented in Figure 2.2a), not the presence or absence of a whole collective.

How should we thus interpret the idea of conflict between levels in situations where particles and collectives refer to the same physical substrate? I propose that conflict between levels in such situations actually corresponds to cases where fitness at each level is estimated over different timescales and for which the environment over these different timescales cannot be considered to be the same on average (Bourrat 2015a, 2015b, 2021, 2023a, 2023b; Black et al. 2020; Bourrat et al. 2022). Particle-level fitness is often estimated over a much shorter timescale than collective-level fitness. For instance, the fitness of a cell is often estimated over a timescale that does not exceed the lifespan of the organism it is part of. In contrast, the fitness of an organism is estimated over at least one organism's generation. Now, if a long-term projection is made with these estimates over different timescales, they might not match, and there might appear to be conflict between levels. However, a measure made over the same timescale at both levels would reveal no such conflict.

To see this point, take the example proposed by Wilson and Sober in the Introduction to this volume. Assume a population of altruistic and selfish individuals organized in groups with a phase during which the group reproduces in a mixing pool following the trait group model (Wilson 1975), one of the simplest models used in the multilevel selection literature to demonstrate the power of group-level selection. Measuring the fitness of both types within a group will show that selfish individuals do better. However, when measured at the level of the group, groups composed of more altruistic individuals do better (produce more offspring) than those composed predominantly of selfish individuals. According to Wilson and Sober, this would demonstrate a conflict between the individual and group levels, as quoted in the Introduction. However, a more accurate description is that estimating the success of an individual within a group does not correspond to its fitness in the long run, as it does not take into account events that are typically described at the group level (e.g., a dissolution of the group, formation of new groups). However, such events can also be accounted for from the perspective of an individual. When this is done, taking the long-term growth rate of an altruistic individual considering events beyond that of the immediate group of this individual leads to no discrepancy or conflict between the two levels.

It follows from this reasoning that discussing conflict between lower and higher levels of selection in situations where the higher levels are composed of lower ones can only be understood in a specific counterfactual sense. In the trait group model, there *would be* a conflict between the lower and the higher level if being selfish (and successful) within a group did not entail being unsuccessful in the long run, as both selfish groups and selfish individuals are unsuccessful when altruism prevails. Accordingly, one might embrace the notion of "counterfactual conflict." I find this problematic due to the oxymoron it creates. An evolutionary conflict either does or does not occur; in the case of the trait group model, no such conflict exists or could even exist without violating the supervenience assumption. Instead, I prefer to refer to such conflicts as metaphorical conflicts or "conflicts."

Thus far, I have argued that when entities at different levels are made of different physical substrates, a genuine evolutionary conflict can only occur between them. However, in such cases, applying a unique level of description would show that such conflicts are not strictly conflicts *between* levels of selection but rather conflict occurring *in multilevel settings*. Second, when the entities are made of the same physical substrate, I argued that due to the constraint that a higher level is constituted from entities of the lower level, separating the evolutionary fate of the higher-level and lower-level entities is not possible (measured by long-term fitness). As a result, the term conflict can only be understood in a metaphorical sense because it relies on counterfactual scenarios.

Subversion from Within and Conflicts Between the Levels of Selection

One context where the foregoing analysis can be useful is cancer, which is often viewed as a case exemplifying multilevel selection, where the levels of the organism and that of the cancerous cells are said to be in conflict (Lean and Plutynski 2016; Shpak and Lu 2016; Okasha 2021). Cancerous cells proliferate in a way that is detrimental to this organism. However, this idea has been contested by Gardner (2015) and Shpak and Lu (2016), who both note that one issue with the idea of conflict between levels of selection in situations of cancer is that cancer cells typically die with the organisms bearing them; in other words, they are an evolutionary dead end. Okasha (2021) calls this the evolutionary dead-end argument (EDA), and I follow suit.

I partly agree with the proponents of the EDA but disagree with some of their claims. First, cancerous cells might be regarded as physically distinct from the rest of the organism. In such cases, one can regard the cancerous cells and the rest of the organisms as being in evolutionary conflict, if the former is interpreted as A and latter as B, as in Figure 2.2a. When the cancer is transmissible, such as in cases of transmissible cancers in Tasmanian devils and dogs or in rare cases of transmission during pregnancy in mammals, both Shpak and Lu and Gardner agree that a conflict exists. However, they argue that this is not a conflict between *levels of selection* but rather a conflict between two *individuals* (albeit one being parasitic on the other). This argument concurs with my analysis; when the higher-level and lower-level entities can be separated, it is a stretch to say that a conflict between levels exists. It is more accurate to refer to conflict between two types of entities, where each is described at a different level of description.

If the cancer is not transmissible, Gardner (2015) argues that cancerous cells have no reproductive value (a measure of fitness related long-term growth rate). This is so because the cancerous cells die with the organism that bears those cells. Thus, when using the notion of "reproductive value," if he requires that it only applies to situations where evolution is indefinite, he is correct that cancerous cells have no reproductive value. However, as pointed out by Okasha (2021), the same argument could be applied for nearly any entity at any level of organization. This is so because, as famously argued by John Maynard Keynes, "in the long run, we are all dead." When estimating the long-term fitness of an organism in a population, it is often assumed that the species to which this organism belongs does not go extinct. However, we know that species extinction is, in the long run, very likely to occur. Using the same tool but applying it at a different scale, we could end up concluding that the long-term fitness of any organism is nil. Thus, if by "indefinite," Gardner means truly indefinite, it is overly restrictive.

There is nothing preventing us from studying within-organism evolutionary dynamics and applying the notion of long-term reproductive success in this context. However, "long term" here refers to a projection in which no deleterious effects of cancer on the well-functioning of the organism are included, rather than "long term" in the absolute sense. The assumption that there is no deleterious effect on the organism is at the basis of the somatic evolution model of cancer, which applies evolutionary principles to study cancer evolution within an organism (e.g., Vogelstein and Kinzler 1993; Burt and Trivers 2006; Lean and Plutynski 2016; Fortunato et al. 2017; Okasha 2021). During the lifetime of an organism, cell lineages exhibit variation, differences in reproductive output, and heritability, three properties that constitute the core of the Darwinian scheme (Lewontin 1985; Godfrey-Smith 2009). Further, because cells within an organism have an asexual mode of reproduction with near-perfect transmission, one can expect that adaptations will occur if mechanisms preventing the suppression of variation are ineffective, as is thought to be the case in cancer (Burt and Trivers 2006; Aktipis et al. 2015; Shpak and Lu 2016). Seen through this lens, the diversity of cancers points to a number of hallmark adaptations within organisms (Gerlinger et al. 2014; Lean and Plutynski 2016; Shpak and Lu 2016; Fortunato et al. 2017). However, it should be noted here that such within-organism conflict (in line with what I argued earlier) could easily be described as conflict arising between different types of cells—that is, as conflict at a single level—which seems to be the interpretation provided by Burt and Trivers (2006, chap. 11) or Frank (2007), among others.

Another more charitable interpretation of Gardner's position is that the context in which he made his remark was one in which he was assuming (along with other protagonists in this literature) that organisms reproduce. Because legitimate fitness comparisons require that they are measured over the same set of events (Bourrat et al. 2022), the reproductive value of cancerous cells over such a timescale is indeed nil. Thus, if cancer is seen in a context where cancerous cells are regarded as an

integral part of the higher-level entity—that is, as part of the organism—
their fitness should be regarded as nil.

However, this argument does not apply to all forms of cancer. To
see this, take the case of inherited cancers: cancers passed on through
the germline, such as breast and ovarian cancer due to mutations in
the *BRCA1* and *BRCA2* genes. Like their somatic counterpart, these
cancers lead to cell proliferation. Despite having reduced fitness, individuals carrying these mutations might still have the opportunity to reproduce. Similarly, consider a case of cancer in an organism for which
there is no germ-soma separation (e.g., plants), where a mutation can be
passed on to the next generation of plants. In such cases, the EDA, which
rests on the existence of a germ/soma separation and the cancer arising
in the soma, would not work.

Notably, according to Shpak and Lu, inherited cancers or cases of
cancer in organisms without germ-soma separation could nonetheless represent true cases of multilevel selection with a conflict between levels, for
they appear to embrace the idea that fitness at the lower level is high while
fitness at the higher level is low, using the multilevel Price equation (Price
1972; Okasha 2006; Bourrat 2021; see also Patten, this volume), a popular
equation in evolutionary theory for apportioning the effect of selection at
multiple levels of organization and where "higher" and "lower" level
selection are operationalized into "between" and "within" collective selection. Following my analysis, however, caching out the distinction between higher and lower levels of selection in terms of between-collective
and within-collective selection suggests a counterfactual notion of conflict
that is not on par with the traditional notion of an actual evolutionary
conflict between two entities (see Shelton and Michod 2014 for more
details about the distinction between these two ways to understand
individual-level selection). When using the between/within distinction and
arguing that the fitness (which could include inclusive fitness effects) of a
cancerous cell is high while reducing the fitness of the organism of which it
is part, one does not answer the question of whether an actual evolutionary
conflict between two entities A and B (as described earlier) exists. Instead,
one provides an answer to the question of whether the fitness of a cancerous cell *would be* in conflict with that of the organism bearing it if,
when measuring cell fitness, the constraints on the growth and survival of
the organism bearing this cell were eliminated.

If no genuine conflict exists between levels of selection, one might ask
what role(s) using this phrase plays in multilevel selection theory. Before

answering this question, it should be noted that one main criticism of the idea that group selection can occur is that a selfish mutant that would not contribute to the production of collective-level benefits could invade the group and ultimately lead to a situation where there is no group benefit to be shared (see Okasha 2006, chap. 6 for an overview of this controversy). Therefore, whether the invasion of selfish mutants is possible, given a particular setting, is an important aspect of multilevel selection theory (Maynard Smith 1964). However, this idea is quite independent from the idea of conflict between levels, where levels are compositional. Nonetheless, the two ideas have not been neatly separated in discussions about levels of selection.

In a hierarchical system, asking about the emergence or maintenance of a higher level of organization has fundamentally to do with whether the invasion of "selfish" entities that would benefit in the short term from disrupting this level is possible. This can be viewed as a "conflict" (i.e., in a metaphorical sense). There *would be* conflict between the levels if these were independent. However, given the many ambiguities surrounding the notion of levels of selection, I prefer referring to this type of situation as a "paradox," as in the expression "the paradox of the organism" (Dawkins 1990; Patten et al. 2023). The existence of higher-level individuals such as organisms appears paradoxical when some factors are neglected or when those factors play out differently in different contexts. However, this paradoxical nature disappears once these factors are fully accounted for.

My proposal is in part compatible with that of Okasha (2021). Following his previous analysis of the levels-of-selection question (see Okasha 2006), he proposes to distinguish a synchronic (i.e., at a time) and a diachronic (i.e., over time) approach to this question. The diachronic approach permits one to ask questions about the origin or emergence of new levels of organization while the synchronic approach permits one to answer questions about whether selection acts at multiple levels at the same time—and, if so, whether they are in conflict. Okasha, following Buss (1987), makes the case that when seen from a diachronic perspective, organismal level adaptations might have been developed against cancer. While I concur with his analysis, I disagree that this shows a conflict between levels other than in a metaphorical sense. There might have been conflict between different lineages of unicellular organisms and the possibility of building higher-level individuals required mechanisms that allowed for the maintenance of these. However, conflict was not occurring *between* compositional levels. It instead occurred between entities at a *single* level and, for the most part, has been solved in modern multicellular organisms through the emergence of anti-cancer adaptations (which sometimes fail). It is correct that a diachronic perspective permits us to explain the existence of such adaptations, but one can make this point without reference to the notion of conflict between levels of selection. In contrast, talking about a "paradox" appears to be more appropriate, as this implies it can be solved and disappear, whereas a "conflict" is grounded in facts.

At that point, some might want to claim that the difference between a paradox or a metaphorical conflict (as I defined them) is merely terminological, that whether one terms the same phenomenon a "conflict" or a "paradox," the same phenomenon has little importance, so long as one is clear about what they mean. However, the problem here precisely lies in the fact that the ideas surrounding multilevel selection have been applied in a wide variety of contexts and to refer to so many phenomena that some terminological hygiene can only be beneficial.

Conclusion

In this chapter, I provide an analysis of the idea of conflict between levels of selection. The upshot is that the idea of conflict between levels of selection is rather ambiguous and can be understood in different ways. When levels are not compositional but merely ways of describing different types of entities, the notion of conflict makes sense. However, discussing conflicts between levels as if more than one process of selection is happening, each occurring at one level of organization, is suboptimal because in many cases one could redescribe the whole setting using a single level of description. Strictly speaking, when levels of selection refer to compositional levels where higher levels are constituted of entities at lower levels, an evolutionary conflict between them cannot exist because long-term evolutionary interests at the different levels are the same. One way to make sense of the idea of conflict in such situations is to refer to counterfactual scenarios in which describing the evolutionary dynamics at the lower level does not impact the evolutionary dynamics at the higher level because the two levels refer implicitly to different scenarios. However, in such cases, any "conflict" is of a metaphorical nature. Finally, I argue that in situations where such "conflict" would occur, such as when an organism develops cancer, it is more appropriate to refer to a paradox standing to be dissolved: Why, despite the short-term benefit of proliferating in the case of cancer or of remaining a free-living particle in the case of evolutionary transitions in individuality, do we observe the maintenance and emergence of collective-level individuality? Once the ecological and evolutionary mechanisms are understood and the paradox is dissolved, so are the conflicts between levels.

Acknowledgments

I thank Arvid Ågren and Manus M. Patten for their invitation to participate in the workshop in Washington, DC, in 2022 that led to this volume, as well as Ellen Clarke and Amy Boddy for their comments on a previous version of this chapter. This research was supported under the Australian Research Council's Discovery Projects funding scheme (Project no. DE210100303). The author gratefully acknowledges the financial support of the John Templeton Foundation (#62220). The opinions expressed in this chapter are those of the authors and not those of the John Templeton Foundation.

References

Aktipis, C. A., Boddy, A. M., Jansen, G., et al. (2015). Cancer across the tree of life: Cooperation and cheating in multicellularity. *Philosophical Transactions of the Royal Society B, 370*(1673), 20140219.

Alizon, S., Luciani, F., and Regoes, R. R. (2011). Epidemiological and clinical consequences of within-host evolution. *Trends in Microbiology, 19*(1), 24–32.

Bedau, M. A. (1997). Weak emergence. *Philosophical Perspectives, 11,* 375–399.

Black, A. J., Bourrat, P., and Rainey, P. B. (2020). Ecological scaffolding and the evolution of individuality. *Nature Ecology and Evolution, 4,* 426–436.

Bourrat, P. (2015a). Levels of selection are artefacts of different fitness temporal measures. *Ratio, 28*(1), 40–50.

Bourrat, P. (2015b). Levels, time and fitness in evolutionary transitions in individuality. *Philosophy and Theory in Biology, 7,* 1–17.

Bourrat, P. (2021). *Facts, Conventions, and the Levels of Selection.* Cambridge University Press.

Bourrat, P. (2023a). A coarse-graining account of individuality: How the emergence of individuals represents a summary of lower-level evolutionary processes. *Biology and Philosophy, 38*(4), 33.

Bourrat, P. (2023b). Multilevel selection 1, multilevel selection 2, and the Price equation: A reappraisal. *Synthese, 202*(3), 72.

Bourrat, P. (2024). Independence and the levels of selection. *Philosophy, Theory, and Practice in Biology, 16*(3), 11.

Bourrat, P., Doulcier, G., Rose, C. J., Rainey, P. B., and Hammerschmidt, K. (2022). Tradeoff breaking as a model of evolutionary transitions in individuality and limits of the fitnessdecoupling metaphor. *eLife, 11*, e73715.

Brooks, D. S. (2021). *Levels of Organization in the Biological Sciences.* MIT Press.

Burt, A., and Trivers, R. L. (2006). *Genes in Conflict: The Biology of Selfish Genetic Elements.* Harvard University Press.

Buss, L. W. (1987). *The Evolution of Individuality.* Princeton University Press.

Charlesworth, B. (2000). Levels of selection in evolution. *Heredity, 84*(4), 493.

Dawkins, R. (1982). *The Extended Phenotype: The Long Reach of the Gene.* Oxford University Press.

Dawkins, R. (1990). Parasites, desiderata lists and the paradox of the organism. *Parasitology, 100*(S1), S63–S73.

Folse, H. J., and Roughgarden, J. (2010). What is an individual organism? A multilevel selection perspective. *Quarterly Review of Biology, 85*, 447–472.

Fortunato, A., Boddy, A., Mallo, D., Aktipis, A., Maley, C. C., and Pepper, J. W. (2017). Natural selection in cancer biology: From molecular snowflakes to trait hallmarks. *Cold Spring Harbor Perspectives in Medicine, 7*(2), a029652.

Foster, K. R., and Bell, T. (2012). Competition, not cooperation, dominates interactions among culturable microbial species. *Current Biology, 22*(19), 1845–1850.

Frank, S. A. (2007). *Dynamics of Cancer: Incidence, Inheritance, and Evolution.* Princeton University Press.

Gardner, A. (2015). The genetical theory of multilevel selection. *Journal of Evolutionary Biology, 28*(2), 305–319.

Gause, G. F. (1934). *The Struggle for Existence.* Williams & Wilkins.

Gerlinger, M., McGranahan, N., Dewhurst, S. M., Burrell, R. A., Tomlinson, I., and Swanton, C. (2014). Cancer: Evolution within a lifetime. *Annual Review of Genetics, 48*(1), 215–236.

Godfrey-Smith, P. (2001). Three kinds of adaptationism. In *Adaptationism and Optimality,* edited by S. H. Orzack and E. Sober, 335–357. Cambridge University Press.

Godfrey-Smith, P. (2009). *Darwinian Populations and Natural Selection.* Oxford University Press.

Godfrey-Smith, P., and Kerr, B. (2013). Gestalt-switching and the evolutionary transitions. *British Journal for the Philosophy of Science, 64*(1), 205–222.

Haig, D. (2012). The strategic gene. *Biology and Philosophy, 27*(4), 461–479.

Haldane, J. B. S. (1932). *The Causes of Evolution.* Harper & Brothers.

Joseph, S. B., and Kirkpatrick, M. (2004). Haploid selection in animals. *Trends in Ecology and Evolution, 19*(11), 592–597.

Kim, J. (1988). Explanatory realism, causal realism, and explanatory exclusion. *Midwest Studies in Philosophy, 12*(1), 225–239.

Lean, C., and Plutynski, A. (2016). The evolution of failure: Explaining cancer as an evolutionary process. *Biology and Philosophy, 31*(1), 39–57.

Lewontin, R. C. (1985). Adaptation. In *Dialectics and Reductionism in Ecology,* edited by R. Levins and R. C. Lewontin, 65–84. Harvard University Press.

Maynard Smith, J. (1964). Group selection and kin selection. *Nature, 201*(4924), 1145–1147.

Maynard Smith, J. (1988). Evolutionary progress and levels of selection. In *Evolutionary Progress,* vol. 219, edited by M. H. Nitecki, 219–230. University of Chicago Press.

Maynard Smith, J., and Szathmáry, E. (1995). *The Major Transitions in Evolution.* Oxford University Press.

Michod, R. E., and Roze, D. (2001). Cooperation and conflict in the evolution of multicellularity. *Heredity, 86*(1), 1–7.

Moore, J., and Moore, J. (2002). *Parasites and the Behavior of Animals.* Oxford University Press.

Okasha, S. (2006). *Evolution and the Levels of Selection.* Clarendon Press.

Okasha, S. (2021). Cancer and the levels of selection. *British Journal for the Philosophy of Science,* 716178.

Patten, M. M., Schenkel, M. A., and Ågren, J. A. (2023). Adaptation in the face of internal conflict: The paradox of the organism revisited. *Biological Reviews* 98(5), 1796–1811.

Price, G. R. (1972). Extension of covariance selection mathematics. *Annals of Human Genetics, 35,* 485–490.

Queller, D. C., and Strassmann, J. E. (2018). Evolutionary conflict. *Annual Review of Ecology, Evolution, and Systematics, 49*(1), 73–93.

Rainey, P. B., and Kerr, B. (2010). Cheats as first propagules: A new hypothesis for the evolution of individuality during the transition from single cells to multicellularity. *BioEssays, 32,* 872–880.

Rainey, P. B., and Kerr, B. (2011). Conflicts among levels of selection as fuel for the evolution of individuality. In *The Major Transitions in Evolution Revisited,* edited by B. Calcott and K. Sterelny, 141–162. MIT Press.

Ratcliff, W. C., Herron, M., Conlin, P. L., and Libby, E. (2017). Nascent life cycles and the emergence of higher-level individuality. *Philosophical Transactions of the Royal Society B, 372*(1735), 20160420.

Shelton, D. E., and Michod, R. E. (2014). Group selection and group adaptation during a major evolutionary transition: Insights from the evolution of multicellularity in the volvocine algae. *Biological Theory, 9*(4), 452–469.

Shpak, M., and Lu, J. (2016). An evolutionary genetic perspective on cancer biology. *Annual Review of Ecology, Evolution, and Systematics, 47*(1), 25–49.

Taylor, D. R., Zeyl, C., and Cooke, E. (2002). Conflicting levels of selection in the accumulation of mitochondrial defects in *Saccharomyces cerevisiae*. *Proceedings of the National Academy of Sciences of the United States of America, 99*(6), 3690–3694.

Tsuji, K. (1995). Reproductive conflicts and levels of selection in the ant *Pristomyrmex pungens:* Contextual analysis and partitioning of covariance. *The American Naturalist, 146*(4), 586–607.

Vogelstein, B., and Kinzler, K. W. (1993). The multistep nature of cancer. *Trends in Genetics, 9*(4), 138–141.

Wade, M. J. (2016). *Adaptation in Metapopulations: How Interaction Changes Evolution.* University of Chicago Press.

Wilson, D. S. (1975). A theory of group selection. *Proceedings of the National Academy of Sciences of the United States of America, 72*(1), 143–146.

Wilson, D. S., and Sober, E. (1994). Reintroducing group selection to the human behavioral sciences. *Behavioral and Brain Sciences, 17*(4), 585–608.

CHAPTER THREE

Internal Conflicts That Organisms Can Live With

MANUS M. PATTEN

ABSTRACT   Most organism concepts are rooted in cooperation. For example, Okasha's stance that a unity of purpose is required for organismal agency, Grafen's notion of the individual as a fitness-maximizing agent, and the major-evolutionary-transitions-in-individuality paradigm each depend on high levels of cooperation within the organism and an absence of internal conflicts. But internal conflicts abound in organisms. In this chapter I ask whether these organism concepts, which portray organisms as agential individuals, can survive in the face of such internal conflicts. I argue that there are two fundamentally different kinds of internal conflicts, and only one of them is a genuine threat to these organism concepts. Conflicts that take place within the organism over the relative success of its parts can be harmful to an organism's fitness, but such disputes have no bearing on the organism's goals otherwise, leaving these organism concepts intact. In contrast, internal conflicts over an organism's development—where the dispute is over the organism's eventual traits—challenge the notion of an organism as an agent and may hinder transitions in individuality.

Introduction

Organisms are made of parts, but not in the same way that machines are made of parts (Riskin 2016). With a machine, the parts are inanimate objects. These parts, just like the machine itself, cannot be thought to possess any kind of agency. The only agents in the vicinity of a machine are the machine maker and the machine user. But organisms are assembled differently. A metazoan, for example, is made of thousands or millions of cellular parts. If we were to travel back in time almost one billion years we would see how homologues of these parts were once free-living cells. Unlike the parts of a machine, we have reason to attribute agency both to the metazoan and to the formerly free-living parts of which it is now composed. Similarly, an organism's genome comprises thousands of genes, but a case can be made for the separate agency of some of them, specifically selfish genetic elements (Okasha 2018).

A machine's designer is likely to ensure that all the parts work together to bring about the best-functioning machine. This should be easy, insofar as the only interests that are being served are those of the designer. The parts of the machine, as noted, don't have interests of their own. But for organisms, there is no such guarantee. The fact that multicellular organisms comprise agential or once-agential parts opens up the possibility for internal conflict.

And because every major evolutionary transition in individuality involves the coming together of parts that were once agential (Maynard Smith and Szathmary 1997; Queller 1997; West et al. 2015), these internal conflicts should be ever present in organisms. Tempting as it may be to analogize organisms and machines, there is danger in that analogy, for it forecloses the possibility of internal conflict and its several consequences. Many organism concepts likewise overlook or downplay the threat from internal conflicts.

In stark contrast, Dawkins's (1990) paradox of the organism confronts this reality. He identified the risk from internal conflicts and selfish elements and asked how it was that organisms—such seeming paragons of cooperation—could persist despite the threat of being torn apart from within. While some chapters in this volume (e.g., Scott and West, chap. 1; Haig, chap. 6) address organismal persistence in developmental time— that is, how token organisms survive and cope with the potential harm from their selfish elements—here I'll be interested in the longer-term persistence of the organism as a type of organization, the very concept of the organism. In what follows I offer what I think represents at least a partial solution to this latter view of the paradox by highlighting that not all internal conflicts are a threat to organisms in this sense.

Two Kinds of Internal Conflicts

A first step is to appreciate that internal conflicts are not a single kind of thing. Recently, Patten et al. (2023) elaborated a fundamental distinction first highlighted by Scott and West (2019) between selfish elements that they termed "trait distorters" and "transmission distorters." Both kinds of elements fail to contribute to the organism's overall goal of fitness maximization, hence their selfishness, but they do so in distinct ways.

"Transmission distorters" were described by Patten et al. (2023) as analogous to stowaways on an airplane. Stowaways benefit from being aboard the plane, but they haven't paid for their ticket. Some selfish elements have this quality. A prime example of a transmission distorter is a homing endonuclease gene (HEG) (Stoddard 2005). When in a heterozygous individual, the homing allele of a HEG causes a double-strand break in the DNA sequence of the non-homing allele. The breakage of that non-homing allele is certainly not one of the goals of the organism. But upon breaking, the DNA repair machinery seeks out a template for repair, which it finds in the homing allele. The homing allele's DNA sequence is then copied over onto what had been the non-homing allele, and the formerly heterozygous genotype becomes homozygous for the homing allele. It's a devilishly simple trick, and the homing allele is able to spread to high frequency on the strength of this trickery. In nature, HEGs tend not to harm their host's fitness (Burt and Koufopanou 2004). They cost the organism a few ATP and nothing more (though synthetic HEGs, or gene drives, are being designed in the lab to deliver harm to their hosts and possibly even extinction of entire populations, e.g., Burt and Crisanti 2018). But, like stowaways, they do not pay their fair share. It is in this sense that they do not contribute to their organism's goals and are therefore selfish.

"Trait distorters" were analogized to hijackers on an airplane. Whereas a stowaway is perfectly happy to accept a ride to wherever the plane is going, hijackers have a different destination in mind. Trait distorters influence the organism's phenotype, its developmental destination, in such a way that the element benefits and the organism suffers. A HEG does no such thing. The organism has no interest in whether the homing or non-homing allele is transmitted. That is of no concern to the organism, and swapping the genotype at that locus leaves organismal phenotype and fitness virtually unchanged. In contrast, an imprinted gene in a mammal or plant may have a different developmental outcome in mind from an unimprinted gene. Indeed, the best explanation for why there is imprinting at all (Haig 2000) posits that imprinting arises as a resolution to a conflict between the two haplotypes over what the eventual phenotype of the organism ought to be. Thus, imprinted genes, like hijackers, attempt to divert the organism to a different developmental destination. It is in this sense that trait distorters do not contribute to the organism's goals; trait distorters' goals run counter to the goals of the organism writ large, hence their selfishness.

Another way to clarify this distinction between trait and transmission distortion is to pay attention to who disagrees with whom. In transmission distortion, it is a matter of part versus part. In the HEG example, the conflict resides within the individual and between the two alleles within the HEG locus. The homing allele and the non-homing allele disagree over what fraction of gametes ought to contain the homing allele. The organism, note, is disinterested, for its fitness is entirely unaffected by this within-individual battle for transmission. Trait distortion, on the other hand, boils down to part versus organism. Here the interests of the part, an imprinted gene in our example, and the interests of the organism, taken to be a reference gene (*sensu* Fromhage and Jennions 2019) or a non-imprinted gene, are not aligned. Trait distortion is also considered an internal conflict, but note that it does not entail any within-individual selection.

To make this last point clearer, I'll introduce a formalism that has been heavily used in discussions of conflict, levels of selection, agency, and the major transitions in evolution—all topics that will be visited later in this chapter. The Price (1970) equation (Box 3.1) is a mathematical device that equates total evolutionary change of some quantity with the sum of two components, the effects of what are typically viewed as "between-individual" selection and "within-individual" selection (but see Okasha and Otsuka 2020, who question whether such a neat partition actually exists). The way Price arrived at his equation was by recognizing first that individuals may vary in how many descendants they leave, owing to the traits they exhibit, and second that these descendants might vary systematically from their ancestors in the traits they inherit (Frank 1995; Price 1995). The first kind of variation is captured in the first term on the right side of the Price equation, the covariance term, which associates an ancestor's phenotype with its fitness, or number of descendants. The second kind of variation is captured in the expectation term, which measures how much, on average, descendants differ from their ancestors owing to a variety of factors that have been labeled "transmission." One way that these transmission effects may manifest is that the parts of the organism engage in a competition for passage to the next generation.

Not all internal conflicts stem from selfish *genetic* elements. We can also decompose organisms into cell lineages, which may similarly harbor qualities of trait and transmission distortion. Consider the kind of organism that regularly develops as a chimera of unrelated cell types. A transmission-distorting cell lineage in this chimera may aim to over-replicate within the organism and become more numerous, especially in the germline, such that the descendants are enriched relative to the ancestor for that cell lineage. This will be registered in the Price equation treatment as a positive expectation term. Such jockeying for numerical

BOX 3.1

The Price Equation and Internal Conflict

The Price equation partitions the total evolutionary change of the average value of some quantity, z, into what is typically interpreted as between-individual selection and within-individual selection. A complete derivation of the Price equation, which can be found in Gardner (2008), is beyond the scope of this chapter. My sole interest here is in showing how the Price equation captures our two kinds of internal conflict: trait distortion and transmission distortion. The Price equation looks like this:

$$\Delta \bar{z} = Cov(w,z) + E(w\Delta z)$$

where w represents the fitness of an individual taken relative to the population average individual fitness, and z is measured on each individual. The Δz of the expectation term captures a change that takes place between an ancestor and its descendants.

For example, if we take z to measure the concentration of a homing allele in an individual's genotype (taking values of 0%, 50%, and 100%), then $\bar{z}$ gives the population allele frequency. The expectation term would be non-zero in this case, because homing alleles effectively convert heterozygous parents ($z = 50\%$) to homozygotes, such that all their offspring inherit the homing allele ($z = 100\%$). The homing alleles can be thought to have higher genic fitness within those heterozygous bodies. Had z instead been taken as the concentration of an allele of an imprinted gene, this expectation term would be zero. Parents with a given concentration of that imprinted allele transmit that allele in the exact same proportion, and the Δz term is therefore zero.

The covariance term will be non-zero provided there is some connection between z, a trait measured on an individual, and individual fitness. If, for some reason, being homozygous for the homing allele reduces fitness for such individuals, we would see a negative covariance. In the case of naturally occurring homing endonucleases, we often find virtually no fitness effect at the individual level and so a zero covariance.

The same may not be true of other sorts of transmission distorters, such as the *Segregation Distorter* complex of *Drosophila*, which drives during male spermatogenesis (Larracuente and Presgraves 2012). A Price equation partition would show the expectation term to be positive, as the driving allele is overtransmitted to offspring of heterozygous males, and the covariance term would be negative, as males who carry the driving allele have reduced fertility and viability.

representation may involve trade-offs with organismal fitness (e.g., cancer) such that the covariance term is at the same time negative, but this isn't a necessary feature of this kind of transmission distortion. Organismal fitness—that is, the number of descendants the organism leaves—may be entirely unaffected by competition within (i.e., a zero for the covariance term), and may even be enhanced if this competition leaves the organism with fitter, healthier cells (i.e., a positive covariance term—in which case we would want to avoid calling this a "selfish" element, as this is more like the *Kampf der Theile*—the struggle of parts—that Roux (1881; in Haig's [2024] translation) described as being of benefit to the organism).

A trait-distorting cell lineage in a chimera takes a different approach in achieving its goals. For example, if a cell lineage knows that it is the minor cell lineage in a chimera, it may achieve a higher inclusive fitness by nudging its bearer toward a more altruistic stance, generously conferring benefits on copies of itself in other, neighboring bodies. This doesn't affect the transmission of that cell lineage through its bearer, and so the expectation term of the Price equation for this example would be zero. But this is no doubt an internal conflict, as the major cell lineage of that generous chimera suffers the costs of such altruism.

The trait- versus transmission-distortion distinction shows that not all internal conflicts are reducible to within-individual selection. As we will see, the Price equation has been put to heavy use in discussions of agency and major evolutionary transitions in individuality, and internal conflicts loom large in all these discussions. Too often, though, internal conflict has been equated with within-individual selection (i.e., a non-zero expectation term).

Organisms as Agents: Threats from Internal Conflicts

Let us agree, at least for argument's sake, that organisms appear as though they have a goal: to maximize their fitness. Taking that as the observation raises some questions. First, is it appropriate to talk agentially like this about organisms, and if so, is it always appropriate? Second, is organismal agency the product of natural selection? That is, when we observe an organism appearing to pursue the goal of fitness maximization, can we say that its goal-directedness was installed there by natural selection?

The first of these two questions has been tackled most prominently by Samir Okasha in his 2018 book *Agents and Goals in Evolution,* and the second of these by Alan Grafen in his Formal Darwinism Project (Grafen 1999, 2006, 2007, 2014). As we will see, both authors accept the possibility of organismal agency, but only when internal conflicts are absent. In the section "Different Kinds of Internal Conflict and Their Implications for Agency and Major Transitions in Individuality" I will clarify what they might mean by "internal conflict" and examine the consequences for their notions of agency, but first I want to show how they arrive at their positions.

Okasha on Organismal Agency

Okasha observes that practicing biologists regularly use agential language in talking about both the evolutionary process (e.g., "mother nature wants . . ."), which he dismisses as illegitimate, and about organisms (e.g., "the meerkat wants . . ."), which he deems both legitimate and useful. He argues that although parts of the organism have functions, it is only whole organisms that have an ultimate goal—namely, fitness maximization (for a formal justification of this, see the section below on Grafen's Formal Darwinism Project)—thus establishing the legitimacy and utility of agential thinking about organisms. Put another way, because an organism is made of many parts with different jobs, but because all of these different jobs contribute to the organism's ultimate goal, it is both acceptable *and useful* to talk about organisms as agents.

A key element of this argument is that all the parts agree on the organism's ultimate goals. There is what Okasha calls a "unity of purpose" to the parts, without which agency breaks down. For comparison, consider a hardware store. The parts of a hardware store, the tools for sale, all have functions, but there is no sense in which they have a unity of purpose. The various tools don't all further the store's ultimate goal of maximum handiness or some such thing. These tools did not evolve in service of any single ultimate goal. Hence, hardware stores are not agents. However, the parts of an organism are different, in that all (or almost all) have been adapted for serving the organism's ultimate goal. This unity of purpose is what gives organisms their agency.

The sticking point comes when internal conflicts are taken into consideration. For not all parts of an organism conduce to the organism's goal, most notably selfish genetic elements. Okasha (2018, 32) declares that talk of organisms as agents is threatened by the presence of internal conflicts and points to two specific examples: cytoplasmic male sterility in plants and spermatogenic meiotic drive in fruit flies. Here, the parts of an organism, selfish genetic elements in both cases, do not contribute to the organism's goal of fitness maximization. Internal conflicts brought on by such genes are a problem for organismal agency in Okasha's view. In the final section, I will show that his two example cases are, in fact, different kinds of internal conflicts, and these different kinds of internal conflicts pose different threats to organisms and organismal agency.

Grafen on the Individual-as-Maximizing-Agent Analogy

If natural selection is the process, then what is its product? Several options suggest themselves as answers. For one, an increase in the average fitness of individuals. This is Fisher's (1930/1999) fundamental theorem, and despite some uncertainty over what Fisher meant exactly (Plutynski 2005), we now have formal demonstrations that natural selection does indeed have an improving tendency on mean fitness (Price 1972; Ewens 1989; Grafen 2021). Second, natural selection makes functional traits well fit to their environments. This is how the leopard gets its spots and the zebra its stripes. And it is simple enough to model a process whereby a population goes from being spotless to spotted or plain to striped. Third, natural selection produces organisms that function agentially. That is, it molds organisms in such a way that they have goals and purposes and they confront their world and do things in an attempt to maximize some "objective function."

This third option is taken as a given by most behavioral ecologists and likely by most practicing evolutionary biologists as they go about their work. And, in fairness, one can be forgiven for thinking that we have formal proof that this is what natural selection produces, because so much of evolutionary biological research is motivated by this view. But until Grafen (1999) raised awareness of the fact, it actually had no formal justification. We had no basis to claim that natural selection would give us individuals that behave as though they are trying to maximize something. Natural selection delivers the first two outcomes, and there were formal proofs in the language of population genetics to demonstrate these results. But the third was nowhere to be found in population genetics. The Formal Darwinism Project was an attempt to clarify how a process like natural selection, which we model and understand with population genetics, may lead to organisms that behave as maximizing agents (Grafen 2014).

To arrive at the conclusion that organisms should behave as maximizing agents, Grafen specifies some conditions that must be met if there is to be a correspondence between individual optimization (i.e., agency) and population genetics. These conditions are summarized by Okasha (2018) as:

1. If all individuals are optimal, then selective equilibrium obtains.
2. If all individuals are equally suboptimal, there is no scope for selection (because of the lack of variation) but there is potential for selection.
3. If individuals vary in their optimality, the change in the frequency of any gene equals its covariance with the individuals' attained value of the objective function (i.e., fitness).
4. If selective equilibrium obtains, then all individuals in the population are optimal.

Grafen points out that the correspondence may fail if there is internal conflict, specifically in the form of within-individual selection. Grafen relies on a Price equation formulation of population genetics, which partitions the overall change in some quantity, taken by Grafen to be the allele frequency, into components due to between-individual selection and within-individual selection (the covariance term and the expectation term of the Price equation, respectively). The canonical example of a gene that experiences selection in both ways is a meiotic driver, such as the *Segregation Distorter* complex in fruit flies (Larracuente and Presgraves 2012). Having a genotype with one of these driving alleles leads to reduced fitness for its bearer, but it also means that the gametes are overrepresented for driving alleles. The individual fails to be a maximizing agent in that both conditions 3 and 4 are not met.

Thus, Okasha and Grafen arrive at the same conclusion: If organisms are to be agents, they must be free (or virtually free) of internal conflicts. But as we found with Okasha's conclusion before, care was not taken in defining what exactly "internal conflict" referred to.

Organisms as Products of Evolution: The Major Evolutionary Transitions in Individuality

From the Modern Synthesis until the 1980s, roughly speaking, evolutionary theory encompassed a body of work devoted to understanding how organisms change. Questions about how leopards get more spotted and how zebras get more striped became tractable after the rediscovery of Mendel, and so much of the Modern Synthesis and what followed reads like a simple vindication of Darwin's ideas (Boomsma 2023). But starting in the 1980s, first with Buss (1987) and later with Maynard Smith and Szathmary (1995), the question of how organisms came to be in the first place became a focus. (Boomsma and Gawne [2018] detail some of the antecedents of these ideas from the early twentieth century, a period in the history of evolutionary thought where talk of superorganisms and hierarchical selection first gained traction.) The individuals of the earlier questions, leopards and zebras, were taken as a given. But, of course, their kind of individuality—multicellularity—is a derived state, a relatively recent invention in the nearly four-billion-year history of life. The question of how single-celled organisms evolved to become multicelled organisms, and more generally of how new kinds of individuality evolve, started to enter the field.

Several decades later we have a general sense of what needs to happen during these so-called major evolutionary transitions in individuality. Lower-level individuals (e.g., cells) must cooperate in forming collectives (e.g., multicellular organisms) that pursue the collective's goals. An impediment to achieving this coordination is that the parts, as they initially come together, are agents in their own right, set on maximizing their own fitness, and have no responsibilities to one another and no sense of a collective goal. A major evolutionary transition in individuality therefore entails the aligning of the parts' goals with that of the collective's.

This sounds quite similar to what Okasha has described as a unity of purpose. Once a major transition has taken place it no longer makes sense to treat each part as having an ultimate goal, for all of the component parts now conduce to the same ultimate goal of the collective; it makes more sense to speak of the collective as having the ultimate goal and each part as helping to pursue some intermediate goal along the way toward it.

Michod (2000) has described this process as an export of fitness from the parts to the collective. In effect, the parts forgo pursuing their own fitness maximization and instead assign themselves the task of contributing to the fitness maximization of the collective. If this were not to happen, the collective would consist of parts that competed with one another for representation within the collective. This competition would be of no benefit to the collective's goals, presumably.

The Price equation is again helpful in thinking one's way around this topic. Effectively, what a major evolutionary transition in individuality has been thought to achieve is the diminution of the expectation term—that is, the elimination of within-individual selection—in favor of an enhancement of the covariance term (Michod and Roze 1997). This would seem to be equivalent to the statement that internal conflicts must be absent.

But, as discussed already, internal conflicts come in a variety of forms, only some of which entail within-individual selection. Past discussion of major evolutionary transitions in individuality has too hastily equated internal conflict with within-individual selection. In the next section, we'll see the progress in understanding major evolutionary transitions in individuality that can come from recognizing the distinction between trait and transmission distortion.

Different Kinds of Internal Conflict and Their Implications for Agency and Major Transitions in Individuality

In this section, I want to show how an appreciation for the different kinds of internal conflicts—transmission and trait distortion—can clarify concepts of agency and major evolutionary transitions in individuality. Two questions will be addressed:

1. Is the unity of purpose undone by both kinds of internal conflict? Is it undone to different extents? Which form of internal conflict is the greater threat to Okasha's and Grafen's notions of organismal agency?
2. What kind of internal conflicts can impede the major evolutionary transitions in individuality? If a major transition takes place, what then are the threats to the newly formed individuals? Which form of internal conflict is more likely to spur "major reversions" in individuality?

Internal Conflicts and Agency

The claim I'll advance here is that pure transmission distortion, in an idealized form, is never a threat to organismal agency. But trait distortion always is.

First, addressing Okasha's unity of purpose criterion, we can evaluate whether a transmission distorter qualifies as a threat to the unity of purpose. Let's take one of the examples that Okasha himself considers: a spermatogenic meiotic driver. In this case we have an allele that causes heterozygotes to kill or scuttle the half of sperm that fail to inherit it after meiosis. This typically leaves males less fertile, and further, many such spermatogenic drivers are accompanied by linked deleterious effects that are unrelated to their sperm killing. Undeniably, these are harmful to the males that carry them and some would point to the differing signs of the two terms of the Price equation as evidence that there is an internal conflict here, just the sort of thing that might threaten agency.

But I do not think a spermatogenic driver should qualify as something that confers a *disunity* of purpose. First, what segregation ratio at this locus is in the organism's interest? (Note that this is different from asking whether in general a 50:50 segregation ratio is adaptive for organisms.) That is, what percent of sperm carrying the driving allele maximizes the organism's fitness? The answer is . . . ? The organism does not in fact "care" what segregation ratio is achieved. Selection operates at a level below the organism, and consideration of its outcome is, so to speak, *beneath* the organism. If anything, this trait is not really a trait of the organism so much as a trait of the part. The organism neither benefits nor suffers under any particular segregation ratio; the two alleles at the locus are entirely fungible from the organism's perspective. The driving allele, however, maximizes its fitness when it appears in 100% of sperm. That is the element's ultimate goal. But because the organism lacks any clear goal for this, there can be no disalignment of interests and therefore no disunity of purpose.

Of course, one might say that I'm ignoring another aspect of drive: subfertility. A separate question then is how fertile the organism ought to be. If we were to ask the selfish element how fertile the organism ought to be, it would undoubtedly respond, "maximally fertile." That spermatogenic drivers often leave their carriers subfertile is simply a constraint of how so many of them operate. If they could influence the segregation ratio and leave fertility unaffected, they would fare better than if they harmed fertility. The harm to fertility is not part of the element's strategy per se. So, for this trait, which is a trait measurable on the organism itself, there is unity of purpose. Both the part and the collective agree on what the optimal value of the trait is, even if the element struggles to achieve that goal. In Grafen's Formal Darwinism terminology, if we were to let the organism or the element choose the value of its instrument with an eye to maximizing its objective function, the choice would be the same for both the element and the organism. Grafen and colleagues have overlooked the possibility of mutations arising that simultaneously maintain the organism's optimum but that still permit within-individual selection. A pure transmission distorter—that is, a selfish element that spreads by transmission distortion alone and leaves organismal traits and fitness unaffected—may be difficult to obtain, given all the many trade-offs possible in development. But the hypotheticals described above are meant to show that pure transmission distortion is no problem for Okasha's unity of purpose or for Grafen's correspondences.

I should underline that in the foregoing I have not claimed that spermatogenic drivers are not harmful to fitness, nor that we should not view them as a potent force in evolution. They are absolutely both of these things. But for them to undermine the agency of the organism, they have to do something additional: have conflicting phenotypic optima and then wrest control of these phenotypes from the organism. In other words, they need to be trait distorters.

Another way to make the case that transmission distorters are no threat to agency is to conjure up some examples inspired by natural selfish elements, and then use these to test whether we think the organism's unity of purpose has been disrupted or whether Grafen's correspondences fail.

In the stalk-eyed fly there is a driving X chromosome that achieves its overtransmission by sabotaging Y-bearing sperm in males, much like the aforementioned spermatogenic driver (Presgraves et al. 1997). One might reasonably expect that a male with the driving X should be half as fertile as a male with an ordinary X. But, rather interestingly, the driving X either causes or is associated with genes that confer larger testes, and so more sperm, to compensate for the sperm killed by drive (Meade et al. 2020; Bradshaw et al. 2022). Consequently, these males are no less fertile than ordinary males.

What is notable about this case is that there is no automatic trade-off between selection within and selection between males. My suspicion is that built-in trade-offs—which are common in the sorts of selfish elements we find in nature (because such trade-offs may keep them from reaching fixation)—has led to the notion that transmission distortion *necessarily* entails changes to the organism's phenotype or fitness. (In fairness, the stalk-eyed fly with the driving X may be as fertile as a standard male, but it appears that compensating for lost sperm comes at the expense of eye span, a sexually selected trait (Meade et al. 2020). Homing endonuclease genes achieve their transmission distortion for the price of just a few ATP, a cost that is so minimal as to likely fall below the detection of between-organism selection. Similarly, meiotic drive in females, where the driving gene's strategy is to find its way to the egg and to avoid the polar body, is achievable without bringing any associated cost to the female's fertility, at least in principle, and without any additional expenditure of resources. And even some spermatogenic drivers, as the stalk-eyed fly example shows, may nonetheless avoid bringing harm to their bearer's overall fitness. The view this suggests is that the dream of all transmission distorters is to be meticulous and precise: to achieve biased transmission, but to do so without any harmful effects at the organismal level.

In contrast, any trait distorter, regardless of how subtle its effect on the organism's traits, is a genuine threat to agency. Take for example an imprinted gene. According to Haig's kinship theory (1997, 2000), an imprinted gene is selected for its effects from one parental origin only, and as such should evolve toward that parental origin's optimum. Genes of paternal and maternal origin are expected to have slightly different optima for traits under parent-offspring conflict (e.g., the allocation of resources from maternal bodies or from other kin). If the parental-origin optima for the amount of resource transferred differ by a single gram of resource, we can expect that maternally expressed imprinted genes are selected to extract one less gram than what the paternally expressed imprinted genes are selected to extract. For unimprinted genes influencing resource transfer, we expect them to be selected for something in between. Thus we have three different goals, differing by no more than a gram of resource, all wrapped up in the same genome.

Such genes may be expressed in different proportions in different tissues of the body. If some tissue that can affect resource transfer is dominated by paternally expressed imprinted genes, then it would appear as though its goal was to extract more resource than a tissue dominated by maternally expressed imprinted genes. Consequently, we could have tissues or organs with different apparent goals.

Though subtle, this is undeniably a disunity of purpose. Using Grafen's terminology, the different genes of the genome and the different tissues of the body they underpin choose different values of the instrument to maximize their objective functions. That which maximizes a paternally expressed imprinted gene's fitness is different from that for a maternally expressed imprinted gene, which both differ from that of an unimprinted gene.

However, the conflict between these parties does not reduce to within-organism selection. Imprinted genes follow Mendel's laws. The expectation term in the Price equation for the situation just described is zero. The conflict in this case is taking place entirely at the organismal level, and not at a level below.

Calling back to another of Okasha's chosen examples of internal conflicts, another trait that may be distorted for the benefit of some parts but not others is the sex of the organism. Consider a cytoplasmic element, such as a mitochondrion or some other endosymbiont, that is transmitted through female reproduction only. The other genes of the organism that hosts this endosymbiont are transmitted biparentally. The cytoplasmic element is in conflict with these other genes over sex determination, for although the other genes may be selected to deliver a 1:1 sex ratio, the cytoplasmic element gains an advantage when the organism develops as a female. Thus, the cytoplasmic element has as its goal to turn its host female. Other genes should have no such goal and should, in the long run, even be selected to avoid assigning their bearers to the more common female fate (Leigh 1977; Gardner 2022). These sex-ratio distorters gain no transmission advantage within the organism, for this would require something we don't have here: heteroplasmy. They gain their selfish advantage by wresting control of development from the rest of the genes in the organism.

Internal Conflicts and Major Evolutionary Transitions in Individuality

In this section, I'll advance the claim that major evolutionary transitions in individuality do not necessarily need to eliminate or overcome within-individual selection. I see no reason why transmission distortion can't accompany a major transition.

I suspect that the fear of transmission distorters stems from an implicit assumption that what is good for the part must be bad for the collective, and vice versa (a collective action problem). The difficulty of evolving cooperative behaviors is often framed this way: the incentive within a collective is to cheat or free-ride, but the best thing for the collective is for its constituent parts to cooperate. It's as though there is a single dial that governs both how a constituent part will behave and how the group will fare. One such trait that fits this single-dial model is motility. Because the centriole can be used either for flagellae, which are needed for motility, or for the spindle apparatus, which is needed for dividing the cell, each cell of a collective faces a choice between differentiating as a motile cell or as a replicating cell (Koufopanou 1994).

Replicating is in the cell's interest, but motility is in the interest of the collective. A selfish cell will opt for replication over motility.

But note that the situation here is much like the situation for segregation distorters mentioned before. If we could ask a single cell residing in one of these collectives what amount of motility is best for the collective, it would likely agree with the collective that a high level of motility is best. And if asked what ratio of yourself relative to others it would like to see within the collective, it would give the selfish answer: 100% of self. Only because there is a mechanistic trade-off between motility and replication do we think that over-replication is a threat to the major transition. But there is nothing about over-replication per se that is threatening to a major evolutionary transition in individuality. If cells could compete for representation within the collective but leave the collective's goals untouched, they would. There is no disagreement between the cells and the collective over the optimal value of the collective trait: both agree that the collective should be motile.

What couldn't possibly work logically, though, would be for some part of the collective to prevent the collective from achieving *the collective's* fitness goal. If trait distorters accompany a major transition, and if they're potent enough to shift the collective away from its fitness goal, then there is no point to the collective's existence and there is no hope that the collective will have fitness enough to maintain itself over evolutionary time. For example, if cells within the collective could experience higher fitness if the collective were less motile, then these cells would be selected to achieve that goal, which runs counter to the goal of the collective. This trait distorter would represent a genuine case of disunity of purpose.

In the discussion of major evolutionary transitions in individuality, many of the transmission distorters that come to mind (e.g., a selfish cell that never differentiates as a motile cell) also imply changes to collective-level traits. The point I have made here is that their transmission-distorting effects are not the real threat to achieving major transitions. It is only their effects on collective-level traits (i.e., trait distortion) that can thwart a major evolutionary transition in individuality.

The Paradox of the Organism

The title of this volume is taken from a paper by Richard Dawkins:

> The paradox of the organism is that it is not torn apart by its conflicting replicators but stays together and works as a purposeful

entity, apparently on behalf of all of them. Not only is it not torn apart; it functions as such a convincingly unified whole that biologists in general have not seen that there is a paradox at all! (Dawkins 1990, S64)

In my view, Dawkins's concern for the organism was perhaps overstated. Not all of the organism's conflicting replicators have the potential to tear it apart. But to be clear, I am not saying that selfish replicators aren't harmful in a general sense. Spermatogenic drivers and cancerous tumors are, almost without exception, harmful to their bearers' fitness, and many selfish elements are known to reduce fertility and viability. But a loss of fitness is not the same as a loss of agency or individuality. Plenty of things external to the organism can harm an organism's fitness. But we wouldn't worry that an organism wasn't an agent or an individual on account of a harsh environment or an infection. The criterion for determining whether organisms lose their agency and individuality is that they lack (or lose) a unity of purpose.

Can organisms ever have such a disunity of purpose that they lose their grip on agency? And can major evolutionary transitions in individuality ever be reversed? I think there is a positive answer to this question, and the place to look is most certainly trait distortion. Pure transmission distorters cannot on their own threaten either of these things. Transmission distorters may bring all manner of harm to an organism's fitness, but their designs for transmission success affect traits that sit below the level of the organism's concern, and as such represent nothing more than a distraction from understanding organismal agency and major evolutionary transitions in individuality.

A separate and related question that Dawkins poses in his paradox is why there aren't more selfish elements. One potential answer is that all transmission distorters secretly long to be as precise as HEGs, and if they could achieve this, they would have done so already. And in that case they would fall below our notice. That is, although they would contribute nothing to the organism, they would at least not cause any noticeable harm to fitness. They would be well-behaved stowaways. Transposable elements come to mind here. Despite being a majority of some genomes by volume, they have fairly low fitness effects, and we tend not to think of organisms that harbor a large transposable-element content as less organismal. It's only the transmission distorters that cause some negative effect on organismal fitness, the badly behaved stowaways, that get hung up in polymorphism and rise to our notice. It is quite possible that the vast majority of transmission distorters reach fixation without detection. The kinds of transmission distorters that harm organismal fitness cannot spread far and wide. And the kinds of transmission distorters that likely do spread far and wide—perhaps more than we yet appreciate—don't tear organisms apart from within (Scott and West 2019).

So then why aren't there more trait distorters? Or, more crucially, why aren't there more *very harmful* trait distorters? There is a null resolution to this: If the internal conflict from such elements were so severe, you wouldn't have an organism within which to observe such conflict. In other words, organisms that harbor severe trait distortion get torn apart, and by a process of species selection, all that remains are those organisms with at worst mild trait distortion. If ever there were an organism that experienced this tearing apart, we would want to label it a "major reversion" in individuality. Such major reversions might be detectable with a comparative approach, where for example, one may find a unicellular species nested among a clade of multicellular organisms or a solitary species of insect nested in a clade of otherwise eusocial insects. It would seem these patterns are rare in nature, which suggests there are mechanisms in place to quash trait distortion before it ever rises to the level of threatening individuality (Scott and West 2019).

On this view, there may be many organisms that simply reflect a compromise between the conflicting agendas of the parts and the collective writ large (Haig 2014). Provided their goals and optima aren't too far apart, the organism muddles through—not wholly unified in its purposes and not exactly achieving its optimum, but not completely torn apart at least.

Conclusion

Internal conflicts take two forms: transmission distortion (i.e., part vs. part); and trait distortion (i.e., part vs. collective). Concerns about transmission distortion (i.e., within-individual selection) stripping agency from organisms or preventing major evolutionary transitions in individuality are overblown. Much has been made of the expectation term of the Price equation in these matters, but selection below the level of the organism is truly *beneath* the organism—that is, of no concern to it. Only when there is trait distortion, when parts and collectives disagree on the goals of the collective, do we have the sort of conflict that can threaten agency, thwart a major transition, or even cause a major reversion in individuality.

References

Bradshaw, S. L., Meade, L., Tarlton-Weatherall, J., and Pomiankowski, A. (2022). Meiotic drive adaptive testes enlargement during early development in the stalk-eyed fly. *Biology Letters, 18*(11), 20220352.

Burt, A., and Crisanti, A. (2018). Gene drive: Evolved and synthetic. *ACS Chemical Biology, 13*(2), 343–346.

Burt, A., and Koufopanou, V. (2004). Homing endonuclease genes: The rise and fall and rise again of a selfish element. *Current Opinion in Genetics and Development, 14*(6), 609–615.

Buss, L. W. (1987). *The Evolution of Individuality*. Princeton University Press.

Dawkins, R. (1990). Parasites, desiderata lists and the paradox of the organism. *Parasitology, 100*(S1), S63–S73.

Ewens, W. J. (1989). An interpretation and proof of the fundamental theorem of natural selection. *Theoretical Population Biology, 36*(2), 167–180.

Fisher, R. A. (1930/1999). *The Genetical Theory of Natural Selection: A Complete Variorum Edition*. Oxford University Press.

Frank, S. A. (1995). George Price's contributions to evolutionary genetics. *Journal of Theoretical Biology, 175*(3), 373–388.

Fromhage, L., and Jennions, M. D. (2019). The strategic reference gene: An organismal theory of inclusive fitness. *Proceedings of the Royal Society B, 286*(1904), 20190459.

Gardner, A. (2008). The Price equation. *Current Biology, 18*(5), R198–R202.

Gardner, A. (2023). The rarer-sex effect. *Philosophical Transactions of the Royal Society B, 378*(1876), 20210500.

Grafen, A. (1999). Formal Darwinism, the individual-as-maximizing-agent analogy and bet-hedging. *Proceedings of the Royal Society B, 266*(1421), 799–803.

Grafen, A. (2006). Optimization of inclusive fitness. *Journal of Theoretical Biology, 238*(3), 541–563.

Grafen, A. (2007). The Formal Darwinism project: A mid-term report. *Journal of Evolutionary Biology, 20*(4), 1243–1254.

Grafen, A. (2014). The Formal Darwinism project in outline. *Biology and Philosophy, 29*, 155–174.

Grafen, A. (2021). A simple completion of Fisher's fundamental theorem of natural selection. *Ecology and Evolution, 11*(2), 735–742.

Haig, D. (1997). Parental antagonism, relatedness asymmetries, and genomic imprinting. *Proceedings of the Royal Society B, 264*(1388), 1657–1662.

Haig, D. (2000). The kinship theory of genomic imprinting. *Annual Review of Ecology and Systematics, 31*(1), 9–32.

Haig, D. (2014). Genetic dissent and individual compromise. *Biology and Philosophy, 29*, 233–239.

Koufopanou, V. (1994). The evolution of soma in the Volvocales. *The American Naturalist, 143*(5), 907–931.

Larracuente, A. M., and Presgraves, D. C. (2012). The selfish Segregation Distorter gene complex of *Drosophila melanogaster*. *Genetics, 192*(1), 33–53.

Leigh, E. G., Jr. (1977). How does selection reconcile individual advantage with the good of the group? *Proceedings of the National Academy of Sciences of the United States of America, 74*(10), 4542–4546.

Meade, L., Finnegan, S. R., Kad, R., Fowler, K., and Pomiankowski, A. (2020). Maintenance of fertility in the face of meiotic drive. *The American Naturalist, 195*(4), 743–751.

Michod, R. E. (2000). *Darwinian Dynamics: Evolutionary Transitions in Fitness and Individuality*. Princeton University Press.

Michod, R. E., and Roze, D. (1997). Transitions in individuality. *Proceedings of the Royal Society B, 264*(1383), 853–857.

Okasha, S. (2018). *Agents and Goals in Evolution*. Oxford University Press.

Okasha, S., and Otsuka, J. (2020). The Price equation and the causal analysis of evolutionary change. *Philosophical Transactions of the Royal Society B, 375*(1797), 20190365.

Plutynski, A. (2006). What was Fisher's fundamental theorem of natural selection and what was it for? *Studies in History and Philosophy of Science Part C: Studies in History and Philosophy of Biological and Biomedical Sciences, 37*(1), 59–82.

Presgraves, D. C., Severance, E., and Wilkinson, G. S. (1997). Sex chromosome meiotic drive in stalk-eyed flies. *Genetics, 147*(3), 1169–1180.

Price, G. R. (1970). Selection and covariance. *Nature, 227,* 520–521.

Price, G. R. (1972). Fisher's "fundamental theorem" made clear. *Annals of Human Genetics, 36*(2), 129–140.

Price, G. R. (1995). The nature of selection. *Journal of Theoretical Biology, 175*(3), 389–396.

Queller, D. C. (1997). Cooperators since life began. *Quarterly Review of Biology, 72,* 184–188.

Riskin, J. (2016). *The Restless Clock: A History of the Centuries-Long Argument over What Makes Living Things Tick*. University of Chicago Press.

Smith, J. M., and Szathmary, E. (1997). *The Major Transitions in Evolution*. Oxford University Press.

Stoddard, B. L. (2005). Homing endonuclease structure and function. *Quarterly Reviews of Biophysics, 38*(1), 49–95.

West, S. A., Fisher, R. M., Gardner, A., and Kiers, E. T. (2015). Major evolutionary transitions in individuality. *Proceedings of the National Academy of Sciences of the United States of America, 112*(33), 10112–10119.

PART II

IMPLICATIONS OF CONFLICT

A Metaphysical Paradox of Organismal Identity

ELLEN CLARKE AND WILL MORGAN

ABSTRACT Dawkins's paradox of the organism asks how organisms can be harmonious units, given the scope for conflict among their parts, such as that caused by selfish genetic elements (SGEs). We argue that in order to even make sense of Dawkins's question, we need to have some way of determining what the relevant unit is. Where are the organism's spatial and temporal boundaries? What things are parts of the organism, and why? We link Dawkins's problem with a much older paradox in metaphysics—the ancient Greek paradox of change—that asks how it is possible for a thing to change its properties over time, and yet remain one and the same thing. The paradox of change is relevant to biology because it seems to suggest that there can be no such things as organisms—at least not in the sense of objects that persist through time. Aristotle blocked the paradox of change by arguing that at least some objects have an essence that allows them to persist through change. An organismal essence would make it possible for organisms to be born, to change over time, to reproduce, and to die. But what could an organismal essence be? This chapter suggests taking inspiration from Dawkins's resolution to the paradox of the organism by looking forward into the future. In this view, an organism's individual essence can be defined by the set of genes that share what Dawkins called a "stochastic expectation of the future." This view offers a helpful distinction between SGEs that compromise organismal persistence by pursuing a future at odds with the rest of the DNA, and elements that merely lower organismal fitness without undermining the shared future. SGEs of the first kind might be better understood as parasitic individuals in their own right rather than as true parts of the organism.

Dawkins's Paradox

In his 1990 paper, Dawkins invoked the metaphor of a river to express a biological problem of change:

> In the perspective of evolutionary time, genes inhabit not a pool but a river, flowing through time, a broad-fronted turbulent river down which the Mendelian particles zig-zag their way, changing partners at every generation, some increasing in frequency, others decreasing. . . . It is possible to do the mathematics of natural selection completely forgetting that, as a matter of fact, the genes are

not swishing about in a liquid pool or river but bound up in solid chunks—organisms—and colossal chunks at that. . . . Their existence as cohesive units of function, in the teeth of potential conflict among the true replicators that they contain, constitutes a paradox which I shall call the Paradox of the Organism. The Paradox of the Organism is that it is not torn apart by its conflicting replicators but stays together and works as a purposeful entity, apparently on behalf of them all. (Dawkins 1990, 100)

Dawkins conceived of organisms as persisting objects in possession of multiple parts—genes—and sought to explain how those parts are able to coexist harmoniously, when the default is for genes to compete against one another. Part of the issue here is explaining how agents can cooperate instead of interacting in a conflictual manner. This problem of cooperation has been very well explored. We address a different, more general issue, about organisms and their parts: How does an organism keep itself together, both at a time and through time? More specifically, we are concerned with two distinct but connected questions:

   1. In virtue of what do distinct parts compose an organism?

and

   2. In virtue of what can such wholes persist over time?

   Question (1) is a more specific version of a general metaphysical issue—that is, the "special composition question" (Van Inwagen 1990), which asks, "Under what conditions do things compose something?" There is a common intuition that, say, sticks of wood may compose a chair in some circumstances, while to Pluto, my left arm and the glass on my table collectively compose nothing. The special composition question asks us to explain why. Applied to biology, we may ask, "What does it take for multiple different things to make up a single organism?" Organisms seem to include or be made up of lots of different stuff—proteins, genes, cells, and tissues of various kinds. What makes it the case that all of these distinct things can combine to make up a single bigger thing?

   Question (2) is asking what it takes for one and the same organism to exist at different instants in time. For example, I existed ten years ago, and I exist now. What makes this the case? While questions (1) and (2) are different, and one answer does not logically provide an answer to the other, they are plausibly connected. First, they are both about the unity of organisms; question (1) is about the unity of the parts of an organism at one time, and question (2) is about the unity of an organism over different times. Second, an answer to one question often strongly suggests an answer to the other. A satisfying account of the unity of the organism should be able to provide an answer to both questions.

These issues are more general than the cooperation problem because they are not limited to cases in which the relevant parts are expected to compete with one another, or to interact with one another in an agential manner, or even to interact with one another at all. For example, we can ask for a justification of the idea that ingredients in some circumstances compose a cake, or stars in the sky compose a constellation. Setting aside the concerns about genetic competition, these deeper questions ask why Dawkins thinks of genes as composing persisting objects—what he calls "solid chunks"—at all? We look for a way to make progress in regard to these questions by reflecting on an ancient Greek paradox—the paradox of change.

The Paradox of Change

A paradox occurs, by definition, when a list of two or more apparently reasonable claims or assumptions are combined to give us a completely *un*reasonable conclusion. We are left feeling like we must abandon one of the claims, but they are each so precious to us that it is hard to know what to do.

The paradox of change is often associated with Heraclitus who expressed it, conveniently for our purposes, via the example of a river. The famous line attributed to him is "No man ever steps into the same river twice, for it is not the same river, and he is not the same man." There are two main premises in the paradox. First is that there are objects such as rivers and men that seem to persist over time. For instance, a river existed ten years ago, and the same river exists now. Insofar as this is possible, the river has persisted through time.

The second premise is that change seems to occur. A river may run deeper or drier at different points in time, depending on rainfall. Its course may even change shape over time, as its banks are eroded by the passage of water. Similarly, one of us owns a black cat called Kiki who was acquired when she was six months and is currently four years old. It is easy to list things about Kiki that have changed in that time—her size, adventurousness, the length of her tail. Given the facts about metabolism and cell turnover, she won't be made of the same stuff—the same atoms or cells—as she was when she was just a kitten. She has always remained black, but this doesn't seem crucial. I can at least imagine dying her fur, or having her turn gray perhaps, without this meaning she isn't Kiki any more. Most of us would agree that Kiki could survive at least some of these changes.

Both of these premises—that objects persist, and that change happens—seem reasonable and widely held. And note again that there is no special issue about the droplets in a river fighting with one another and needing to keep peace. Whatever motivates us to think of different droplets as parts of a single river, it isn't that they are being friendly to one another or sharing a common goal. Nonetheless, it is much harder to reconcile the two premises than we ordinarily realize. In fact, persistence through change has been a central problem in philosophy since the ancient Greeks and continues to figure in contemporary philosophical debates. We can identify four main flavors of solution (Boulter 2013). One gives up on the reality of persisting objects. A second gives up, instead, on the reality of change. A third abandons the standard logic used to combine the premises. And a fourth does some fancy metaphysical footwork in order to keep both premises and standard logic intact. We will primarily be exploring the last option, which has become so widespread in contemporary thinking that the fanciness of its footwork has largely become invisible. But first, let's make clear why the two premises are difficult to reconcile.

> Premise 1: If objects persist through time, then an object existing at time t^1 is identical to—the same as—an object existing at time t^2.
> Premise 2: If change occurs, then an object at time t^1 is not the same as an object at time t^2.

To combine these—to suppose that objects can both persist and change over time—seems to require that an object at time t^1 be both the same as—and not the same as—an object at time t^2. But that is a contradiction.

To invoke the river example again, we seem to want to say that the Euphrates River in 1985 is the same as the Euphrates River in 2023. But it also seems plausible that it isn't exactly the same—it will have changed in loads of ways. But how can it be both different and the same? The paradox, therefore, threatens to show that it is impossible for objects to change over time.

Heraclitus concluded that the appearance of persisting objects must be an illusion. The river when I first step into it has a particular set of atoms rushing along it, a particular color and temperature and shape. The second time I step in, all these properties have changed and so it is a different river. Note that it doesn't actually matter what time interval I choose or how much change has occurred. Even if just one atom in the river changes, this would mean that a new river takes the place of the old. This conclusion does not just apply to the river but to all supposedly persisting objects, such as human beings. Were you to fall over and scrape your knee, thereby losing a few skin cells, a new, numerically distinct organism would take your place.

Some contemporary philosophers have favored the conclusion that reality consists in a rapid succession of different objects, rather than in stable persistence of objects over time (Armstrong 1980; Hawley 2001). Instead of moving through a world of changing objects, this view holds that reality is a kaleidoscopic succession of time slices each occupied by numerically different rivers, people, and objects. But scientific theory as well as ordinary human language seems to presuppose that objects *do* persist over time and survive change. Organisms are thought of as having lifecycles, in which a single individual undergoes a predictable series of changes. Biologists track organisms through time, so they understand one and the same amphibian as regenerating a limb, one and the same embryo as developing in time. Dawkins visualizes organisms as chunks flowing along some sort of evolutionary river. He questioned what prevents such chunks from being blown apart by internal conflict but sidestepped deeper questions. Given the seemingly constant change that characterizes reality—movement, decay, chemical processes, and a perpetual cycling of atoms—why think there are persisting chunks at all?

An Essentialist Solution: Accidental and Essential Properties

One school of thought, in response to the problem of change, has been enormously influential in Western philosophy: essentialism (see Robertson Ishii and Atkins 2023 for a review). Aristotle is usually credited with developing it in this context. In brief, Aristotelian essentialism avoids choosing between persisting objects or change by allowing that while some of the properties of an object persist through change, and thus underpin its numerical identity over time, other properties genuinely change. Rather than saying that for A and B to be identical, they must have all of the same properties, it says that A and B are identical if they have the same essence, or essential properties. Essential properties are such that if an object was to lose them, the object would cease to exist. My having a particular skin cell is not an essential property of me, so I can lose a skin cell without ceasing to exist.

People occasionally find this solution so obvious that they think the whole paradox trivial, as if it is obvious that being numerically the same person is different from being exactly, qualitatively, the same in every respect. But it is worth getting clear that some substantive metaphysical commitments are required to sustain the solution. In particular, we become obliged to a distinction between two sorts of properties—accidental properties, whose possession by an object is contingent, so that they may come and go—and essential properties, whose possession is obligatory in a certain sense. In order to preserve the idea that an object persists over time, we need to invoke an individual essence—that is, a set of properties that are essential to being *that particular object*. We can distinguish this from a "kind essence," which is necessary for being a particular kind of thing.

Examples might make this easier to grasp. Suppose a banana is sitting atop a fruit bowl. It is easy enough to understand that some of the banana's properties might change over time, without jeopardizing the fruit's continuation. If I pick it up then its location changes, for one. Some people think it is less obvious how intrinsic properties—those that the banana has all by itself, rather than in virtue of its relation to other objects—can change. But we are familiar with the idea that a banana will ripen over time, changing not only its color, but also its internal texture and eventually shape. So let's say the banana's location, color, texture, and shape are all accidental properties of the banana. The distinction will be object-specific, in that shape may seem plausibly accidental to a banana or a tissue, but not to a chair. We might decide that a chair necessarily has a certain shape, so that if it is bent or broken it no longer qualifies as that chair, or indeed as any chair at all. Such a chair would have lost both its individual essence, necessary for its numerical identity as that particular chair, and its kind essence, because it ceases to qualify as a chair.

With the right choice of example, then, this distinction between accidental and essential properties can be grasped intuitively. At the same time, it is worth being clear that it is quite a radical and problematic commitment. For example, it is difficult to say how the distinction could ever move beyond an intuitive sense and become subject to empirical evidence. Necessity and possibility, more generally, go beyond what we can observe, and strike some philosophers as elements that a proper metaphysic ought to leave behind. Even someone less opposed to the general idea of necessity as a natural phenomenon will often find themselves hard pushed to motivate the selection of some properties rather than others as the necessary ones. Of the Euphrates River, we might suggest its geographic location is essential, or its shape, or the point at which it meets the sea, or perhaps a characteristic odor caused by the rocks across which it flows. But all of these seem arguable. How could we ever access knowledge about which of the Euphrates River's properties are essential?

So every route out of the paradox is at least awkward. Either go with Aristotle and make a case for some properties or other being essential. Or go with Heraclitus and deny that objects persist, and understand the appearance of a persisting banana or river as akin to the appearance of a moving image on a cinema screen. This framework might seem counterintuitive and disorientating, but it has gained recent support among metaphysicians and philosophers of biology as a potentially coherent alternative to Aristotelian metaphysics. For example, according to what they are calling process biology, "What we identify as things are no more than transient patterns of stability in the surrounding flux, temporary eddies in the continuous flow of process" (Dupré and Nicholson 2018, 13). However, challenges remain in spelling out exactly how a process differs from an object, how to understand the relation between the objects at different time slices, and so on (DiFrisco 2018).

Another ancient Greek philosopher, Democritus, suggests a further option. He argued that complex objects are not real, but are illusions generated by the interaction of underlying particles—what he called atoms. He located persisting objects at a lower compositional level, in effect, by positing that atoms persist and are unchanging, and the appearance of ordinary persisting objects such as rivers and people is produced by the way those unchanging atoms interact. Modern-day proponents of this sometimes call themselves "nihilists." They argue that only the fundamental particles of physics are real, and all larger composites are illusory, just particles "arranged river-wise" or person-wise (van Inwagen 1990; Rosen and Dorr 2003). Concerning organisms, a nihilist might allow that there are things such as cells and genes but deny that they ever compose organisms. We might be tempted to draw parallels here with the way that genetic reductionists sometimes want to say that genes are more real than organisms. What it means to say that some objects are "more real" than others is unclear, but perhaps they have something like nihilism in mind. Recall that Dawkins thinks of evolutionary lineages as a river of flowing genes. In that sense, we might avoid having to make sense of the existence of stable chunks at all and claim that only genes are persisting objects. But this is costly, as Dawkins points out, because we have good reasons to want to understand how higher-level properties, such as cohesion and purpose, are generated by the way genes interact in their particular chunks. The nihilist needs to be able to explain all of the properties of composites in terms only of their fundamental particles. Many commentators have been skeptical that this would be possible, let alone desirable (Fodor 1974).

A discussion of the general merits and flaws of essentialism and its rivals goes beyond the scope of this chapter. We are instead going to attempt to articulate a decent candidate for an organismal essence, such that Aristotelian essentialism can be applied to organisms. This would preserve the commonsense idea that organisms are persisting objects, composed of parts. Finally, we will explore the consequences of applying this idea to how we think about SGEs.

Organismal Identity

Applying essentialism to the problem of organismal change requires us to think of organisms as having individual essences—properties or sets of properties that are necessary for a particular organism to continue being that particular object. Can we find some property of an organism that plausibly never changes—and furthermore, *may not change* without resulting in death or replacement of that organism by some other object? Going back to Kiki the cat, what, if anything, makes her Kiki? Why don't we think I have had a succession of similar but different cats? What have all her parts got in common that make them her parts, rather than being part of her collar or part of her home? Why is her claw part of her but a flea is not?

Here is another case: Consider a bacterial cell that is in the process of being infected by a horizontally transmitted plasmid. This is an event in which a small packet of DNA enters the cell from outside, and then persists within that cell, possibly giving the cell new functionality, possibly being transmitted vertically to offspring cells, and certainly hanging around and interacting with other stuff inside the cell without being thrown out.

What do we want to say about the bacterial cell, about its identity, about its parts? Something has changed, certainly. There is DNA within its boundaries that was not there before. Do we want to say that a new bacterial cell has taken the place of the old? Or that the initial cell survived this change in some way? Is it the same cell, just with a larger amount of DNA, or has the addition altered its identity? Has the plasmid become a new part, or is it something foreign that is merely inside the cell?

We hope it is clear that there is a conceptual choice to be made here. None of the possibilities seem illogical or ruled out by empirical evidence. So which one is right? What does it depend on?

Some Existing Responses

A number of quite different suggestions have been made about what to think of as essential to organisms such as cats.

The "Same Life" Solution

According to van Inwagen (1990), who draws on John Locke, an organism persists in virtue of the continuation of its life, where a life is a self-maintaining biological event or activity that maintains an organism's structure despite the constant turnover of matter. In response to the paradox of change, we can take an organism's life to be essential to it—an organism persists through change, even quite drastic change, so long as it has the same life. Despite Kiki's change in tail length, cells, and personality, she still has the same life. In this sense, the "same life" view provides a unified account of the unity of an organism—an account of how it is unified at a time, and over time.

While there is nothing obviously false about this solution, it is too superficial. For how can we determine whether two objects have the same life? Do Kiki's cells have the same life as her? What about her gut microbes? If Kiki became pregnant, would the fetus developing inside her have the same life? Why? What about the plasmid-infected bacterium? The "same life" answer does not seem to take us much further than we started.

The Genetic Solution

For biologists, it is perhaps natural to look to DNA to define an organism's essence. In some ways, this is perhaps the obvious modern update to Aristotle's solution, which was to think an object's essence consisted in its "form." The canonical illustration of the difference between form and matter is a statue, where the matter is the clay and the shape is the form. The intuition is supposed to be that the statue persists only if it retains its form. Contemporary essentialists say that a carbon atom persists so long as it retains its atomic number (Ellis 2001). An essence is something about the object's microstructure, something inside it, something that can be copied and instantiated by multiple different physical bases. DNA gives us the intuitive answer here, when it comes to the living world.[1]

On this view, what makes Kiki *Kiki* is her particular genotype. Mapping this back to Dawkins's metaphor, we might say that what gives a particular chunk its identity is just the identity of the particular genes it contains. The chunk flowing along the river is the same chunk because of what's in it—it has the same genetic parts. So, the genetic solution tells us that things belong to the same organism as parts if they have the same DNA as the rest of the organism, and it tells us that an organism persists so long as it contains the same genome.

Appealing though this might sound, it cannot be the whole answer. One reason is that genotypes are not always unique to different organisms. For example, Kiki might have an identical twin sister, and if that were so then we'd seem to have to say that both cats are Kiki, because they'd have matching DNA. Another reason is that bits of DNA can be added to and deleted from organisms over time. For example, transposable elements are bits of DNA that insert copies of themselves into the genome, while viruses of various kinds are adept at inserting their DNA into host genomes. Pregnancy often causes the mother to acquire fetal cells via the placenta, which in some cases have been detected hanging

---

[1] It is worth mentioning here that essentialism has picked up such a bad name in the context of biology that it is somewhat out of vogue to even discuss it. Our view is that this reputation is ill-deserved because it was gained in two problematic contexts. Species essentialism names the flawed idea that species can be defined by possession of genes shared among their members. Genetic essentialism names the flawed idea that genes tell you all you need to know about an organism, and that they fix all the traits of organisms in rigid ways. It is perfectly possible to reject both of those false views while continuing to appeal to essentialism in other ways.

around in her body many years later (Gammill and Nelson 2010). Gene deletions occur either by accident during transcription, or as a consequence of cellular machinery there to combat viruses and transposable elements. Finally, genes change when they mutate. Most organisms will also experience flux over their lifetimes in their hosting of endosymbionts. If some of these qualify as true parts, then change of endosymbionts is another way that organismal genetic change might occur.

So there seem to be at least some contexts in which, to continue Dawkins's analogy, chunks may gain or lose different genetic parts as they flow down the river over time, which threatens the simple idea of understanding a chunk's identity as settled by its inclusion of particular genetic parts. Or at least it would imply that every time some mutation, addition, or deletion occurred, the chunk/organism in question would cease to exist and be replaced with a brand new, numerically different organism. That isn't impossible—it is a coherent conceptual scheme, at least—but it is not the way that we typically think about organisms.

However, there is yet a deeper problem with the foregoing thought, which is that we are still presupposing something about the imagined chunks. Dawkins's metaphor is quite visual—we can easily picture lumps in a river. But we are just imagining genes as stuck together to form these chunks in the way that bits of Lego might be stuck together. Real organisms are not bits of glued-together DNA, of course. It is only a metaphor, but it is hiding a serious question: In virtue of what are different parts held together to form an organism anyway? And if we are serious about giving primacy to genes as the essential properties that fix an organism's identity over time and ability to change, then which genes are we talking about? The ones in the organism, you might say. But which ones are they?

It is presupposed that we have some way to decide where an organism's boundaries are. Its parts will probably be all stuck together within some kind of skin or wall. But there will be other things within that wall that are not parts, such as prey, transient symbiotes, pathogens, and nonliving matter such as balls of cotton forgotten by a surgeon (Kaiser 2018). In some cases we might consider as organismal an entity whose parts are not permanently stuck together—insects in a bee colony, perhaps. We might say that Kiki's DNA is all the DNA in or inside her skin, assuming we don't count DNA in dead cells such as in her fur. But what about her mitochondrial DNA? And her intestinal bacteria? Perhaps we don't count the digestive passages as properly inside a cat. But what about if she develops a tumor or is infected by a virus?

Just to note, again, that these problems are logically prior to any concerns about conflict between parts. Dawkins's puzzle about how unity is maintained among genes presupposes that we know which genes we are talking about. Intuitive though it may be, a simple genetic solution will not work to settle organismal identity. The problem is that there are various ways in which DNA changes in things we usually want to call organisms. Bits of DNA mutate or get deleted. And different bits are gained because of mutation or infection or symbiosis. But as we've already suggested, even if DNA never changed, this solution would be incomplete. We can't point to DNA as fixing the identity of an organism unless we can first determine *which* DNA is concerned. A vicious circularity threatens if we try to say "the DNA in the organism," because which DNA is that?

The Bottleneck Solution

Several philosophers have defended the idea that an organism's origin settles its individual essence. In this view Kiki is Kiki by virtue of the fact that all her parts developed from a particular zygote; for example, if her father had been different, or she had been conceived on a different day via a different sperm, then she would have been a different cat (Forbes 1985). Some biologists have encouraged us to decide which bits of stuff qualify as organismal parts in terms of an origin point, too (Bonner 1974; Maynard Smith and Szathmary 1995). In answer to the question, *Which DNA defines an organism's essence?* we can say *All the DNA that is copied faithfully from a shared template in the zygote.* Appeal to a bottleneck origin point allows us to get clear about exactly which DNA is essential to an organism, and to say that any DNA that is added later on is not part of the organism. Returning to the case of plasmid uptake, on this view we might say that the bacterial cell's essence is fixed by the form of its DNA immediately following its birth by fission. The plasmid never becomes a part of the cell, but the cell continues to exist (because only a new bottleneck event forms a new cellular individual). Similarly, in this view Kiki's hypothetical tumor wouldn't cause her to cease to exist but would be best conceived as a sort of parasite living within her body rather than as a part of her, in the same way that the plastic bead that Kiki swallowed is inside of her but is not one of her parts. However, some sorts of changes might be harder to accommodate in this way. If a genetic mutation occurred early in Kiki's development, then it might become widespread in her adult body. All of those subsequent parts would have to be counted as not belonging to her, but as parts of a separate organism that is somehow physiologically entwined with her.

A more general problem with this view is that it feels a bit flat-footed. Why prioritize the past? We aren't really given an explanation of why the state of things at one point in time is so significant for determining the state of things at later times. The bottleneck view doesn't really explain persistence because it isn't obvious why having a shared origin would make various different bits of DNA hold together through time in a way that grounds explanations of the properties of the whole. One might respond that the significance of the past lies in its consequences for the future. Dawkins argued that development from a single-celled bottleneck is significant because it raises the extent to which the parts of the organism will later be genetically identical to one another, and that's important because it reduces the risk of selective competition among them. The problem is that bottlenecks aren't actually sufficient to eliminate the risks of conflict in the future, for the reasons of mutation and deletion and addition already described.

But this opens up another possibility. If what matters is really the way that the future is constrained, then perhaps we could focus on that more directly.

A New Response: Look to the Future?

Dawkins writes, "An organism is an entity all of whose genes share the same stochastic expectations of the distant future" (Dawkins 1990, S63). This is a pretty opaque phrase. Dawkins is thinking of the way that fair meiosis gives all the genes in the genome equal access to the gametes— an equal 50/50 chance of having their form copied into each offspring (Leigh 1971). It is fair insofar as all genes are in the same position, and it is theorized to prevent conflict among the genes because they all have a high chance of finding themselves in the same lifeboat, as it were. Or rather, their descendants are highly likely to have to work together in future genomes, so if they were to damage each other's prospects, they'd be shooting themselves in the foot, if you'll excuse the mixed metaphors.

Various authors have used this idea of a shared evolutionary fate as something that holds the parts of an organism together and inhibits conflict. Leo Buss (1987) talked about "fate-coupling" by the evolution of germ-soma separation. His argument was that the separate germline evolved in order to circumvent conflicts among cells as groups grew increasingly large and complex during the evolution of multicellularity. Sober and Wilson gave perhaps the most explicit account of shared fate in 1994, describing it as a component of Dawkins's vehicle concept:

> Using one of Dawkins's own metaphors, we can say that genes in an individual are like members of a rowing crew competing with other crews in a race. The only way to win the race is to cooperate fully with the other crew members. Similarly, genes are "trapped" in the same individual with other genes and can usually replicate only by causing the entire collective to survive and reproduce. It is this property of shared fate that causes "selfish genes" to coalesce into individual organisms. (Wilson and Sober 1994, 590)

Sober and Wilson argued here that having a shared fate is not merely about having the same fitness—there must be a causal relationship tying the fitnesses together. Others have called the phenomenon "fitness alignment" (Folse and Roughgarden 2010) and talked about its role in securing "harmonious design" (Gardner and Grafen 2009).

The idea of the current chapter is to understand organismal persistence as underpinned by the mechanisms that secure shared fate, mechanisms that include but are not limited to the developmental bottleneck, the separation of germ and soma, and fair meiosis. In this view, we would define an organism's individual essence as comprised by all the DNA whose fitness is aligned. More generally, looking to the future allows us to recognize that conflict—or rather, its resolution—is relevant to organismal boundaries, and to recognize that work has to be done to integrate DNA into chunks. They don't simply find themselves stuck together, and thus there is no theoretical motivation for stipulating genetic stasis relative to an arbitrary time point. What matters—what makes the relevant genes in the evolutionary river compose an organism—is that their fitness is aligned.

We will say that what is essential to an organism is the shared future of its parts. To be a part of an organism is to share its evolutionary future. By this account, organisms persist over time by virtue of the fact that their DNA at a given timepoint expects the same future as expected by the DNA at some future timepoint.

More precisely (because organisms are not only composed of DNA!):

Definition: An organism's spatial and temporal boundaries include all cells whose genes share a common future, and cells immediately derived from cells containing such DNA, where the future is shared because of individuating mechanisms.

Of course, this is still very metaphorical and there are many details that will need to be ironed out. For example, it might make a difference how far into the future we look. It would be helpful to be able to identify moments at which a new, separate individual comes into existence by separating from its mother, or fragmentation of a parent. We will want to equate these with moments when genes acquire independent futures because the causal coupling between them is released. But there may be no precise point at which this happens. It might be gradual and context-dependent.

We might wonder how persistence and identity can be defined in terms of events that lie in the future. After all, the future hasn't happened yet! The thought is that we avoid appealing to the future directly if we make the evolved mechanisms of conflict resolution essential. They exist in the here and now and explain why parts of organisms have a shared future (Clarke 2013). So for example, the vertebrate immune system keeps Kiki's parts committed to a shared future by eliminating cellular mutants that start to proliferate in ways that threaten the goals of the rest. We can say that any parts rejected by the immune system are not parts of the organism. One benefit of this view is that the Kiki's boundaries become more than a matter of observer's perspective, because it is features of Kiki herself—her immune system, her developmental system—that are doing the work. The idea of sharing a future is ultimately about prediction—what we expect to evolve, and whether the fitness of the parts is maximized by the same strategy.

Can we really use this to explain why Kiki's tail belongs to her but her twin and her lice do not? What does it mean to say that Kiki's three-week-old DNA has the same future as her four-year-old DNA? When Kiki was a tiny kitten, her DNA "expected" to be passed onto future generations via cells in Kiki's ovaries. As an adult, those cells retain the route via which her evolutionary heritage is transmitted, and the route is shared by Kiki's mitochondria. However, it isn't shared by the lice in her fur, because the lice do not expect to transmit their genes to future lice offspring along with Kiki's genes. Even if this view seems promising for distinguishing what is internal and what is external in regard to neat cases where there is clear alignment of genes, there is further work to do concerning trickier cases where there is internal conflict, such as SGEs and horizontally transmitted bacterial plasmids.

Selfish Genetic Elements

SGEs are a class of genes that are able to enhance their own transmission at the expense of other genes in the genome. In this sense, they fail to be properly aligned with the other genes in an organism. There is an internal conflict, in that the SGE's goal (maximize own transmission) looks different from the organism's goal (give each gene an equal chance of being transmitted). So if organismal boundaries include all the cells whose genes share a common future, as suggested above, this may mean that SGEs undermine organismal identity quite radically. For when SGEs are present in the genome, this means the genes don't share a common future—at least, they don't have equal stakes in that future. But then no organism could have cells containing SGEs as parts. Given the ubiquity of SGEs this would be a serious blow to our intuitions.

One possibility is that we view the genome as containing different factions whose strategies mostly overlap, but where those strategies diverge in particular cases, or in respect of particular organismal traits. We might view the "shared fate" of the whole genome then as depending on the extent to which there is successful suppression of divergent strategies, such as those attempted by SGEs. Maybe suppressed SGEs are normal parts of a unified genome. But what of successful cheats? Do these become separate parasitical parts? Or does their success spell bad news for the parthood of their rival genes perhaps, those who their success displaces? Or could we say that SGEs are still parts so long as there are mechanisms in place with the function of suppressing conflict, even if those mechanisms are failing some of the time?

Patten (chapter 3 in this volume) argues that we need to distinguish two different kinds of SGEs in thinking about these questions. Trait distorters, such as imprinted genes, influence the phenotype of the organism, altering it from what the normal unimprinted developmental outcome would have been. This often leaves the organism less fit. Conversely, transmission distorters, such as homing endonuclease genes, increase their representation in offspring by knocking a rival gene out and replacing it with a copy of themselves. But there is often no fitness cost for the organism as a whole—it doesn't care which of the two rivals occupies that locus on the genome, because it doesn't affect the phenotype.

Patten argues that although these can both be called internal conflict, only the trait distorters undermine organismal identity by undermining the unity of purpose. Continuing the boat analogy, we might imagine that in transmission distortion, most passengers look away while one of their number quietly pushes their neighbor off the boat and pulls their friend on instead, while the boat calmly continues on its way. Intuitively, it feels right to say that the boat's identity isn't threatened here. What is threatened is the parthood of the gene that got pushed out. Although we might think the other passengers will start to become interested if too much of that behavior is permitted. The unity and identity of the boat as a whole will depend on mechanisms that are in place to suppress transmission distortion, even if successful distortion does occasionally occur.

Conversely, trait distorters, such as imprinted genes, are not in the business of pushing anyone on or off—changing the membership of the boat. Rather, they are grabbing the rudder and attempting to change the destination of the boat. In this case there is a clear conflict among the passengers on the boat—a difference of strategy, of opinion about the optimal destination. What's the right thing to say here, about the identity of the boat? One possibility is that the fighting gets so bad that the whole thing sinks, or that it keeps switching between two alternative destinations such that it never gets anywhere. This is the case when imprinted genes seek to steer the phenotype in a direction that lowers organismal fitness, such as if paternally imprinted genes bring about preeclampsia in pregnant women.

Another possibility is that one side wins and reaches its favored destination. Talk of rival cabals does seem right here, and intuitively the different parts are acting as different evolutionary individuals. But which team is best characterized as the parasite and which as the host seems determined only by which wins. And matters are further complicated by the fact that this struggle might only concern a single trait of the organism. In that case, the rival factions might be fighting to control which station the radio is tuned to, or how fast the vehicle goes, as opposed to its final destination. And that really does set up a situation with the possibility of multiple intersecting factions whose members have mixed allegiances that are, metaphysically speaking, quite far from how we ordinarily think about composition and identity.

It isn't a million miles from how we think about identity in social ontology. Consider the Republican Party in the United States. It seems unproblematic to say that the Republican Party has persisted over many years. But as for whether that persistence is better understood as being underpinned by continuities of membership, or by its ruling ideas, I wouldn't like to say. Certainly there will always be factions and cabals, and arguments about which are the true inheritors of the party. Some of these will threaten the success and even survival of the party. What seems right then, is that in the case of the Republican Party and in the case of trait distorters, these very complexities will make it difficult to predict what will happen next, compared to the case of an object such as a carbon atom, whose persistence over time is much simpler.

Consider Pando, a.k.a. the trembling giant, a huge clone of quaking aspen in Utah. Because the trees grow up clonally from a shared root web, they are similar to each other genetically. But because the clone is very old, there has been lots of time for mutations to accumulate. As time passes, different areas of the clone are gradually diverging from one another genetically. This is a real-life case of another ancient Greek puzzle—the Ship of Theseus (Horvarth 1997). According to Patten's distinction, if the spreading mutations don't make any difference to the clone's fitness, as is plausible, then they are not generating any disunity of purpose. So perhaps we can rule that this sort of slow, cumulative change to Pando's genotype doesn't undermine Pando's identity. Similarly, it doesn't make any difference to the identity of the lifeboat if some passengers are switched out by replacements, as long as the way the boat behaves doesn't change. Only conflicts, in which different parts of the clone try to steer the phenotype in different directions, undermine Pando's individuality.

Conclusion

There are more questions here than answers. But one conclusion that comes through is that Dawkins's original problem was ill-formed. If the parts of an organism are all those that share an expectation of the future—where there are mechanisms guaranteeing no evolutionary conflict between them—then there can be no conflict within organisms. Any entities that failed to entirely share the goals of the organism would simply not qualify as parts but would be better conceived as more like foreign parasites. We don't find it helpful to define within-organism conflict as impossible by definition.

Options to avoid that conclusion include the following:

1. Understand parthood as applying as long as conflict-control mechanisms are in place, whether or not they are fully effective.
2. Understand parthood as applying more or less.
3. Understand parthood as surviving some sorts of conflict, but not others.
4. Abandon conflict-control as settling organismal identity/persistence and find some other essence.
5. Abandon essentialism about organisms and adopt a temporal parts view or some such.
6. Abandon organisms as persisting objects.

The overall collection has the agenda of taking conflict seriously in order to better understand organisms. This chapter began by arguing that before we even get to the conflict issue, there is a logically prior issue concerning the identification of parts as composing an organism, and as potential sites of conflict. Taking the problem of change seriously means we need to work out why we treat different bits of DNA as composing a single organism, as well as how to conceptualize conflicts between those bits.

We then argued that we can appeal to conflict (or rather its avoidance/resolution) in order to settle the first issue, and to determine the nature of an organism's identity both at a specific time and over time. In order to develop this idea, we suggested taking inspiration from Dawkins's observation that organismal parts have a shared expectation of the future. We can say that different parts compose an organism by virtue of being committed to a shared future, where this commitment is maintained by multiply realized conflict-control mechanisms. This answer allows for organisms to change over time, and even allows for the membership—the parties to the commitment—to change over time. But it rules against the inclusion as a part of anything that fails to be committed. Elements that do not share the commitment to the future are better conceived as separate life forms, more like parasites or interaction partners than parts.

This suggestion needs more development and leaves many issues unclear, especially what it means precisely to share a commitment to the future, how to ascertain which parts have a shared future when, and exactly which sorts of conflict or divergent interest are fatal to a shared future.

References

Armstrong, D. M. (1980). Identity through time. In *Time and Cause,* edited by P. van Inwagen, 67–78. D. Reidel.

Bonner, J. (1974). *On Development.* Harvard University Press.

Boulter, S. (2013). *Metaphysics from a Biological Point of View.* Springer.

Buss, L. W. (1987). *The Evolution of Individuality.* Princeton University Press.

Clarke, E. (2013). The multiple realizability of biological individuals. *Journal of Philosophy, 110*(8), 413–435.

Dawkins, R. (1990). Parasites, desiderata lists and the paradox of the organism. *Parasitology, 100*(S1), S63–S73.

DiFrisco, J. (2018). Biological processes. In *Everything Flows: Towards a Processual Philosophy of Biology,* edited by D. J. Nicholson and J. Dupré, 76–95. Oxford University Press.

Dupré, J., and Nicholson, D. (2018). A manifesto for a processual philosophy of biology. In *Everything Flows: Towards a Processual Philosophy of Biology,* edited by D. J. Nicholson and J. Dupré, 1–46. Oxford University Press.

Ellis, B. D. (2001). *Scientific Essentialism.* Cambridge University Press.

Fodor, J. (1974). Special sciences, or the disunity of science as a working hypothesis. *Synthese, 28,* 97–115.

Folse, H. J., III, and Roughgarden, J. (2010). What is an individual organism? A multilevel selection perspective. *Quarterly Review of Biology, 85*(4), 447–472.

Forbes, G. (1985). *The Metaphysics of Modality.* Oxford University Press.

Gammill, H. S., and Nelson, J. L. (2010). Naturally acquired microchimerism. *International Journal of Developmental Biology, 54*(2–3), 531.

Gardner, A., and Grafen, A. (2009). Capturing the superorganism: A formal theory of group adaptation. *Journal of Evolutionary Biology, 22*(4), 659–671.

Hawley, K. (2001). *How Things Persist.* Oxford University Press.

Horvarth, C. D. (1997). Some questions about identifying individuals: Failed intuitions about organisms and species. *Philosophy of Science, 64*(4), 654–668.

Kaiser, M. I. (2018). Individuating part-whole relations in the biological world. In *Individuation, Process, and Scientific Practices,* edited by O. Bueno, R. L. Chen, and M. B. Fagan. Oxford University Press.

Leigh, E. G. (1971). *Adaptation and Diversity.* Freeman Cooper.

Maynard Smith, J., and Szathmáry, E. (1995). *The Major Transitions in Evolution.* Oxford University Press.

Robertson Ishii, T., and Atkins, P. (2023). Essential vs. accidental properties. In *The Stanford Encyclopedia of Philosophy,* Spring, edited by E. N. Zalta and

U. Nodelman. https://plato.stanford.edu/archives/spr2023/entries/essential-ac
cidental/.

Rosen, G., and Dorr, C. (2002). Composition as a fiction. In *The Blackwell Guide to Metaphysics,* edited by R. Gale. Blackwell.

Van Inwagen, P. (1990). *Material Beings.* Cornell University Press.

Wilson, D. S., and Sober, E. (1994). Reintroducing group selection to the human behavioral sciences. *Behavioral and Brain Sciences, 17*(4), 585–608.

CHAPTER FIVE

The Conflict Behind the Conflicts

Organisms as Agents and as Ecosystems

PHILIPPE HUNEMAN

ABSTRACT   Current theoretical biologists and philosophers appeal to two very different schemes to investigate and explain organisms. For some evolutionary biologists, an organism is analogous to an agent maximizing its utility (namely, its inclusive fitness); for others, agency is needed to make sense of plasticity or niche construction. But other research programs emphasize the fact that all organisms are made up of entities of various genomes (e.g., a host and its microbiota) that render them like ecosystems. The former case uses an economic understanding of organisms; the latter appeals to an ecological conceptual apparatus. Both schemes seek to make sense of conflicts internal to organisms. In this chapter I describe these two schemes, explain their respective rationales, and argue that they capture heterogeneous aspects of the organism. In the last section, I explore the tension between them. Finally, considering two phenomena at the ontogenetic and the phylogenetic level—aging and major evolutionary transitions in individuality—I argue that the two schemes can be compatible within a diachronic approach.

Introduction: The Agential View and the Ecosystem View

While evolutionary and molecular biology hasn't always made room for organisms, we are witnessing a "return of the whole organism" (Bateson 2005). Some disagree on the nature and intensity of this return. But it is undeniable that systems biology, which integrates complex genomic and epigenomic pathways, and evolutionary developmental biology (evo-devo) have put organisms in the theoretical foreground by emphasizing the role of developing organisms within evolution. Two concepts—phenotypic plasticity and heterochrony (Raff 1996; West-Eberhard 2003)—are key in this theoretical transition. Adopting these concepts, Pepper and Herron (2008) argue for the organism's explanatory indispensability, though Huneman (2010) distinguishes specific conditions under which evolutionary biology can bracket the developing organism in its models and explanations. Queller and Strassmann (2009) and Strassmann and Queller

(2010) propose that organisms should be part of an evolutionary space of organismality, while Walsh (2013) prefers organismal biology to the suborganismal biology still dominant among molecular biologists and evolutionists. These ideas raise the deeper question of how organisms should be conceived.

Historically, evolutionary biology has provided a general theoretical framework for thinking about biological individuality, based on grades of selective processes. This framework (Maynard-Smith and Szathmary 1995; Michod 1999) unravels the general processes leading collectives of entities (e.g., cells) to become a cohesive individual that reproduces and behaves as a single entity. Multicellular organisms are conceived as a type of individual, yet the ontological conclusions of this program hold for any individual and therefore can't fully characterize what the term "organism" means. As Godfrey-Smith (2009, 2013) argues, "Darwinian individuals" and "organisms" are different concepts whose extensions often overlap.

Biologists and philosophers alike have considered a conceptual scheme for making sense of organisms' agency. It is striking that across very different conceptions of what evolution, evolutionary processes, and evolutionary biology are, some views concur that organisms are agents. Two major philosophy books have followed Rob Wilson's book *Genes and the Agents of Life* (2006): Samir Okasha's *Agents and Goals in Evolution* (2018) and D. M. Walsh's *Organisms, Agency, and Evolution* (2015). Both books explore the idea that organisms are agents. Yet they disagree on what agents are, aside from the fact that agents instantiate a pole of self-standing action. Okasha mostly considers a tradition that began with R. A. Fisher, which views organisms as the seat of a fitness-maximizing process because they have been shaped by natural selection. Even though he endorses neither Fisher's nor Alan Grafen's more recent view of agency as a legitimate concept (Grafen 2006, 2007, 2014), he subscribes to their idea that agency is a scheme justified by the fact that natural selection causes organismal traits. For Okasha, agency is an explanatory tool used by modelers, but not part of the "furniture of the world," as Bertrand Russell used to say. By contrast, Walsh, building on the views of biologists like Sultan (2017) or West-Eberhard (2003) on phenotypic plasticity, argues that "agency" should be seen as an irreducible property of organisms.

Seemingly incompatible with the agency view, another view of organisms has emerged recently. It sees organisms as ecosystems. According to this view, no multicellular organism is functional only through its genetically homogenous cells, which stem from the initial zygote. For instance, many bacteria are involved in the essential functions of an organism. Why should the members of this ecosystem be considered conceptually different? A set of individuals of various species undergoing mutualistic, sometimes parasitic, but also competitive and predatory interactions, is exactly what ecologists call a community—and it is an ecosystem, when considering it with all abiotic materials it involves (water, minerals etc.). This way of thinking appears in the study of cancer and physiology (Costello et al. 2012). It also appears when authors use concepts such as "niche" or "niche construction" to study cell development (Scadden 2006).

Both the agency and the ecosystem concepts seem imported from outside of biology—agency from the rational agents studied by economists, and ecosystem from ecology, which is still independent and distinct from evolutionary biology, as demonstrated by the various unfulfilled attempts to unify the two fields (Huneman 2019). The tension between the two is thus partially a tension between disciplinary approaches. This chapter investigates this epistemic tension, which appears to raise the issue of pluralism—namely, the coexistence of two or several explanatorily or predictively equivalent epistemic schemes. An important instance of pluralism in evolutionary biology concerns the units of selection issue. Researchers disagree on the proper target of natural selection (genes, organisms, groups?). For example, biological altruism can be explained by an appeal to multilevel selection (occurring at several levels, often including genes and organisms) or by using kin selection (occurring at the level of genes). Pluralism is a way to acknowledge that, while different, the two views are both illuminating and justified. I will argue that each organism scheme is pervaded by theoretical conflicts but that each also offers a way to make sense of a range of conflicts. I will also address the major evolutionary transitions in individuality (ETI), a research program aimed at unraveling the general processes through which what was once a group of "separate" individuals becomes an individual proper (Maynard Smith and Szathmary 1995). While visible at different levels of biological organization, the canonical transition is the one toward multicellular organisms. This transition raises several questions: Should newly derived multicellular organisms be conceived of as agents? Are the major transitions simply journeys toward agency? And in this case, how would the ecosystem concept of the organism be applied?

The first section of this chapter explores the rationale and internal epistemic conflicts of the agency view. The second section considers the ecosystem view. The third addresses the scope and limits of each view. The fourth analyzes the conflict between these two views as a conflict between two logics—economic and ecological. Finally, the fifth section examines the way this conflict actually occurs within the organism itself, from both synchronic and diachronic viewpoints. The first subsection focuses on aging, the second on the ETI program and what the ecosystem/agency conflict can tell us about these transitions.

Organisms as Agents

There are many concepts of agency; given the philosophical controversies that have surrounded the word "agency" I won't pretend to give a definitive concept of the word. Still, we can begin a rough characterization with which everybody should probably agree, before disagreeing on the explication of all its elements: An agent is a system that initiates a sequence of events, in relation to properties of its environment, which is likely to be interpreted as leading toward a "goal" or a preferential state of the world.

Wolves hunt; peacocks court peahens; foxes run after rabbits; beavers build dams; and so on. All of this intuitively counts as action—these behaviors are directed toward a goal, and the direction is explanatory. As philosophers since Wittgenstein and Anscombe have emphasized, even when such behavior misses its goal, it can still be described as goal-directed or as fulfilling a function (a hunt, a courting behavior). An empty-handed hunter is still a hunter, and a broken steering wheel retains its function of governing the car's direction. Finally, the thing that does the actions should be seen as an agent. Teleological terms like actions and functions are normative to the extent that they introduce two non-symmetrical states for the world (goal reached, or not; function fulfilled, or not).

Because biology tends to avoid causes that are not nomothetic or efficient, agency therefore is often considered an anthropomorphic description rather than an explanation of what goes on. Its legitimacy in the biological literature is much weaker than the legitimacy of functional concepts, for instance. I won't delve into the huge philosophical literature on teleology, functions, and goal-directedness. Suffice it to say that there are several philosophical ways to make sense of these notions without expelling the agent. But the recent trend toward taking agency seriously stems from two independent and even contradictory sets of justifications.

Two Approaches to Biological Agency

The first justification for conceiving of organisms as agents is that they have been shaped by natural selection, which is in essence an economic process. The connection between economics and evolution has been widely accepted but rarely investigated, with the exception of Okasha (2018) and André et al. (2022). The core idea is that natural selection and rationality—the key processes in evolutionary biology and in economics, respectively—are somehow parallel, or even (for some writers) identical. Explaining a phenotypic trait by natural selection means that it exists because it maximizes fitness, and therefore it is an adaptation; explaining an action with rationality means that an action is chosen because it maximizes utility. Both explanations share a maximizing logic. Not everyone considers natural selection a maximizing process (Birch 2016). However, under this perspective, organisms can be seen as agents the same way economics considers humans economic agents. Granted, natural selection is a population-level process while rationality compares actions that are in the mind of one individual. Furthermore, the utility function is subject to and specific to the individual, while fitness, as John Maynard Smith (1982) points out, is objective because it is measured in relation to number of offspring. But in any case, the two processes work the same way.

Alan Grafen, in his "Formal Darwinism" project, created a sophisticated framework within which the population-level process modeled by population genetics can be isomorphic to a rational choice of optimal phenotype by organisms (Grafen 2002, 2007). He describes the "maximizing agent analog," emphasizing that this account is analogical: Amoebas or nightingales are not by themselves agents in the same sense that they have a mass or are composed of cells. But this analogy nonetheless allows the use of economic concepts such as rationality and utility maximization to make sense of the data regarding what organisms do. It avoids the difficult issue of generality. While many of us are ready to ascribe agency to gorillas or sheep, we might be reluctant to do the same with lice or oysters because the latter seem to lack cognitive abilities, at least in the colloquial sense of the term. (Granted, there is an ongoing discussion about ascribing cognition or even consciousness to animals devoid of comparable brain functions, such as insects.) Yet natural selection still allows us to consider them agents. Hence one can apply the category of "agency" to all organisms, even to those that lack cognitive ability.

As Okasha (2018) shows, while likening selection to a rational agent is contentious, the argument that organisms are agents is quite plausible. He argues that this is a result of natural selection, since organisms have options for maximizing fitness and seem to choose one. When it comes to species capable of learning and therefore of "choice," we might ascribe to them a minimal degree of rationality—the instrumental rationality of economics (Kacelnik 2006), which requires transitivity of preferences and immunity to a changing third option (or constancy of preferred pairs): An agent who prefers C to B and B to A must prefer C to A, if she is rational in this sense. In other words, she must exhibit constancy in her preference for B over A. If this requirement is not met, any explanation or prediction of her actions is not possible. However, we seem to violate this principle of transitivity all the time in our daily lives. For example, we might prefer a $3 chocolate bar to a $4 chocolate bar when presented with chocolate bars at $3, $4, and $5, but then we might prefer the $4 chocolate bar when presented with chocolate bars at $3, $4, and $22, thus violating the constancy of our preference for $3 over $4 chocolate bars. Here, the agent might think that, after all, a $1 difference for having better chocolate is nothing as compared to the difference with the most expensive chocolate.

The fact that some animals fail to behave rationally might be seen as a falsification of the rationality hypothesis in biology. However, experiments in which individuals seem to violate the transitivity of preferences can always be redescribed in a way that does not force us to see these animals as irrational, in the sense that selection would have favored irrationality (Huneman and Martens 2017). An easy way to deal with the third-option immunity problem, such as for chocolate bars, entails redescribing options so that the counterexamples disappear. For instance, in the latter examples, the objects of our preferences are not only chocolate bars, but the tastes of those bars joined with a feeling of saving more or less money, and in this case our ranking appears constant across choices.

However, some biologists and philosophers refuse to think of agency in analogical terms (e.g., Sultan et al. 2022). They argue in favor of a genuine instantiation of agency by organisms. It is not the fact of selection that justifies agency; on the contrary, it is the limits of natural selection that do so. According to them, natural selection can't be a cause of adaptation. Walsh (2010) makes this case forcefully, drawing on a view of natural selection as a statistical rather than causal explanation. He bolsters this epistemological analysis—which some might find unconvincing—by drawing on West-Eberhard (2003), who sees phenotypic plasticity as initiating adaptive evolution rather than recording it (the "genes as followers" as opposed to the "genes as leaders" view); on Sonia Sultan's (2015) work on plasticity in plants; and on Odling-Smee et al.'s (2003) theory of "niche construction," which stresses the causal impact of organisms on their environment and the selective pressures around them. For Walsh, natural selection is a statistical aggregate recording effects of differences in fitness. But the cause of these differences, he claims, should be organisms' activities in the struggle for life—an idea that resonates with Darwin's original conception. Thus, agency for him is presupposed by a selectionist explanation. And since agency is distributed across all species, including plants, this account minimizes the requisites for being an agent. Cognitive capabilities are no longer necessary. But an agent still possesses a behavioral repertoire, correlated to several possible states of the world, which it ranks according to its preferences.

Thus, there are two rival approaches to organisms as agents, but they concur at least in the description of what an organism does. Interestingly, their cleavage reflects a broader cleavage in the philosophy of science, between instrumentalism and realism. For instrumentalists, one can forge any model of the focal system, provided the model allows robust predictions and causal inferences. This is what "maximizing agent analogies" do. Realists believe that science intends to unravel the furniture of the world. For them, this is exactly why agency should be ascribed to organisms, since it appears as a key property of the living. A drawback of the latter stance is that it uses "agency" in a counterintuitive way. A drawback of the former is that it erases the differences between species since it treats them as agents in the same way.

Challenges

One of the major concerns for Grafen and proponents of the maximizing agent analogy is establishing the conditions under which this analogy is valid, given that organisms are made of organs, parts, cells, and other elements such as plasmids, or commensal bacteria, and genes. Why should we consider the organism the agent and not the genes or the cells?

One answer stems from what Okasha (2018) calls "unity of purpose," which I consider more a spectrum than a binary concept. Okasha

argues that organisms display goal-directed behavior; that they are analogous to rational agents because they were mostly shaped by natural
selection, a fitness-maximizing process; and that both of these factors
produce decision-makers oriented toward the maximization of utility,
which is a proxy for fitness. However, these agents are made up of parts
that will disrupt the goal-seeking behavior of the whole if their preferences contradict the preferences of the whole. (Think of a cancer cell
preferring lots of sugar, or bacteria in the dental film, which show this
preference too.) Thus, organisms are agents only when their parts are
significantly aligned in their fitness-maximizing interests. The economic
notion of alignment of interests is key to the concept of agency. This
alignment should be understood as more or less realized. Some lack of
alignment between cells or within a genome is possible within an organism, and its "decisions" may precisely proceed from this weak alignment, as in the case of genomic imprinting. But unity of purpose still
holds to the extent that no part or gene silences all the others. Any
individual gene may have an interest in enhancing its probability of inclusion in the gamete, but because all genes have an interest in being
represented their interests can be aligned under fair meiosis, which
equalizes chances of inclusion. Granted, there is potential conflict.
Hence, genes like segregation distorters evolve. But the overall alignment of interests for all genes, which ensures that none enjoys a high
probability of representation, establishes fair meiosis against segregation
distorters. Models showing how this alignment of interests is established
and maintained have been designed by researchers studying the evolutionary transitions of individuality. But whatever model we prefer, it is
clear that alignment, or "unity of purpose," has a key role in specifying
what we mean when we talk about agency in evolutionary biology.

The ETI paradigm has often been characterized as a research program
about the emergence of organism-like entities—for example, the origin of
multicellular individuals from single cells (Grossberg and Strathmann
2007). It therefore has a natural connection with the agency view of organisms because these transitions appear to bring unity of purpose to a
group of interacting living entities, a condition of agency. But what about
cases in which this transition yields organisms but the alignment of interests is not complete? The agency framework may fail to be relevant, and in
a subsequent section ("The Paradox Within the Organism") I'll say a word
about what can frame an investigation of collectives of particles not likely
to reach full individuality but still likely to play evolutionary roles.

Metaphysical Stances

We can now posit that an organism is an agent because, unlike cells or genes, it is the larger set of entities that displays unity of purpose. One could ascribe agency to each cell (in the eighteenth century, van Helmont talked about the "soul" of each organ), but of course, it remains more parsimonious to see the organism as the agent instead of multiplying agencies.[1]

However, for realists about agency, things go the other way around: Unity of purpose, instead of being a requisite for ascribing agency, is a consequence of organisms being real agents. Because the organisms are genuine agents—as our phenotypic plasticity studies, evo-devo models, and niche construction accounts taught us—they should not be torn apart by conflict. Hence, when conflict does arise, for instance between cancer cells and normal cells, the realist viewpoint considers it a pathological anomaly—just as economics considers irrational behavior an anomalous deviation from the utility-maximizing norm. Yet the same caveat holds: Minor irrationalities, like our preference for the $4 chocolate bar, are not marks of pathological insanity but routine occurrences. Likewise, minor conflict should not be seen as pathological even in the realist viewpoint. (Notice also that the agency realist, unlike the instrumentalist, faces an additional issue in explaining the origin of agency.)

Even though the connections between the alignment of interests and agential talk run in opposite directions for realists and instrumentalists, both stances agree that organisms as agents have somehow achieved a unity that allows them to behave as a whole and therefore to be seen as transparent wholes. I call this strong individuality, and its discussion is key to the ecosystem view, to which we turn now.

Organisms as Ecosystems

A Shift Within Organicism

Until the 1950s, many ecologists tended to see ecosystems as organisms. In 1916, Frederick Clements famously defended the idea that ecological communities—namely, sets of species interacting at a local scale—display

---

[1] I leave aside here the question of the timescales between processes at the level of the parts and at the level of organisms themselves. See Bourrat (2023) and Pocheville (2018). This difference, besides parsimony, justifies ascribing agency to the organism. Thanks to Pierrick Bourrat for pointing this out.

development and metabolism, and hence go through stages of infancy, adulthood, maturity, and decay. Later ecologists integrated this view into a general Darwinian account based on the idea that selection for communities would optimize and integrate communities in the same way that selection at the level of organisms is responsible for their integration and vital functions (Allee et al. 1949; see also Huneman 2025). For many reasons, including the dismissal of group selection by Williams (1966), this view faded away (Mitman 1988; Borrello 2003). But the inverse idea, that organisms can be seen as ecosystem-like, has gained popularity. As in the former conception, this is often an analogy. It may be pushed to the extreme with the Gaia account proposed by Lovelock and Margulis, which has been carried over by Earth System Sciences (Dutreuil 2024). Such an analogy strongly contrasts with our previous "agent-based analogy."

But it is sometimes more than an analogy. Genuine ecological schemes of thought—namely, the analysis of ecological interactions of predation and competition or mutualism—are often employed to pursue physiological research questions. Costello et al. (2012) use "ecological theory" to model the microbiome. The fact that vertebrate organisms have many commensal bacteria that allow them to perform vital functions is a justification for this scheme. This is clear to cancer researchers because the competition between cancer cells and healthy cells for resources is key to understanding tumors. Featherstone and Durand (2012), for example, see cancer progression as an ecological process.

Nevertheless, ecosystem talk is often motivated by other considerations. For instance, explaining the dynamics of stem cells may require appealing to the ecological notion of niche, a key concept of community ecology as Hutchinson (1957) formulated it (Scadden 2006). Kupiec (2009) intends to substitute the model of competing and predating cells underpinning the development of an organism for the classical notion of a genetic program underlying the cells' overall functioning. Along these lines and in a more orthodox fashion, Lecointre et al. (2020) question the notion of a difference between evolution and development. They view development as an evolutionary trajectory that relies on the ecological interactions of cells and bacteria, in the same way evolutionary change relies on the ecological interactions of organisms.

Ecosystem Individuality

If organisms are indeed ecosystems, what is the nature of these ecosystems? Ecological organicism, such as Clements's, is not a satisfying account here, because if ecosystems are organisms, we would face a logical circle between being an organism and being an ecosystem. However, if one follows the view of Gleason, Clements's fierce and foremost critic, there is no ontological robustness of ecosystems: Species that come to live in a community are there because of stochastic processes (Gleason 1926, 1939). But then we would lose the integrated aspect of organisms when we want to think of them as communities or ecosystems.

In fact, ecologists themselves have debated the ontological character of communities and ecosystems, since Gleason's view makes them an artifact of the ecologists' modeling instead of parts of the "furniture of the world" (see Sterelny 2006 and Huneman 2014b for an overview). At stake here is the availability of an account of ecosystems and organisms as real individuals that avoids organicism while not falling into Gleason's methodological individualism.

One powerful account of biological individuality comes from evolutionary biology—more precisely, from the philosopher David Hull's (1980) claim that "to be an individual" means "to be a unit of selection." Most of the more recent accounts of biological individuality subscribe to this general idea, providing complex specifications (e.g., Sober and Wilson 1998; Godfrey-Smith 2009; Bouchard 2010; Folse and Roughgarden 2010; Clarke 2013; Goodnight 2013). However, this is not a solution for ecosystems or communities, since they do not display any uncontroversial heredity. Moreover, ecosystems can't show a response to selection (because they include abiotic elements), even though organisms can be said to inherit a territory (Odling-Smee et al. 2003). While test-tube situations have been interpreted as evidence in favor of ecosystem selection (Swenson et al. 2000), this is not a promising pathway for addressing the question of the individuality of ecosystems in a manner familiar to practicing ecologists.

I have previously proposed conceptualizing ecosystems as "weak individuals," as subsets of interacting entities that emerge among a huge set of interactions and are characterized by the fact that subset-internal interactions are somehow much stronger than the other interactions—using a sense of "strong" that has to be specified by a focal theory (namely, the one pertaining to the research questions) (Huneman 2014a, 2014b, 2020). In a nutshell, there is a decoupling between the pattern of interactions proper to the putative individual and the overall set of interactions. The detail of this account—which includes a probabilistic scheme for detecting individuals—is not relevant here. The point is that there is a way to ascribe a theory-based, objective individuality to some ecosystems, even though this individuality is weaker than the individuality of the "unit of se-lection." It is weaker in two distinct senses. First, determining the indi-viduality of ecosystems depends on a theory of interactions. Therefore, these individuals don't stand by themselves in the same way that units of selection do, because the latter inherit the ontological character of se-lection. Second, individual ecosystems do not reproduce by themselves.

In this sense, one can conclude that organisms are indeed ecosystems because the many species entities that participate in their functioning, de-velopment, and reproduction fulfill the general scheme of interaction-decoupling that defines weak individuality. In virtue of this, organisms have the same kind of individuality as ecosystems. In effect, an ecosystem often amounts to a set of organisms or bacterial colonies that compete, predate, or are commensal or parasitic upon other organisms in a way that is decoupled from all other interactions.

I began this section by considering metaphorical views of ecosystems as organisms that are used when one wants to investigate a specific issue, allowing for this metaphor. This seems to assume that organisms are individuals in a stronger sense than ecosystems, because their individu-ality (in terms of cohesion and integration) is arguably not metaphorical. But adopting the previously mentioned weak notion of individuality induces a shift in the ontological status of individuals; they are not un-derstood as substances but as a kind of decoupling pattern, assuming a set of interactions and a model allow for this decoupling. If organisms are individuals in this weak sense, then they are on a par with ecosys-tems. Let's briefly explore this parity.

The fact that an organism's parts may conflict is now an element of the picture since ecosystems, by definition, include antagonistic inter-actions such as predation and competition. Hence, while the agent scheme requires some alignment in fitness between all cells and parts, the ecosystem scheme is more liberal because conflicts are an inherent part of the picture and may still resolve into stability, equilibria, and organism maintenance. Ecosystems do show stability properties as a re-sult of predation or competition, as ecologists have documented (see Kingsland 1995, Cooper 2003, and Justus 2008 for analyses). Organ-isms and species in ecosystems may not be aligned in fitness or interests, yet ecosystems display some cohesion and some individuality—namely, weak individuality. Indeed, one of the key long-standing questions of ecology concerns the existence of population regulation, which means that the abundance of a species fluctuates in a way that is not chaotic and is determined from within the population. This can be due to competition, yielding what is called density-dependent regulation (see Lack 1954), or due to predation. Conflict is part of the existence of an organism, but ecological logics provide us with ways to understand its presence because ecological dynamics promote some regulation and stability, even through antagonistic interactions (i.e., competition, predation, parasitism).

The weak-individuality idea might seem like a tiny metaphysical quibble. However, it lends legitimacy to the ecosystem-analogy approach to physiology and cell biology that I highlighted above. And just as the agency approach is torn between realistic and analogy-based variants, the organisms-as-ecosystems approach is split between methodological individualism and holism. The former equates weak individuality with artifact individuality, because weak individuals are strongly theory-dependent. The holistic view, for its part, ultimately equates ecosystems with organisms and ascribes organism-like properties to them, such as metabolism, development, and the capacity for health and disease (Dussault 2019). But in the present case, I would argue that weak individuality is neither holistic (as all brands of ecological organicism are) nor purely artifactual; theory-dependence cannot determine all aspects of the interaction-decoupling pattern constituting a weak individual because the fact of these interactions themselves is not theory-dependent.

Now that we have covered two possible schemes accounting for organisms, let's question their respective scopes.

Agents, Ecosystems: Their Rationales and Their Conflict

A Pluralism

Agents and ecosystems offer two powerful ways to account for important features of organisms. The agent analogy comes from the logic of natural selection, which parallels rationality. The ecosystem analogy makes sense of the fact that each multicellular organism hosts a variety of microorganisms, as does an ecosystem. And in both cases, there are views that drop the analogy dimension altogether and view organisms as agents or ecosystems, sensu stricto.

The rationale of both views is different, and they pull organisms in two different directions: economics on the one hand and ecology on the other. This may reflect the evolutionary process itself, since this process starts with organisms preying on and competing with one another—engaging in ecological interactions—and gives rise to natural selection; it is akin to rationality, as it is a trend toward maximization (Fisher 1930; Okasha 2006). Sober (1984) explains this as a duality between "source laws" (laws of ecology and development, which constitute the fitness value of an organism) and "consequence laws" (population and quantitative genetics, which model the allele frequency dynamics in virtue of the various fitness functions, hence supervening on the former level of source laws). Thus, the duality between the organism schemes reflects the dual nature of evolutionary dynamics.

Different as they are, the two schemes I analyze here are not exclusive, and in principle not competing. For example, take the Hawaiian bobtail squid, whose luminescent appendages are composed of a society of bacteria, allowing it to shine and therefore hunt without being betrayed by its own shadow (Bouchard 2010, 2013). The squid without its bacteria may be seen as the agent in the "principal-agent" model (Edwards et al. 2006; Weyl et al. 2010), according to which it chooses bacteria strains daily. The bacteria it swallows each morning allow its phosphorescent appendage to project a light beam onto the benthic floor, concealing its own shadow to the potential prey. Alternatively, it can be seen as an ecosystem when considering how the squid organizes the turnover of the bacteria strains. Here, several bacteria are competing for resources, constituted by the squid appendix, while the squid through its mutualistic association with some of the bacteria of a given strain increases its fitness; mutualism and competition therefore combine as ecological interactions to yield a dynamic that maintains the bobtail squid's existence.

Another example is the termite mound. Turner (2013) makes a strong case for seeing it as a cognitive agent, which is a strong type of agent. Why? Because the mound itself seems to be aware of its internal state and triggers maintenance by the termites when it is altered. But one can just as plausibly study the mound as an ecosystem, where termites, fungi, and bacteria interact and allow for its persistence. Thus, we seem to have a clear case of explanatory pluralism (see Massimi 2022 and Huneman 2023 as applied to aging and death): Each scheme captures one dimension of what it is to be an organism and gives rise to legitimate models. The organism-as-agent scheme captures the adaptive nature of organisms. They are adapted because they are (conceivable as) rational decision-makers choosing the fittest phenotypes. Inversely, the organism-as-ecosystem scheme recognizes that the abundance of cells and other suborganismal parts fluctuates and therefore grasps the maintenance of organismal integration and its co-existence with an intrinsic diversity.

Evolutionary biologists have become used to the pluralism issue because they have worked through the conflict between multilevel selection and the gene's-eye view. Many have realized that kin selection and multilevel selection are in principle equivalent, even though biologists favor one or the other (Kerr and Godfrey-Smith 2007; Gardner et al. 2011). Yet, pluralism as an epistemological category is equivocal. Some pluralisms are about making pragmatic choices between two equivalent and compatible stances—let's call this "conciliative pluralism." Other pluralisms involve processes that are equally possible and sound, but objectively conflicting. Only one can take place at a time in a given system—let's call this "conflicting pluralism." Applied to the case of kin versus multilevel selection, a conflicting pluralist account might argue that altruism in the case of vervet monkeys or antelope stotting behavior is due to kin selection, while altruism in the case of bats is due to multilevel selection. "Conventionalism" is an appropriate label for conciliative pluralism because we set conventions to use one or the other view. The question raised by the coexistence of agent and ecosystem schemes for making sense of organisms is therefore whether we should adopt conventionalism (as in Kerr and Godfrey-Smith 2007 or Gardner et al. 2011) regarding multilevel selection), or whether conflicting pluralism is more reasonable.[2] This conventionalism has been favored in the case of units of selection, especially by Bourrat (2025), who argues that biological individuality with respect to natural selection has to be seen as mostly justified conventions.

The duality or coexistence of stances also determines the scopes of each. The agent scheme has to assume some integration, which implies that many systems such as aging organisms losing functional integrations wouldn't be counted as agents (and I will return to this point), while one intends a scheme of organismality to hold for all organisms and constantly. On the other hand, the ecosystem scheme hardly makes sense of the fact that organisms are adapted. The principled privilege of adapta-

---

[2] I thank Pierrick Bourrat for having raised this suggestion, which puts my case for pluralism in a much more general light. And to some extent, as he says, the ecosystem view sees organisms in a bottom-up way, from their most elementary components and based on the interactions; however, the agent scheme considers things from the top down, like multilevel selection, which starts from groups and not elementary genetic units. See also Huneman (2021) on the relations between kin and multilevel selection.

tions among all traits of organisms, and their decisive character for understanding the normative aspect of organic life, is brushed aside here. Thus, each scheme incurs a cost in terms of the scope of its capacity to make sense of all the many facets of organisms' existence.

Different Contenders

The ecosystem scheme and the agent scheme are both rooted in evolutionary thinking. The agent scheme appears to come from the applicability of natural selection to living organisms. It is a logical consequence of the fact that the organism evolved through natural selection, as the "evolutionary transitions" program argues (Maynard Smith and Szathmary 1995; Michod 1999). The ecosystem scheme's connections with evolutionary thinking are less direct. But considering the entanglement between metazoan organisms and bacteria, the ultimate rationale for the ecosystem scheme, we can see that an evolutionary process is underpinning all the relevant states of affairs. Let's explore this point in more detail.

Consider the microbiome, which allows mammals—and even some nonmammals, like the Hawaiian bobtail squid—to digest food. In these cases, organisms evolved through the intertwining of host and bacteria. They exist at the crossroads between two distinct and distant evolutionary pathways: vertebrates or cephalopods on one hand, bacteria on the other. Several accounts of this intertwining have been proposed. Is it the result of coevolution, in which each species evolves traits in reaction to traits other species have evolved (Bordenstein and Theis 2015)? In other words, is it like the arms race that endowed carnivorous mammals with efficient eyesight and claws or herbivores with acute motion detectors and speed? Or is it more than that? Instead of two evolutionary dynamics mirroring each other, could the host-bacteria pair constitute a unit of selection, as Bouchard (2010, 2013) suggested? According to the idea of individuality sketched by Hull, this would make the pair a biological individual. Such a "holobiont," as Zilber-Rosenberg and Rosenberg (2008) call it, adds another kind of entity to the furniture of the biological world.

Not all biologists and philosophers have accepted this idea. To some, the holobiont concept seems to ascribe too much to the microbiome, which can be conceived of as a mere result of coevolution. Others point out that associations and collectives occur throughout the ecological hierarchy, from cells and organisms to colonies and ecosystems, as well as throughout the genealogical hierarchy, from genes to organisms to species (see Eldredge 1985 on these notions). In the former case, the concept of the holobiont is rejected for reasons of parsimony; in the latter case, it appears almost too trivial. Doolittle and Inkpen (2018) have developed the "It's the singer not the song" theory to emphasize the prevalence and pervasiveness of associations that play some role in Hutchinson's "ecological theater," even though the members in these associations change into other functionally equivalent species. Bapteste and Huneman (2018) argue that the evolutionary units should be seen as living networks that include abiotic materials in addition to cells and bacteria and that operate at various time scales.

The jury is still out on the views that emphasize collectives over separate entities. Yet it is clear that evolutionary dynamics, both logical and psychological, are at the root of both the agent and ecosystem schemes. However, their extensions are not comparable. While many organisms—even those comprising a combination of homogeneous cells (descending from the same zygote) and bacteria, connected to each other through complex ecological relationships—may be counted as ecosystems, agency is more restricting. Its unity-of-purpose condition may prevent some purported organisms from counting as agents because their parts experience too much internal conflict. An example here would be the slime molds in the genus *Dictyostelium*. When these are in a multicellular phase, they behave like organisms. Given the division of labor between the slug, the stalk, and the spores, they display the alignment of fitness required for agency. However, when trophic stress (or any stress) decreases and the cells come apart, it is hard to see the system as an agent. But the system can be seen as an ecosystem, emphasizing how other unicellular species are entangled with the population of *Dictyostelium* cells. Thus, the overlap between the agent and ecosystem schemes are partial. Each scheme captures distinct aspects of being an organism. Epistemologically speaking, it means that the coexistence of these two schemes instantiates a specific case of explanatory pluralism. Nonetheless, the simultaneous existence and use of these two schemes might lead to conflicts, some of which I discuss in the next section.

The Ecologico-Economical Paradox

Although we haven't yet seen an obvious contradiction between the agent and ecosystem schemes, one might still doubt that they are compatible and suspect that we are compelled to choose one or the other.

Conciliations?

As long as we assume our theoretical frameworks are only descriptive, there need not be a major conflict between the agent and ecosystem views. One can endorse the two stances simultaneously in order to cast different light on the same organism, while recognizing that the schemes are only meant to be an approximation and not reality itself. The explanatory conciliative pluralism described above should be the default.

Before proceeding further, however, we should clarify the distinction between realist and more relativistic notions of organisms as ecosystems. Most nominalist or relativistic notions of ecosystems—in line with what Sterelny (2006) calls "indexical communities," that is, communities that exist in the eye of the beholder—are not suited to make sense of organisms. This is because they give up the idea that the integration of parts, like the tree species in a Gleasonian forest, can be real, independent of the modeling intent of the researcher. But organisms apparently exist and are not mere theoretical constructs. Inversely, the ontologically strongest notion of ecosystem is a holism of Clements's kind. Here, the ecosystem itself is conceptualized along the lines of an organism, which reduces the epistemic significance of the ecosystem-organism analogy and makes it less useful for understanding organisms. Remember, the ecosystem stance on organisms intends to make sense of some features of organisms, so it cannot presuppose what organisms are. Hence, ecological organicism is excluded here.

The weak-individuality approach to organisms aims at a notion of ecosystem that is neither purely individualistic, nominalistic, and relativistic, à la Gleason, nor organicist (Huneman 2014a, 2014b, 2020). It is faithful to the notion Hutchinson (1961) had in mind when he thought about genuine ecological communities. Regarding agency, Hugh Desmond and I have defended a Kantian position that seems to be a middle ground between realism and analogy—namely, the thesis that agency is explanatorily indispensable (pace Grafen) but not ontologically weighted (i.e., animals are not sensu stricto agents, pace Walsh) (Desmond and Huneman 2020). Bourrat (2025) defends an analogical view of individuality, viewing it as an explanatorily indispensable concept that nonetheless does not explanatorily belong to the furniture of the world. Thus agents, according to our Kantian account, and ecosystems, according to the weak-individuality view, both occupy a middle ground between extreme ontological views and might be perfectly compatible for this reason.

Leaving this aside, if we treat agency in an ontological sense, implying that amoebae, sponges, or sunflowers are agents (like me, a human, minus some cognitive abilities), then we cannot see these organisms as ecosystems, which lack the agential capacity for acting as a whole. Therefore, we face a conflict between the economic logic, which bases its explanations on rationality, and the ecological one, which encompasses several distinct explanatory strategies (ecological networks, niche models of biodiversity patterns, neutral models, etc.). However, this epistemic conflict covers another more interesting dimension of within-organism conflict, a case for conflicting pluralism as indicated above. The rest of this chapter will explore it.

Internal Conflicts

Recall that both the agent scheme and the ecosystem scheme can make sense of a specific intra-organismal situation of conflict and provide distinct clues about its resolution. The agent scheme implies that conflict exists but has been superseded by the alignment of fitness between all cells. Often, genetic homogeneity provides such alignment, as in the case of metazoa, in which the bottleneck constituted by the one-celled zygotic initial stage ensures that all cells will have the same genome. On the other hand, the ecosystem approach can address internal conflict with the idea that ecosystem interactions yield population regulation. As I previously noted, population ecology struggled for decades with the question of the regulation of species abundances. The cleavage between density-dependent and density-independent views of regulation proved especially divisive (Kingsland 1995). Hutchinson (1957) established a truce in this controversy by showing how both approaches can be valid, though for distinct species and at different temporal and spatial scales. For my purpose here, it is enough to say that the ecosystem scheme can make sense of persistent conflicts within the system by showing that they produce regulation by themselves, through antagonistic interactions such as competition. Thus, the paradox of a system that emerged from entities whose interests may diverge—to put it in language familiar to evolutionists—can be resolved through the most general idea of ecosystems' dynamics.

To summarize, conflicts within organisms are ubiquitous, and being an organism means being capable of containing these conflicts at an imperceptible or nonconsequential level. Agents and ecosystems are two ways contemporary theorists account for the resources that some organisms use for monitoring and defusing such conflict. Therefore, a basic internal tension that underpins an organism's dynamics stands between kinds of conflict and their potential resolution, because the ecosystem stance and the agent stance are rivals in the establishment and explanation of a pacified or buffered conflict. However, this notion of conflict is very abstract. In the next sections, I will consider how such conflict emerges within the organism, diachronically,[3] at two privileged timescales: the ontogenetic (with the question of aging) and the phylogenetic (with the issue of evolutionary transitions). I will then explore how a paradox of continuing conflict proper to organisms can be understood within either the ecosystem scheme or the agent scheme.

The Paradox Within the Organism: When It Becomes Impossible to Be an Agent and an Ecosystem

We have seen that our two schemes for understanding organisms can conflict. An organism can't always be both, even though as epistemic stances the ecosystem scheme and the agent scheme offer an explanatory pluralism. Ecosystems don't act. And an agent's requirement of unity of purpose means they can't be as heterogeneous (constituted by heterospecific entities) as ecosystems are. The two schemes are both valuable and sometimes overlap. However, when an organism-as-ecosystem is affected by intense internal conflicts, it becomes hard to characterize it as an agent. This issue affects the realist view of agency. But even the weak view, which treats agency analogically along the lines of Grafen, is ruled out because unity of purpose, a necessary condition for its applicability, is absent. The solution to this dilemma, I propose, is to accept the validity of both explanations but to recognize that they coexist at distinct timescales. On this view, ecosystem and agency are two poles of the biological changes that can occur to the kind of being we call an organism.

Aging

Let's consider the scale of the individual. Individuals develop, as we usually say, and species evolve. Showing all the ways in which agency-thinking and ecosystem-thinking co-occur or rather, alternatively occur, during

---

[3] Philosophers often distinguish synchronic and diachronic emergence, the former being the concomitant facts of two properties, one being emergent upon the other like thought and brain states; and the former being a process through which the emergent item is yielded. Humphreys (2016) proposed that diachronic emergence is the only genuine emergence.

the life cycle, would be a long task. I will therefore focus on one specific stage—aging—and then sketch a more general argument.

Aging can be described as the increase in probability of death after the reproductive peak. Although patterns of aging are distinct from species to species, a regular pattern has been identified, called the Gompertz curve. One major question in the evolutionary study of aging is why this curve is so pervasive (Huneman 2023, chap. 8). There are several theories, but one of the most general evolutionary explanations rely on what Haldane (1941) called the "selection shadow"—namely, the fact that the intensity of selection decreases with time because the pool of competitors naturally shrinks. This lets genes that are deleterious later in life accumulate in gene pools (Medawar 1957). Or it makes the evolutionary advantages of later-expressed phenotypic traits, such as life extension, much less significant than advantages occurring earlier; thus, even when these early-fitness advantages come with a cost to longevity, they will be favored by natural selection (Williams 1957).[4]

Due to the shadow of selection late in life cycles, life-threatening conflicts between parts of an organism can emerge long after the reproductive peak because, in general, they will be much less likely to be counterselected. Thus, the requisite of agency—namely, the unity of purpose—becomes less respected. Eventually, agency may no longer be a legitimate scheme to apply to an organism. Alternatively, if we take a realist approach, we would have to view these agents as abnormal, living in a subnormal state.[5] But these late-in-life organisms can still easily be conceived as ecosystems: Competition exists between parts, though it can be more intense than earlier, and hence can fail to produce as much regulation as it used to do.

Cancer, whose frequency tends to increase with age, offers a useful illustration of how aging can be conceptualized as a transition from agency to ecosystem. Broadly, cancers are rogue cells that multiply faster than normal cells, create a kind of niche in the interstitial tissue, and then live at the expense of normal cells. The immune system attempts to control emerging cancers (Pradeu 2013). The distribution of cancers at different ages shows that the efficacy of this control declines with time, which can be seen as a sign of senescence. Because cancers disrupt the

---

[4] See Huneman (2023) for a general overview and critical examination of these views.
[5] This latter view conflicts with the common idea that aging is "normal," and I won't enter into this complex discussion of natural norms.

unity of purpose, this is a case of aging fostering a failure of agency as an epistemic stance suited to organisms. However, the competition between cancer cells and wild type cells is an ecological process and legitimately falls under the ecosystem scheme of organismality.

A second example comes from recent literature on aging. Bapteste and colleagues (Teulière et al. 2021) have shown that some bacteria can be defined as "age distorters." Their interaction with hosts distorts the aging pattern of the host in a way that is beneficial to the bacteria. Teulière, Bernard, and Bapteste hypothesize in their studies that age distorters are common and that some lifespans can be understood as a result of compromises between the gene-based lifespan of an organism and then the age-distorting effect of a symbiont. Here too, the unity of purpose is disturbed. Likewise, what we see in an aging organism is the result of an ecological interaction between age distorters and the host. Aging seems to be a decline in agency, which renders manifest the existence of organisms as ecosystems.

Evolutionary Transitions

Evolution presents us with a general pattern. Living individuals often constitute groups, which, through evolutionary time, may cohere into a new kind of individual at a higher level of composition or complexity. Between more than four billion and one billion years ago, only unicellular organisms existed. Then multicellular organisms appeared. And then ways of living collectively evolved, culminating with organism-like social insect colonies that throughout their own social organization display a division of reproductive labor (queen/workers) akin to what characterizes most metazoa—namely, the sequestration of the germline (Buss 1987).

The major evolutionary transitions in individuality (METI) paradigm aims to explain the general processes that combine previously independent individuals into new, higher-level individuals at various timescales and levels. The most common approach uses the idea of selection at multiple levels, or multilevel selection (Michod 1999). In this view, individual-level selection in a population that is divided into groups favors "selfish" individuals that reproduce better or faster than others. Meanwhile, at the group level, another selection process favors groups containing the highest number of "altruist" individuals working for the group at the cost of their own fitness (see West et al. 2007). This very general scheme, which has been refined several times, has provided many insights on both the generality of evolutionary transitions and the specific case of the evolution of multicellularity (see Strathmann and Grossberg 2008). For us, the lesson is that the cohesive character of multicellular organisms results from evolution by natural selection, which means that this cohesiveness was not there all along.

The METI program aims at explaining the emergence of organisms. Hence, it focuses on the emergence of altruistic behaviors and their maintenance, as well as the possibility of collectives reproducing as a single entity. This property obviously constitutes a second stage in the process of transition toward individuality. Ratnieks et al. (2006) distinguish between complete and component resolution of the famous "tragedy of the commons." Along these lines, I have distinguished complete from component transitions (Huneman 2013). When a transition is complete, we get unity of purpose, enforced by mechanisms such as policing (e.g., worker bees killing a newly arising queen in the hive) that prevent the complex individuals from separating into a collection of elementary individuals. Component transitions are steps toward this state.[6]

In the case of egalitarian transitions, I propose viewing this distinction between component and complete transitions as one between ecosystems and agents. When a transition is complete, some unity of purpose is reached between low-level individuals that make up the whole, often ensured by policing mechanisms and spatial settings (e.g., containing membranes) that reinforce an alignment of fitness among them. Hence, the novel individual is an agent, or at least it satisfies the maximizing agent analogy. But before that, at an earlier stage of the evolutionary transition process, we find cohesive groups whose members are more likely to interact with each other than to interact collectively with other entities. In other words, we find weak individuals. And because entities from various species interact ecologically (both facilitating and competing) within these weak individuals, they could be seen as ecosystems.

Thus, when an evolutionary transition is complete, the last stage of the process is a transition from being an ecosystem—which owes its stability and self-maintenance to an internal ecological dynamic—to being an agent. Component transitions yield individuals that correspond to the

---

[6] According to Queller (2009), fraternal transitions characterize those that involve genetically homogeneous entities, while egalitarian transitions involve entities that are radically different, as in the case of the endosymbiosis that gave rise to chloroplasts, or to eukaryotes (Margulis 1978). I combine this distinction with the complete/component transitions distinctions to partition the possible cases of transition into four types (Huneman 2013).

ecosystem scheme; complete transitions yield individuals that correspond to the agent scheme. The exact structure of this process and the issues raised by its modeling can be left for another book.

From the viewpoint of transitions, one can therefore expect more unified organisms to be less like ecosystems, and more conflicted individuals to be more like ecosystems. Unity of purpose yields agency; inversely, ecological processes may yield some unity but themselves include forms of conflict—namely, competition and predation. Thus, the present account predicts that we will find stronger within-organism ecological processes when the organism's constituent parts conflict more severely, so that the presence and prominence of ecological processes could be taken as a proxy for, or a measure of, the severity of internal conflict. Likewise, the present account would be falsified if individuals displaying a high degree of conflict are less likely than other, less internally conflicted organisms to feature ecological or ecological-like processes.

The two major conclusions to be drawn from this short examination of the evolutionary transitions are as follows:

1. While the ecosystem and agent schemes seem to offer conflicting accounts of organisms—and, more precisely, of the conflict-resolving dynamics proper to organisms—they can coexist across time.
2. If we map the space of individuality, including all component and complete transitions, we find that most of the component transitions will be ecosystems. Hence the ecosystem scheme is much more general. But complete transition is a transformation through which some of these ecosystems are endowed with agency.

Hence the duality between the ecosystem scheme and the agency scheme is neither a rivalry between competing explanations nor simply a case in which two complementary accounts of the same reality figure within an explanatorily pluralist epistemic framework. They are a pair of poles that determine the structure of transformations in the latest stages of evolutionary transitions in individuality.

Conclusion

In this chapter I have tried to make sense of the coexistence of two pervasive schemes used by evolutionary biologists and ecologists to make sense of the fact of organisms. Rather than deciding for or against one of these views, I have explained how they constitute distinct stances fulfilling distinct explanatory goals and therefore may coexist. I have presented each of them as a way to make sense of the conflicts intrinsic to organisms, and of the ways that organisms use their internal resources to overcome them. I argued in favor of an epistemic pluralism that allows for potential conflicts between the schemes.

But the duality between agents and ecosystems is not only epistemic, with each account being equally legitimate and capable of capturing one dimension of the nature of organisms. The duality is also ontological. No matter our metaphysical interpretation—realist, instrumental, or the sort of Kantian middle ground sketched in Desmond and Huneman (2020)—agency is a category into which some organisms can be arranged. The same goes for ecosystems. But the former and the latter may not be the same organisms. We saw that the ecosystem stance is much more liberal. Finally, the joint validity of the ecosystem scheme and the agency scheme—which might be puzzling for anyone who is a realist in terms of philosophy of science—becomes less puzzling when understood diachronically. To this extent, agency and ecosystem are not only a set of powerful analogies to interpret organisms but are also the poles that structure evolutionary transitions. Hence this concept pair provides us with a novel interpretive tool to model evolutionary transitions in a very general manner (as a transition from "ecosystemality," so to speak, toward agency) and to cast a light on their universality.

References

Allee, W. C., Park, O., Emerson, A. E., Park, T., and Schmidt, K. P. (1949). *Principles of Animal Ecology*. W. B. Saunders.

André, J. B., Cozic M., DeMonte S., et al. (2022). *From Evolutionary Biology to Economics and Back: Parallels and Crossings between Economics and Evolution*. Springer.

Bapteste, E., and Huneman, P. (2018). Towards a dynamic interaction network of life to unify and expand the evolutionary theory. *BMC Biology, 16,* 56.

Bateson, P. (2005). The return of the whole organism. *Journal of Biosciences, 30,* 31–39.

Bordenstein, S. R., and Theis, K. R. (2015). Host biology in light of the microbiome: Ten principles of holobionts and hologenomes. *PLoS Biology, 13*(8), e1002226.

Borrello, M. E. (2003). Synthesis and selection: Wynne-Edwards' challenge to David Lack. *Journal of the History of Biology, 36,* 531–566.

Bouchard, F. (2010). Symbiosis, lateral function transfer and the (many) saplings of life. *Biology and Philosophy, 25,* 623–641.

Bouchard, F. (2013). What is a symbiotic superindividual and how do you measure its fitness? In *From Groups to Individuals,* edited by F. Bouchard and P. Huneman, 243–264. MIT Press.

Bouchard, F., and Huneman, P., eds. (2013). *From Groups to Individuals.* MIT Press.

Bourrat, P. (2023). A coarse-graining account of individuality: How the emergence of individuals represents a summary of lower-level evolutionary processes. *Biology and Philosophy, 38*(4), 33.

Bourrat, P. (2025). Moving past conventionalism about multilevel selection. *Erkenntnis, 90,* 1363–1376.

Buss, L. (1987). *The Evolution of Individuality.* Princeton University Press.

Clarke, E. (2013). The multiple realizability of biological individuals. *Journal of Philosophy, 110*(8), 413–435.

Clements, F. (1916). *Plant Succession: An Analysis of the Development of Vegetation.* Carnegie Institution.

Cooper, G. (2003). *The Science of the Struggle for Existence.* Cambridge University Press.

Costello, E., Stagaman, K., Dethlefsen, L., Bohannan, B. J., and Relman, D. A. (2012). The application of ecological theory toward an understanding of the human microbiome. *Science, 336,* 1255–1262.

Desmond, H., and Huneman, P. (2020). The ontology of organismic agency: A Kantian approach. In *Natural Born Monads: On the Metaphysics of Organisms and Human Individuals,* edited by A. Altobrando and P. Biasetti, 33–64. De Gruyter.

Doolittle, W. F., and Inkpen, S. A. (2018). Processes and patterns of interaction as units of selection: An introduction to ITSNTS thinking. *Proceedings of the National Academy of Sciences of the United States of America, 115*(16), 4006–4014.

Dutreuil, S. (2024). *Gaia Terre Vivante.* La Découverte.

Edwards, D. P., Hassall, M., Sutherland, W. J., and Yu, D. W. (2006). Selection for protection in an ant-plant mutualism: Host sanctions, host modularity, and the principal-agent game. *Proceedings of the Royal Society B, 273*(1586), 595–602

Eldredge, N. (1985). *Unfinished Synthesis: Biological Hierarchies and Modern Evolutionary Thought.* Oxford University Press.

Fisher, R. A. (1930). *The Genetical Theory of Natural Selection.* Clarendon Press.

Folse, H. J., III, and Roughgarden, J. (2010). What is an individual organism? A multilevel selection perspective. *Quarterly Review of Biology, 85,* 447–472.

Gardner, A., West, S., and Wild, G. (2011). The genetical theory of kin selection. *Journal of Evolutionary Biology, 24*(5), 1020–1043.

Gleason, H. A. (1926). The individualistic concept of the plant association. *Bull Torrey Bot Club, 53*(1), 7–26.

Gleason, H. A. (1939). The individualistic concept of the plant community. *The American Midland Naturalist, 21*, 92–110.

Godfrey-Smith, P. (2009). *Darwinian Populations and Natural Selection.* Oxford University Press.

Godfrey-Smith, P. (2013). Darwinian individuals. In *From Groups to Individuals,* edited by F. Bouchard and P. Huneman, 1–14. MIT Press.

Goodnight, C. J. (2013). Defining the individual. In *From Groups to Individuals,* edited by F. Bouchard and P. Huneman, 37–54. MIT Press.

Grafen, A. (2002). A first formal link between the price equation and an optimization program. *Journal of Theoretical Biology, 217*(1), 75–91.

Grafen, A. (2006). Optimization of inclusive fitness. *Journal of Theoretical Biology, 238*(3), 541–563.

Grafen, A. (2007). The Formal Darwinism project: a mid-term report. *Journal of Evolutionary Biology, 20*(4), 1243–1254.

Grafen, A. (2014). The Formal Darwinism project in outline. *Biology and Philosophy, 29*, 155–174.

Grosberg, R. K., and Strathmann, R (2007). The evolution of multicellularity: A minor major transition? *Annual Review of Ecology, Evolution, and Systematics, 38*(1), 621–654.

Haldane, J. B. S. (1941). *New Paths in Genetics.* Allen and Unwin.

Hull, D. L. (1980). Individuality and selection. *Annual Review of Ecology and Systematics, 11*, 311–332.

Humphreys, P. (2016). *Emergence: A Philosophical Account.* Oxford University Press.

Huneman, P. (2010). Assessing the prospects for a return of organisms in evolutionary biology. *History and Philosophy of Life Sciences, 32*(2/3), 341–372.

Huneman, P. (2013). Adaptation in transitions. In *From Groups to Individuals,* edited by F. Bouchard and P. Huneman, 141–172. MIT Press.

Huneman, P. (2014a). Individuality as a theoretical scheme. I. Formal and material concepts of individuality. *Biological Theory, 9*(4), 361–373.

Huneman, P. (2014b). Individuality as a theoretical scheme. II. About the weak individuality of organisms and ecosystems. *Biological Theory, 9*(4), 374–381.

Huneman, P. (2019). How the modern synthesis came to ecology. *Journal of the History of Biology, 52*, 635–686.

Huneman, P. (2020). Biological individuality as weak individuality: A tentative study in the metaphysics of science. In *Biological Identity,* edited by A. S. Meincke and J. Dupré, 40–62. Routledge.

Huneman, P. (2023). *Death: Perspectives from the Philosophy of Biology.* Palgrave.

Huneman, P. (2025). The Chicago School of Ecology's evolutionary superorganism and the Clements-Wright connection. *History and Philosophy of Life Sciences.* https://doi.org/10.1007/s40656-024-00652-4.

Huneman, P., and Martens, J. (2017). The behavioural ecology of irrational behaviour. *History and Philosophy of Life Sciences, 39*(3), 23.

Hutchinson, G. E. (1957). Concluding remarks: Cold Spring Harbor symposium. *Quantitative Biology, 22*, 415–427.

Hutchinson, G. E. (1961). The paradox of the plankton. *The American Naturalist, 95*, 137–145.

Justus, J. (2008). Ecological and Lyapunov stability. *Philosophy of Science, 75*(4), 421–436.

Kerr, B., and Godfrey-Smith, P. (2002). Individualist and multi-level perspectives on selection in structured populations. *Biology and Philosophy, 17*(4), 477–517.

Kingsland, S. E. (1995). *Modeling Nature: Episodes in the History of Population Ecology.* 2nd ed. University of Chicago Press.

Kupiec, J. J. (2009). *The Origins of Individuals.* World Scientific.

Lack, D. (1954). *The Natural Regulation of Animal Numbers.* Oxford University Press.

Lecointre, G., Schnell, N. K., and Teletchea, F. (2020). Hierarchical analysis of ontogenetic time to describe heterochrony and taxonomy of developmental stages. *Scientific Reports, 10*(1), 19732.

Massimi, M. (2022). *Perspectival Realism.* Oxford University Press.

Maynard Smith, J., and Szathmáry, E. (1995). *The Major Transitions in Evolution.* Oxford University Press.

Medawar, P. B. (1957). *The Uniqueness of the Individual.* Routledge.

Michod, R. (1999). *Darwinian Dynamics.* Oxford University Press.

Mitman, G. (1988). From the population to society: The cooperative metaphors of W. C. Allee and A. E. Emerson. *Journal of the History of Biology, 21*(2), 173–192.

Odling-Smee, J., Laland, K., and Feldman, M. (2003). *Niche Construction: The Neglected Process in Evolution.* Princeton University Press.

Okasha, S. (2006). *Evolution and the Levels of Selection.* Oxford University Press.

Okasha, S. (2018). *Agents and Goals in Evolution.* Oxford University Press.

Pepper, J. W., and Herron, M. D. (2008). Does biology need an organism concept? *Biological Review of the Cambridge Philosophical Society 83*, 621–627.

Pocheville, A. (2018). Biological information as choice and construction. *Philosophy of Science, 85*(5), 1012–1025.

Queller, D. C., and Strassmann, J. E. (2009). Beyond society: The evolution of organismality. *Philosophical Transactions of the Royal Society B, 364*, 3143–3155.

Raff, R. (1996). *The Shape of Life.* University of Chicago Press.

Ratnieks, F. L., Foster, K. R., and Wenseleers, T. (2006). Conflict resolution in insect societies. *Annual Review of Entomology, 51*, 581–608.

Scadden, D. (2006). The stem-cell niche as an entity of action. *Nature, 441*(7097), 1075–1079.

Sober, E. (1984). *The Nature of Selection: Evolutionary theory in Philosophical Focus*. University of Chicago Press.

Sober, E., and Wilson, D. S. (1998). *Unto Others: The Evolution and Psychology of Unselfish Behavior*. Harvard University Press.

Sterelny, K. (2006). Local ecological communities. *Philosophy of Science, 73,* 215–231.

Strassmann, J. E., and Queller, D. C. (2010). The social organism: Congresses, parties, and committees. *Evolution, 64,* 605–616.

Sultan, S. (2017). *Organism and Environment*. Oxford University Press.

Sultan, S. E., Moczek, A. P., and Walsh, D. (2022). Bridging the explanatory gaps: What can we learn from a biological agency perspective? *BioEssays, 44,* e2100185.

Swenson W., Wilson D. S., and Elias, R. (2000). Artificial ecosystem selection. *Proceedings of the National Academy of Sciences of the United States of America, 97*(16), 9110–9114.

Teulière, J., Bernard, C., and Bapteste, E. (2021). Interspecific interactions that affect ageing: Age-distorters manipulate host ageing to their own evolutionary benefits. *Ageing Research Reviews, 70,* 101375.

Turner, S. (2013). Individuality in a social insect assemblage. In *From Groups to Individuals: Evolution and Emerging Individuality,* edited by P. Huneman and F. Bouchard, 219–242. MIT Press.

Walsh, D. M. (2010). Not a sure thing: Fitness, probability, and causation. *Philosophy of Science, 77*(2), 147–171.

Walsh, D. M. (2013). Mechanism, emergence, and miscibility: The autonomy of evo-devo. In *Functions: Selection and Mechanisms,* edited by P. Huneman, 43–65. Springer.

Walsh, D. M. (2015). *Organisms, Agency, and Evolution*. Cambridge University Press.

West, S. A., Griffin, A. S., and Gardner, A. (2007). Social semantics: Altruism, cooperation, mutualism, strong reciprocity and group selection. *Journal of Evolutionary Biology, 20*(2), 415–432.

West-Eberhard, M. J. (2003). *Developmental Plasticity and Evolution*. Oxford University Press.

Williams, G. (1957). Pleiotropy, natural selection and the evolution of senescence. *Evolution, 11,* 398–411.

Williams, G. C. (1966). *Adaptation and Natural Selection*. Princeton University Press.

Weyl, E. G., Frederickson, M. E., Yu, D. W., and Pierce, N. E. (2010). Economic contract theory tests models of mutualism. *Proceedings of the National Academy of Sciences of the United States of America, 107*(36), 15712–15716.

Zilber-Rosenberg, I., and Rosenberg, E. (2008). Role of microorganisms in the evolution of animals and plants: The hologenome theory of evolution. *FEMS Microbiology Reviews, 32*(5), 723–735.

CHAPTER SIX

Cooperation and Conflict in Metazoan Development

Toward a Neo-Darwinian Embryology

DAVID HAIG

ABSTRACT The sequestration of somatic cells from the germ line resolves most intra-genomic conflicts because all genes in somatic nuclei have a shared interest in the reproduction and survival of the germ line, and no gene in a somatic nucleus can gain an intergenerational benefit from its own preferential replication. Intra-genomic conflicts involving preferential germline replicators and postsegregational distorters are concentrated in nuclei of the germ line. These conflicts include competition among cells to acquire a germline rather than a somatic fate, and can result in evolutionary changes in the mechanisms and locations of germ-cell specification. In many organisms, germline conflicts are dampened by a delay of zygotic genome activation until after the sequestration of somatic cells, followed by supervision of germline proliferation by somatic supporting cells. Postzygotic maternal provisioning in mammalian development destabilized these mechanisms of conflict resolution because it was associated with early activation of the embryonic genome before the differentiation of somatic cells and because imprinted genes of maternal and paternal origin disagree over resource extraction from mothers.

Introduction

"Organism" and "body" are synonyms. Multicellular bodies are the temporary homes of genetic lineages that have resided in unbroken chains of past bodies, all of which survived to reproduce. The raison d'etre of these bodies has been production of sperm that fertilize ova (if the body is male), ova that are fertilized by sperm (if the body is female), or ova and sperm (if the body is hermaphroditic). The path of posterity has passed from zygote to zygote, via ova and sperm, with the vast infrastructure of successive somas constructed as sophisticated life-support systems for the genetic denizens of the germ line. These bodies are the constructed niches of their resident genes, and as such comprise a major component of the environment to which their genes have become adapted (Haig 2012b,

2020). Although complex multicellular bodies have evolved many times, this chapter will focus on the development of metazoan animals.

All important terms are used in multiple senses. Germ line is no exception. Weismann (1892, 242) defined a *Keimbahn* as a lineage of cells connecting a zygote to a reproductive cell. Conversely, Wilson (1925, 311, 313) defined germ lines as being comprised of cells without somatic descendants. The two definitions differ as to whether cells with both somatic and germline descendants are included in the germ line (for further discussion see the appendix to this chapter). Although most embryologists employ a Wilsonian definition of the germ line, this chapter employs a Weismannian definition because Weismann's definition not only has historical priority but is also better suited for addressing questions of adaptive function and the phylogenetic transmission of genetic information. This is because mutations that occur before the differentiation of somatic cells can be inherited. Under the Weismannian definition, if a mutation is inherited, then the cell in which it occurred is necessarily part of the germ line. Statements that somatic mutations in plants, planarians, or corals violate the "Weismann's barrier" misunderstand Weismann (Niklas and Kutschera 2014; Vasquez Kuntz et al. 2022).

Weismannian germ lines can be considered to be comprised of two segments: a *germ stem* of cells with somatic as well as gametic descendants, and a *germ crown* of cells without somatic descendants. This chapter will define a somatic cell as a cell that belongs to neither germ stem nor germ crown (Figure 6.1). These definitions of germ line and soma correspond to a distinction I have made elsewhere between the cooperative calm of the *pax somatica* ("peace of the body") and the internecine strife of the *arena germinalis* (an arena was the site of combat in a Roman ampitheater) (Haig 2016b, 2022). Intra-genomic conflicts can be expressed in both the germ stem and germ crown because genetic changes in both can be inherited.

The performance of most bodily functions by somatic cells reduces intra-genomic conflicts because no gene obtains a multigenerational benefit from preferential replication in cells that do not leave direct genetic descendants. Therefore, all genes of somatic cells have a common interest in the survival and reproduction of the germ line with which they are associated and all are expected to work together for the communal good as components of a well-integrated organism. However, a similar coordination of roles can no longer be assumed in cells of the germ line. Genes

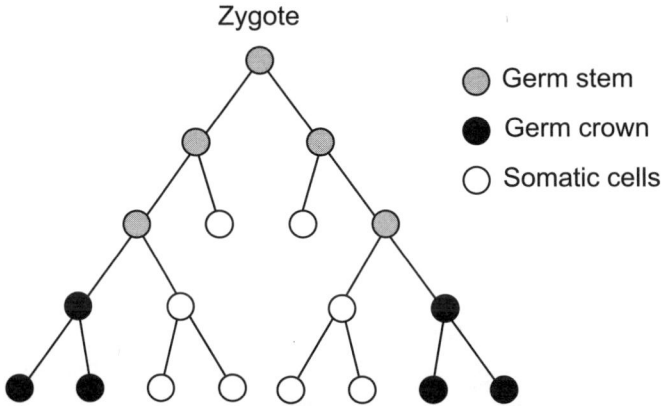

Figure 6.1. Cellular divisions during development of a hypothetical organism. Cells of the germ stem (gray) have both somatic and germ cells as descendants. Somatic cells (white) do not have germ cells as descendants. Germ cells (black) of the germ crown do not have somatic cells as descendants. The Weismannian "germ line" includes cells of the germ stem and germ crown. The Wilsonian "germ line" is restricted to cells of the germ crown and excludes cells of the germ stem.

that coexist peaceably in somatic cells can erupt into rancorous conflicts in the germ line as rogue elements vie for preferential replication or attempt to corrupt the distribution of communal goods for selfish benefit (Haig 2016b, 2022).

The *arena germinalis* avoids civil war because of important restraints on conflict. First, genes that exhibit regular Mendelian inheritance are spectators or umpires rather than combatants in the arena. Second, somatic cells have evolved to manage the germ line for organismal benefit. This management includes the apoptotic elimination of germ cells that step out of line. Germ cells must conform to the edicts of somatic cells that supervise their activities (such supervision is absent for cells of the germ stem before differentiation of somatic cells). Third, transcription is silenced until after the differentiation of somatic cells in the embryos of many species (including *Caenorhabditis*, *Drosophila*, zebrafish, and *Xenopus*).

Early sequestration of the germ line has been presented as a means of protecting delicate germ cells from the vicissitudes of the soma (Strome and Updike 2015). This chapter reverses the narrative. Somatic cells are set aside early to protect bodily functions from the robust conflicts of the germ line and to manage those conflicts for the collective good. Germ lines are less organismal than the somas in which they reside. Somatic selflessness restrains germline selfishness. Bodies function as cohesive wholes.

Genetic Versus Organismal Fitness

Haploid genomes unite at syngamy to do things together they could not achieve on their own. Their diploid union is a marriage of sorts. (*Gamos,* the ancient Greek word for marriage, is the etymological root of mono*gamy,* poly*gamy,* *gam*ete, and syn*gamy.*) Successful unions produce many gametic progeny. Each gamete is associated with a dissolution of a diploid marriage in meiotic divorce. The once harmonious union risks degeneration into acrimonious disputes over the division of conjugal property. The regular dissolution of successful marriages to take a chance with untried partners has been considered one of the paradoxes of sexual reproduction (Otto 2009).

George Williams (1966, 24) wrote, "It is only the meiotically dissociated fragments of the genotype that are transmitted in sexual reproduction, and these fragments are further fragmented by meiosis in the next generation." A gene was a "potentially immortal" chromosomal segment that persisted intact for several generations. Richard Dawkins (1976, 30) adopted Williams's definition and identified such genes as "selfish" in the sense that successful variants possessed attributes that ensured their own propagation whether by collaboration with, or exploitation of, other genes. Organisms were relegated to the role of ephemeral vehicles of immortal genes (Dawkins 1982).

Cosmides and Tooby (1981) divided "the genome into fractions whose defining rule is that the fitness of all genes in the set is maximized in the same way." They called such a set of genes a coreplicon. The most important coreplicon of meiotic organisms is the set of genes present as conjugal pairs in diploid cells for which the two alleles at each locus receive an equal share of marital goods at meiotic divorce. This coreplicon can be called the *conjugal genome*. The genic fitness of its members is usually equated with individual or organismal fitness. Membership of the conjugal genome is defined by the inability of a gene to direct benefits preferentially to offspring that inherit its copies, and away from offspring that do not inherit its copies. Such genes are spectators and umpires of the *arena germinalis* rather than gladiators.

Meiosis is the fundamental source of conflicts within Mendelian genomes. If genomes never recombined, the long-term fates of all genomic parts would be inexorably coupled and a conceptual distinction between organismal and genic selfishness might seem needless pedantry. What was good for one part would be good for all. The genome would evolve as a whole. Meiosis fragments genomes into smaller units with distinct evolutionary trajectories, destabilizing the cohesion of the whole and creating conflicts among the parts that now possess individual goods distinct from the communal good. After meiotic segregation, former allelic partners are now rivals and there is no sentimentality among genes.

But meiosis is double-faced. Not only is it a source of conflicts within genomes but it is also a partial solution to these conflicts provided that genomes fragment into small enough parts (Haig and Grafen 1991). In equitable divorces, each allele receives half of marital resources and shares equally in future opportunities. Recombination reduces the inherent risks of divorce because smaller genomic chunks can exert less agency than larger chunks and are therefore less able to bias the distribution of marital goods. A single high-stakes divorce between selfish genomes is replaced by many smaller divorces between conjugal pairs of selfish alleles. Furthermore, the unpredictable nature of segregation and crossing-over creates uncertainty about which genes will segregate together to which offspring. If this information is hidden before the division of marital resources among offspring, then the best each gene can do is maximize the total number of offspring. Genes of the conjugal genome have been said to make decisions about the allocation of goods among offspring from behind a *meiotic veil of ignorance* (Haig and Bergstrom 1995; Okasha 2012).

The Commonwealth of Genes

The meiotic veil of ignorance is lifted by selfish genetic elements that employ two basic stratagems to subvert the fairness of meiosis and expropriate more than their fair share of conjugal goods at meiotic divorce. *Preferential germline replicators* (PGRs) produce extra copies of themselves in germline nuclei before meiosis. By this means, a sequence that enters a zygote as a single copy transmits its copies to more than half of ensuing gametes. *Postsegregational distorters* (PSDs) withhold goods from cells or embryos that do not inherit their copies for the benefit of cells or embryos that do (Fishman and McIntosh 2019). PGRs typically act in mitotic cells of the germ line, although a PGR could also act in a meiocyte before it divides, whereas PSDs typically exploit meiotic segregation for their selfish advantage, although a PSD could also exploit mitotic divisions of the germ line. PGRs and PSDs are both examples of what Patten et al. (2023) called *transmission distorters*.

James Madison (1787) wrote that however small a republic may be, the number of its representatives "must be raised to a certain number to guard against the cabals of the few" and however large the republic may be, its representatives "must be limited to a certain number, in order to guard against the confusion of the multitude." Madison's prose entered evolutionary biology in Egbert Leigh's (1977) invocation of a "parliament of the genes" that opposed "the cabals of the few." Because each gene segregates independently of most other genes, genetic shuffling ensures that the collective of genes is selected to suppress self-interested behavior of smaller groups of genes.

Recombination has contrasting consequences for PSDs and PGRs. A PSD typically involves a conspiracy between two genes that occur together on a shared haplotype, one encoding a toxin that does not segregate and the other encoding an antitoxin that segregates. The toxin kills cells or offspring that do not inherit the antitoxin. Crossing-over between the genes for the toxin and antitoxin generates "suicide haplotypes" that possess the gene for the toxin without the gene for the antitoxin. Recombination constrains the activities of PSDs, and promotes equitable division of resources, by unpredictably breaking the genome into smaller heritable chunks, thereby separating all but tightly linked pairs of toxins and antitoxins (Haig and Grafen 1991). On the other hand, two copies of a PGR on the same chromosome can be transmitted to different gametes if there is recombination between them. Thus, recombination impedes the spread of PSDs through the gene pool but enhances the spread of PGRs. Perhaps Madison's "confusion of the multitude" can be interpreted as a warning against the shifting chromosomal allegiances of PGRs. Just as a parliament may have too many or two few members, the harmony of the genome may be maximized by some intermediate level of recombination.

Conjugal gene pairs, PGRs, and PSDs are co-residents of genomes in which some have interacted for millions of years. I prefer to avoid the language of "host" and "parasite," with some genes as privileged citizens and others undocumented aliens, and focus instead on the complex interplay of cooperation and conflict within the commonwealth of genes that has had important consequences for the embryonic development of complex multicellular bodies.

The Conjugal Genome

Genes of the conjugal genome are predicted to collectively manage the activities of PGRs and PSDs to maintain fertility and reduce organismal fitness costs for progeny. Evolutionary change in the conjugal genome occurs by mutation. A useful distinction can be made between mutations and innovations. A mutation is a genetic change in the nucleus of a single cell. An innovation is the first appearance of a mutation in a zygotic nucleus. Not all mutations become innovations. Some occur in somatic cells. Others are eliminated by stochastic or selective processes within the germ line in which they occur. Innovations will be a biased set of germline mutations to the extent that some form of selection in the germ line intervenes between mutation and innovation. (Innovation is introduced here as a new technical term.)

Mutations that promote proliferation of germline cells will be disproportionately represented among innovations, whereas mutations that inhibit germline proliferation or disrupt gametogenesis will be underrepesented among innovations. A key question is whether these selective biases favor innovations that benefit or impair bodily fitness. On the one hand, cellular selection in the germ line and organismal selection are likely to work in concert to maintain functions of genes responsible for basic cellular processes (Hastings 1989, 1991; Otto and Hastings 1998). On the other hand, some disease-causing innovations occur at high frequency among live births because the mutation conferred a proliferative advantage on cells of the germ line (Choi et al. 2012; Giannoulatou et al. 2013; Yoon et al. 2013; Shinde et al. 2013).

Development from a single cell contributes to the cohesiveness of metazoan bodies by ensuring genetic homogeneity among a body's cells (Grosberg and Strathmann 2007). After the first generation in which an innovation occurs, subsequent bodies inherit the innovation in all of their cells, or none of their cells, because these bodies develop from single-celled zygotes. An innovation in a gene of the conjugal genome gains no further advantage, or disadvantage, from germline selection after this first generation (Queller 2000). For this reason, innovations in genes of the conjugal genome cannot spread through a population of bodies by germline selection alone. The long-term fates of genetic innovations will be determined by their effects on organismal phenotypes. Germline and organismal selection are most likely to work in concert for basic metabolic processes because what is good for a cell is also good for the organism in which it resides, but cellular selection in the germ line proceeds without regard for effects on phenotypes, such as visual acuity, that are expressed only in somatic cells (Haig 2024b).

Preferential Germline Replicators

A ground-based radar tracks an incoming fighter jet to guide a surface-to-air missile to destroy the intruder. The fighter jet uses the radar signal to guide an air-to-surface missile to destroy the radar and its crew. If one assumed this was a harmonious interaction, one might conclude that the function of the ground-based radar was to guide the air-to-surface missile. Biologists risk making similar errors when they assume that interactions among genes in the germ line are necessarily cooperative. To make sense of a conflict, one must understand the tactical and strategic objectives of the contending parties, but also understand that these objectives can be obscured in the fog of war by incidents of friendly fire and collateral damage.

The genomic signature of a PGR is the presence of multiple copies per haploid genome. Unlike genes of the conjugal genome, PGRs can spread through a population of multicellular bodies by germline selection alone. Because their activities are frequently mutagenic, and thus costly for bodily fitness, natural selection favors PGRs that restrict their activities to the germ line by responding to reliable cues of the difference between germline and somatic cells (Haig 2016b, 2022).

Control of the machineries of DNA replication and RNA transcription will be strongly contested in the *arena germinalis*. Replication of a retrotransposon requires transcription of an mRNA that is translated as a reverse transcriptase and transcription of a "genomic" RNA that is reverse transcribed as DNA and inserted at a new location. Replication of a DNA transposon requires the translation of a transposase that cuts a DNA copy of the transposon from an old location and inserts it at a new location. The number of its copies is increased when a DNA transposon is cut from an already replicated chromosomal location and is inserted at a yet-to-be-replicated site or when the excised copy is restored by homologous recombination using the sister chromatid as a template for repair (Wells and Feschotte 2020).

Retroelements require transcription for their preferential replication, whereas DNA transposons depend on the machinery of DNA replication and homologous repair. Zebrafish and *Xenopus* genomes are dominated by DNA transposons with a relative paucity of retroelements (Chalopin et al. 2015). The absence of embryonic transcription during the critical early cleavage divisions of their embryos may have greatly restricted opportunities for retrotransposition (Kane and Kimmel 1993; Blitz and Cho 2021). By contrast, the large genomes of mammals and salamanders are dominated by remnants of retrotransposons perhaps because of a relative relaxation of controls on germline transcription (Sun et al. 2012; Keinath et al. 2015; Sotero-Caio et al. 2017). Mammalian genomes may be particularly vulnerable to accumulation of retroelements because transcription is activated early in mammalian embryonic development (Vassena et al. 2011). Primordial germ cells of axolotls (a kind of salamander) are the last cells to differentiate (Chatfield et al. 2014). Therefore, the exceptionally large genomes of salamanders may be a maladaptive consequence of their possession of especially long, transcriptionally active, germ stems that provide many opportunities for proliferation of retrotransposons.

Each kind of multicopy sequence will have its own adaptations for preferential replication. Satellite repeats proliferate by the induction of chromosomal breaks that recruit the machinery of homologous recombination to repair the break with a "patch" copied from a homologous template. Successful repeats replace a damaged unit with a patch containing more than one unit as occurs, for example, when a replication fork chases another replication fork or a replication fork chases itself. Biased gene conversion in a germline nucleus allows one allele to replace its partner and be transmitted to all offspring of that cell lineage (Duret and Galtier 2009). Such biases could substantially increase the mutational load in a population (Bengtsson 1990), but have yet to be shown to be a major source of strategic behavior at the genetic level. All of these mechanisms of preferential replication in the germ line will be subject to countermeasures imposed by the rest of the genome.

Constitutive heterochromatin forms at multicopy sequences in somatic cells (Janssen et al. 2018), but is reduced in the germ stem (Meshorer and Misteli 2006; Becker et al. 2016). In somatic cells, constitutive heterochromatin can be considered an expression of the *pax somatica* in which multicopy sequences and *trans*-acting regulators collaborate to silence repeats and minimize their genome-destabilizing properties. Epigenetic inactivation is a collaborative adaptation of both kinds of genes for the enhancement of organismal fitness. In contrast, the decondensation and transcriptional activation of multicopy sequences in germline nuclei can be seen as an adaptation of the repetitive sequences to promote their intranuclear fitness and not of *trans*-activators to promote organismal fitness. Rather, *trans*-activators are being exploited by repetitive sequences for the latter's selfish benefit.

Postsegregational Distorters

Mitotic postsegregational distorters (mitotic PSDs) are a theoretical possibility in which both copies of an original sequence go to one sister cell and the other sister cell undergoes cellular death. In this scenario, copy-number differences arise between sister cells and differential proliferation or survival of the sister cells results in a change of average copy number in surviving cells. Consider, as a hypothetical example, a DNA transposon that has undergone DNA replication before it excises one of its copies from of a pair of sister chomatids and inserts the excised copy at a new chromosomal location leaving cell-lethal damage at the source location. After mitotic segregation of sister chromatids, the cell that receives the damaged chromatid dies but its sister cell survives with at least one, possibly two, copies of the DNA transposon. If such processes occur in the germ line, then they could function as a form of mitotic drive, especially if the death of a germ cell was compensated by an extra division of an undamaged germ cell.

However, PSDs have usually been assumed to act after meiotic segregation. Meiotic PSDs differ in male and female animal germ lines (1) because eggs receive more resources than sperm, (2) because transcription ceases after meiosis in spermatogenesis but before meiosis in oogenesis, and (3) because the completion of female meiosis is usually delayed until after a sperm has entered an oocyte (Longo 1973). Female meiosis is asymmetric unlike symmetric male meiosis. All but one of the meiotic products of an oocyte is eliminated in the first or second polar bodies. This creates opportunities for genetic agents that manipulate meiotic mechanisms to preferentially segregate away from polar bodies (Clark and Akera 2021; Silva and Akera 2023). Such agents have been considered segregation distorters in the strict sense, rather than PSDs, but I include them under the rubric of PSDs because polar bodies can be considered lethal locations causing postsegregational death.

Most meiotic PSDs encode a "toxin" expressed before segregation and an "antitoxin" expressed after segregation. The toxin does not participate in segregation but the antitoxin segregates (Bravo Núñez et al. 2018; Kruger and Mueller 2021). Progeny that do not inherit the antitoxin undergo death or disability whereas progeny that inherit the antitoxin survive unharmed. The toxin selects for progeny that express the antitoxin, with the gene for the toxin preferentially segregating to surviving progeny along with the closely linked gene for the antitoxin. Such haplotypes of conspiratorial genes function as "protection rackets" (Haig 1997) or "spiteful greenbeards" (Gardner 2019).

Haploid spermatids are connected by cytoplasmic bridges with some gene products shared among spermatids and others restricted to the spermatid in which they were expressed (Bhutani et al. 2021). A period of postmeiotic transcription creates opportunities for toxin–antitoxin systems that kill spermatids that do not inherit the antitoxin. A toxin could be expressed before meiosis and be distributed to all products of meiosis or expressed after meiosis but be shared across cytoplasmic bridges. Its antitoxin needs to remain localized to spermatids that inherit the gene encoding the antitoxin. Postmeiotic transcription ceases in haploid spermatids as their nuclei condense and histones are replaced with protamines (Steger 1999).

After male meiosis, sperm from the same ejaculate may compete for ova creating an opportunity for driving alleles that act in heterozygous spermatocytes to sabotage the function of spermatozoa that inherit their former allelic partner (Lyttle 1991). A potential mechanism of a PSD involves complementarity between a sense RNA (as toxin) and its own antisense (as antitoxin). For example, piwi-interacting small RNAs (piRNAs) regulate the expression of other RNAs with which they form double-stranded RNA. Many piRNAs expressed in premeiotic male germ cells target transposable elements and are interpreted as defenses against transposable elements (Aravin et al. 2007), although it is possible that some piRNAs may be adaptations of the transposable elements for self-restraint in germline transposition (Haig 2024b). However, most piRNAs expressed at pachytene in mammalian and avian testes target unique, rather than repetitive, sequences. Some of these piRNAs either eliminate or activate mRNAs in postmeiotic spermatids (Gou et al. 2014; Wang et al. 2022, 2023). Inhibition of the biogenesis of piRNAs at pachytene results in male infertility (Zheng and Wang 2012; Wu et al. 2020; Chen et al. 2021; Choi et al. 2021). These piRNAs have been suggested to promote spiteful segregation (Chen et al. 2020; Aravin 2020; Wu et al. 2020; Choi et al. 2021). One simple mechanism would involve gene products (translated after meiosis) of mRNAs (transcribed before meiosis) functioning as sperm-specific toxins that are inhibited by piRNA antitoxins (transcribed at pachytene or in the haploid phase). Such a mechanism would require that piRNAs are transcribed from the complementary strand to their target and are not shared among spermatids in the haploid phase.

Because transcription occurs in haploid spermatids, a toxin can be transcribed in spermatocytes before segregation with its antitoxin transcribed in spermatids after segregation. This form of postsegregational rescue is unavailable in female germ lines because transcriptional quiescence is established in diploid oocytes before the first meiotic division and transcription does not resume until after a sperm has entered the egg. Therefore, for a toxin–antitoxin system to function in female germ lines, the toxin would need to be expressed before meiosis but lack toxic effects until its antitoxin was expressed after activation of the embryonic genome. I will call such a system an *embryonic PSD* because it kills embryos rather than gametes. A major difference of embryonic PSDs from spermatogenic PSDs, is that an embryo can survive a maternally deposited toxin if it inherits the antitoxin from its mother or its father, but a sperm survives a paternally deposited toxin only if it inherits the antitoxin from the same spermatocyte.

Embryonic PSDs gain greatest benefits when embryos compete with siblings for access to resources because the death of embryos without a PSD then benefits siblings with the PSD (Wade and Beeman 1994). Opportunities for postsegregational killing are greatly constrained in organisms with large yolky eggs and delayed activation of the embryonic genome because prezygotic provisioning restricts the extent to which siblings compete for resources (unless they eat each other). Nevertheless, embryonic PSDs that reduce posthatching competition among siblings have been reported in oviparous invertebrates. *Medea* elements of *Tribolium castaneum* kill embryos that do not inherit their copies. Embryos are rescued that inherit *Medea* either maternally or paternally (Beeman et al. 1992).

Ontogenetic Veils of Ignorance

The separation of somatic cells from germ cells can be considered an *ontogenetic veil of ignorance* by which beneficiaries in the germ line are hidden from decision-makers in the soma. Somatic cells perform the administrative functions of highly efficient but disinterested civil servants. Three additional ontogenetic veils of ignorance limit "selfish" cellular behavior by creating ontogenetic uncertainty about which genes receive benefits (or are harmed). The first feature is the suppression of transcription in cells of the germ stem until after the differentiation of somatic cells. The second feature is the supervision of stem cell proliferation by supporting cells that have a different ontogenetic origin from the cells whose proliferation they regulate. The third feature is the formation of germline syncytia that share gene products among small clones of cells.

Delayed Zygotic Genome Activation

Cells of the germ stem divide without supervision by somatic cells because somatic cells are yet to differentiate. In many taxa, all mitotic divisions in the germ stem occur before transcriptional activation of the embryonic genome using maternal gene products deposited in the oocyte. Therefore, early embryonic development proceeds under the remote control of maternal genes operating behind the meiotic veil of ignorance. I will call these enslaved cells *embryonic drone cells*. Mutations in a drone cell have no effect on its phenotype and are thus not subject to germline selection. In these taxa, transcription in the germ line is suppressed until after somatic cells are present to supervise germline proliferation. The effect is to deny RNA-dependent agency to genes in the germ stem.

Stem Cells and Stem Cell Niches

Most cells of complex multicellular organisms are derived from stem cells that divide symmetrically to produce two new stem cells or asymmetrically to produce a stem cell and a cell of another kind. Feedback from the population of differentiated cells regulates the ratio of asymmetric to symmetric divisions and thus regulates the size of the population of stem cells and the size of bodily parts generated from those stem cells (Morrison and Kimble 2006; Simons and Clevers 2011).

Genetic differences among cells within bodies are limited to new mutations, where mutations are broadly defined to include changes in numbers of multicopy sequences. Mutations in stem cells that cause a higher proportion of symmetric divisions will increase in frequency within a population of stem cells unless countervailing selection operates to eliminates cells with such mutations. Such mutations can be considered *selfish* (for the good of its cellular lineage). Conversely, a higher proportion of asymmetric divisions of somatic stem cells can be considered *altruistic* (for the good of the body) because such divisions do not increase stem cell number but generate cells that perform somatic functions.

Proliferation of stem cells is typically regulated by non-dividing supporting cells that constitute their somatic niches. Stem cells that detach from their niche either differentiate or die (Li and Xie 2015). Thus the number of stem cells is controlled by the limited number of niches and the limited capacity of each niche to accommodate stem cells. Supporting cells of a niche typically have distinct embryonic origins from the stem cells whose division they regulate. In mammalian somatic development, osteoblasts form the niche of hematopoietic stem cells, mesenchymal cells form the niche of intestinal stem cells, and endothelial cells form the niche of neural stem cells (Li and Xie 2015). Therefore, a mutation in a supporting cell that increases the proportion of symmetric divisions by stem cells has no selective advantage within the body because the mutation is absent from the stem cells that undergo increased proliferation.

The distinct embryological origins of stem cells and their supporting cells function as an ontogenetic veil of ignorance because new mutations are less likely to be shared by stem cells and their supporting cells. This is an important defense against cancer. If stem cells were the direct progenitors of their own supporting cells, then mutant stem cells could readily form tumors by generating their own niches.

Selfishness of germline stem cells is more complex because symmetric divisions increase the number of stem cells but asymmetric divisions are required for the production of gametes. A mutation in a germline stem cell that caused all divisions to be symmetric would form a germ-cell tumor but would be an evolutionary deadend. Primordial germ cells (PGCs) typically originate far from the somatic supporting cells of the future gonad and then migrate to populate gonadal niches (Richardson and Lehmann 2010; Barton et al. 2024). As a consequence, germline stem cells in gonads are unlikely to share de novo mutations with the somatic supporting cells that regulate their numbers. Indeed gonads may have evolved as restraints on germ-cell freedom. However, PGCs remain dependent on interactions with somatic cells throughout their migration to the gonads (Cooke and Moris 2021). These shifting interactions with different populations of somatic cells have been described as a "traveling niche" or "dynamic niche" (Gu et al. 2009; Cantú et al. 2016). Ontogenetic separation between germ cells and their somatic niches minimizes opportunites for nepotistic cabals between somatic cells and germ cells that share the same mutations. Therefore, evolutionary lineages in which PGCs have a different embryological origin from their somatic supporting cells are expected to suffer lesser costs from such cabals. Conversely, the migratory movements of germ cells are not without risks in colonial organisms because germline stem cells of one genetic individual can move through somatic tissues to parasitize the gonads of another genetic individual (Rinkevich and Weismann 1987; Magor et al. 1999).

Germline Syncytia

A syncytial stage precedes meiosis during mitotic proliferation of germ cells of both sexes of most animals (Pepling et al. 1999; Lu et al. 2017; Chaigne and Brunet 2022; Gerhold et al. 2022). Syncytia are initiated by incomplete cytokinesis during a series of synchronous mitotic divisions after germ cells exit their gonadal stem cell niche. As a result, a small clone of cells remains interconnected by cytoplasmic bridges through which gene products and organelles can be shared (Greenbaum et al. 2011). In female germ lines, intercellular connections are severed before meiosis resulting in cytoplasmically isolated oocytes. In male germ lines, cytoplasmic bridges are maintained, and continue to be formed, during the meiotic divisions of spermatocytes resulting in a cyst of interconnected spermatids that eventually cellularize as spermatozoa with compacted nuclei.

In *Drosophila* male and female germlines, germline stem cells divide to produce another germline stem cell and a cystoblast. The cystoblast undergoes four rounds of mitosis with incomplete cytokinesis to produce a premeiotic cyst of sixteen cells. In male germlines, each nucleus of the premeiotic cyst undergoes two meiotic divisions with incomplete cytokinesis to form a postmeiotic cyst of sixty-four interconnected spermatids. In female germ lines, one of the members of the premeiotic cyst becomes an oocyte and the other fifteen cells differentiate into endopolyploid nurse cells that supply materials to the oocyte (Slaidina and Lehmann 2014). Nurse cells eventually empty their remaining contents into the oocyte and undergo programmed cell death (Hinnant et al. 2020).[1]

---

[1]Are nurse cells somatic? If they are, then cystocarps would, by strict definition, be part of the germ stem, leaving oocytes as the only cells of the germ crown. But such insistence on definitional purity would obscure the evolutionary origin of nurse cells in differentiation of a new cell type from within an ancestral germ crown. For convenience, I consider nurse cells, and similar cul de sacs from the germ thoroughfare, to be parts of the germ crown.

A cyst formed by n synchronous divisions contain 2^n cells. In mouse ovaries, cysts of eight or sixteen cells fragment into individual oocytes (Pepling and Spradling 2001; Lei and Spradling 2013). Cysts in *Xenopus* ovaries contain sixteen oocytes (Kloc et al. 2004). The number of interconnected spermatids in mammalian testes varies markedly among species. In mice and rats, ten mitotic divisions and two meiotic divisions generate syncytia of 4,096 spermatids whereas, in human testes, three mitotic divisions and two meiotic divisions generate syncytia of thirty-two spermatids (Fayomi and Orwig 2018; Yoshida 2019). The female germ crown of the urochordate *Oikopleura dioica* is a single cell, containing tens to hundreds of thousands of nuclei before simultaneous maturation and spawning of oocytes arrested in meiosis I (Ganot et al. 2007). After exit from their stem cell niche, all premeiotic germ cells in *Caenorhabditis* gonads form a syncytium. Divisions of neighboring cells within this syncytium are asynchronous and cytoplasmic sharing of materials among germ cells is limited (Gumienny et al. 1999; Hubbard and Schedl 2019).

Deep evolutionary conservation of a syncytial phase raises the question of its functional significance. Synchronous division and the sharing of gene products reduces opportunities for cellular selection based on genotypic differences among nuclei within syncytia (Pontecorvo 1944; Nguyen et al. 2015). The sharing of gene products between the four haploid spermatids produced by a diploid spermatocyte probably limits opportunities for meiotic PSDs. However, this does not explain the formation of premeiotic syncytia. One possibility is that connections among oogonia or spermatogonia defend against mitotic PSDs in the divisions after a germ cell leaves its germ cell niche and before the formation of gametes.

In female cysts of *Drosophila,* nurse cells "sacrifice" themselves for the benefit of a single large oocyte. Pepling and Spradling (2001) suggested that syncytial cysts in other taxa contain cryptic nurse cells and that the cystic architecture facilitates the selection of better-quality oocytes rather than obviates such selection, as suggested by Pontecorvo (1944). From this perspective, intercellular transfers of mitochondria allow selection of better-quality mitochondria for incorporation into the functional oocyte (Lei and Spradling 2013, 2016), although it should also be noted that mitochondria have a vested interest in their own transmission to functional oocytes (Haig 2016a). Such hypotheses do not explain the formation of similar syncytia in male germ lines nor in female germ lines that lack nurse cells (Lu et al. 2017).

The Developmental Hourglass

Various ontogenetic veils of ignorance collectively contribute to disinterested supervision of cellular divisions in the germ line. However, the dependence of germline cells on somatic supporting cells is absent during early embryonic divisions of the germ stem. In taxa that possess embryonic drone cells, nuclear divisions in the germ stem take place under the control of stored maternal gene products. However, in mammalian development the embryonic genome is activated before the differentiation of somatic cells. Therefore, the conflicts of the *arena germinalis* are largely unsupervised until differentiation of somatic cells brings the germline melee under greater control. The disorderly nature of early mammalian development (see below) can be considered to be an indirect consequence of the emancipation of early embryonic cells from maternal servitude.

Embryologists have noted high conservation of a phylotypic stage of development around the time of gastrulation with development before and after this stage, exhibiting greater evolutionary lability. This phenomenon has been called the *developmental hourglass* (Raff 1996, 208–209; Piasecka et al. 2013). Divergent development after the phylotypic constriction has been considered unsurprising as the bodies of different evolutionary lineages adapt to their changing environment, but the extensive variation in early development has been considered paradoxical because the initial intuition was that the earliest stages of development should be the most highly conserved (see Raff 1996, 197, and Lee et al. 2014, 601).

I propose that one contributing factor to the unexpected variability of early development is that the *pax somatica* has not yet been established with the consequence that the conflicts of the *arena germinalis* are relatively unfettered. An evolutionary arms race for ontogenetic advantage therefore ensues between the conjugal genome and a menagerie of PGRs and PSDs, which destabilizes early embryonic processes. These conflicts are substantially, but not completely, brought under control with the differentiation of somatic cells that regulate germline development. Furthermore, in organisms with embryonic drone cells the activities of RNA-dependent PGRs and PSDs are highly constrained during the early cleavage divisions before the sequestration of somatic cells. As a corollary, the evolutionary variability of early development should be particularly pronounced in organisms with early activation of embryonic transcription.

Specification of the Germ Line

Weismann's *Keimbahn* was the repository of *Keimplasma* (germ plasm) that resided in nuclear chromatin and contained hereditary "determinants" of form (see the appendix to this chapter). For Weismann, "continuity of the germ plasm" referred to the inheritance of what we would now call genes, but for many embryologists "continuity of the germ line" referred to cytoplasmic determinants that were deposited in eggs and segregated to germ cells. Arguments about "continuity of the germ line" were often framed as a conflict between whether the germ line (by which was meant cells of the germ crown) was specified by *preformation* or *epigenesis*. Under preformation, germ-cell identity was specified by cytoplasmic factors inherited from oocytes. Under epigenesis, germ cells differentiated in response to signals from neighboring cells. In these debates, germ plasm was often identified with microscopically visible germ granules deposited in the oocyte and selectively inherited by the next generation of germ cells (Hegner 1911), and which were interpreted as "determinants" of germ-cell identity. Thus, "continuity of the germ plasm" and "continuity of the germ line" became conflated and confused.

A history of germline hide-and-seek between PGRs and the conjugal genome may have contributed to the paradoxical lack of conservation of mechanisms of germline specification (Johnson et al. 2011; Haig 2016b). Once preferential replication by PGRs was cued to existing indicators of germline status, there may have been short-term benefits for the rest of the genome from changing mechanisms of germline specification so that germ cells no longer expressed these cues or so that germ cells differentiated in previously somatic locations. In the short term, the rest of the genome would have been protected from the mutagenic costs of PGRs until new PGRs evolved to target novel cues of germline status. Conversely, PGRs that had just increased their copy number within a nucleus may have benefited from biasing their cellular lineages toward a germ cell fate.

It is now generally accepted that some animals specify their germ cells by preformation and others by epigenesis (Extavour and Akam 2003; Voronina et al. 2011). Short germ stems comprised of embryonic drone cells are typically associated with preformation of the germ line marked by germ granules deposited in oocytes. This combination of characters appears to have evolved multiple times (Johnson et al. 2003; Kumano

2015). Long germ stems are typically associated with epigenetic determination of the germ crown and absence of germ granules in the germ stem, although granules may subsequently be assembled in the germ crown (Chatfield et al. 2014; Johnson and Alberio 2015).

Posttranscriptional Germline Control

Ribonucleoprotein condensates (known by various names, including nuage, chromatoid bodies, germ granules, and germ plasm) are a conspicuous feature of metazoan germ lines (Eddy 1975; Matova and Cooley 2001; Voronina et al. 2011; Mukherjee and Mukherjee 2021). These condensates are phase-separated liquid droplets containing RNA and RNA-binding proteins (Brangwynne et al. 2009; Strome and Updike 2015; Chiappetta et al. 2022; Westerich et al. 2023). Some granules are assembled in the oocyte and segregate to PGCs of the germ crown during early embryogenesis. By this method, maternal genetic control is extended into the germ crown. Other granules are assembled from embryonic gene products in cells of the germ crown.

The association of germline cells with ribonucleoprotein condensates is one expression of the prominent role of RNA-based mechanisms in germline control (Seydoux and Braun 2006; Strome and Lehmann 2007; Lai and King 2013; Nousch and Eckmann 2013; Lebedeva et al. 2018). Somatic differentiation is mediated by the expression of tissue-specific transcription factors, but attempts to identify specific transcription factors that initiate germline differentiation have failed. Instead, specification of the germ crown is accompanied by transcriptional repression rather than activation (Nakamura and Seydoux 2008). Constraints on germline transcription are usually interpreted as a means of repressing somatic gene expression (Matova and Cooley 2001; Leatherman and Jongens 2003; Cinalli et al. 2008) but they also restrict the activities of PGRs. Posttranscriptional control in the germ line may have been favored because it is less easily subverted by PGRs that depend on transcription for their proliferation (Haig 2016b). As a corollary, restraints on transcription can be relaxed in somatic cells because PGRs have evolved somatic self-restraint.

In those taxa that silence transcription in early embryos, retroelements either must wait until after the resumption of transcription to transpose or must load their RNA into oocytes before the prezygotic cessation of transcription. RNAs deposited in oocytes will be most effective in promoting germline retrotransposition if they segregate together with determinants of germline fate during early cleavage divisions. Therefore, whenever cytoplasmic determinants of germline fate are deposited in oocytes, retroelements will evolve to co-localize their gene products with these determinants. Furthermore, given an association of selfish RNA with determinants of germline fate, defenses against selfish RNA are expected to be targeted to the same locations. This association between preformation of the germ line and deposition of selfish RNA in oocytes can also be expressed the other way around. If the germ crown is determined by epigenesis, then RNA deposited in oocytes cannot be reliably targeted to germ cells but will be distributed during cleavage to somatic as well as germ cells. Therefore, the advantages for retrotransposons of prepackaging RNA in oocytes are reduced in organisms with epigenesis of the germ line.

Different proteins nucleate the germ granules of different species (Kulkarni and Extavour 2015). In *Drosophila* ovaries, *oskar* mRNA is synthesized in nurse cells of ovarioles before its transfer to the posterior end of oocytes where oskar protein nucleates the condensation of multiple RNAs and RNA-binding proteins into germ granules (Ephrussi and Lehmann 1992; Snee and Macdonald 2003; Curnutte et al. 2023). Oskar protein is sufficient to induce the formation of germ cells at ectopic locations (Ephrussi and Lehmann 1992; Lehmann 2015). However, a role for oskar in germline specification is restricted to holometabolous insects (Lynch et al. 2011; Quan and Lynch 2014; Blondel et al. 2021). Germ granules of *Caenorhabditis* are assembled by PGL proteins (Hanazawa et al. 2011; Aoki et al. 2016), although these granules are not required for germline specification (Gallo et al. 2010). Germ granules in zebrafish oocytes are assembled around Bucky ball protein (Bontems et al. 2009). Germ granules need not directly determine germline fate but may precipitate around such determinants.

Under the view advanced here, ribonucleoprotein germ granules may combine determinants of germ-cell identity, RNA-based selfish genetic elements (SGEs) that are targeting those germ cells, and genomic defenses that are targeting those SGEs. The double-faced nature of germline ribonucleoprotein condensates is illustrated by the germ granules of *Drosophila* oocytes that incorporate the machinery of piRNA synthesis, possibly as a defense against transposable elements but also contain RNAs derived from transposable elements (Spichal et al. 2021). For example, *Tahre* retrotransposons are transcribed in nurse cells as RNAs that mimic *oskar* RNA and are incorporated into polar granules using the same transport machinery as *oskar* RNA (Tiwari et al. 2017). Similarly, *I*-elements are transcribed in nurse cells with their RNA transported into oocytes (Wang et al. 2018).

Ontogenetic Variations

The ancestral mode of metazoan development appears to have involved a population of multipotent stem cells capable of producing both somatic and reproductive cells in adult bodies (Extavour and Akam 2003; Juliano et al. 2010). These cells enabled forms of asexual reproduction in which a body divided into two or more fragments and then each fragment regenerated the missing parts. They also enabled forms of colonial development in which new zooids budded from older zooids (Bely and Nyberg 2010). Transient germ stems, with early separation of somatic cells from the germ line, have evolved many times accompanied by loss of the ability to reproduce by budding or bodily fission (Kumano 2015).

Similarities of gene expression between multipotent stem cells and germ cells have been characterized as a "germline multipotency program" (Juliano et al. 2010; Fierro-Constaín et al. 2017; Piccinini and Milani 2023). These similarities include a heavy reliance on posttranscriptional control (Rouhana et al. 2010; Alié et al. 2015). In this resurrection of the Weismannian *Keimbahn,* cells with both somatic and germ-cell descendants have been known by taxon-specific names such as neoblasts in platyhelminths and acoels (Wagner et al. 2011; Srivastava 2022), archeocytes in sponges (Alié et al. 2015; Funayama 2018), and i-cells in cnidarians (Leclère et al. 2012); and by generic names such as multipotent progenitor cells (Juliano et al. 2010), pluripotent adult stem cells (Wagner et al. 2011; Kimura et al. 2022), or primordial stem cells (Solana 2013). In my terminology, these multipotent stem cells belong to germ stems of indefinite length.

Animals vary markedly in the organization of their germ lines. For bodies that produce the same number of meiocytes, the number of mitotic divisions between a zygote, the first cell of the germ stem, and a meiocyte, a terminal cell of the germ crown, is greater for bodies with longer germ stems (Figure 6.2). The population dynamics of PGRs will be strongly influenced by the ontogenetic architecture of the germ line and by the presence versus absence of embryonic drone cells. Other things being equal, PGRs will have fewer opportunities for proliferation in

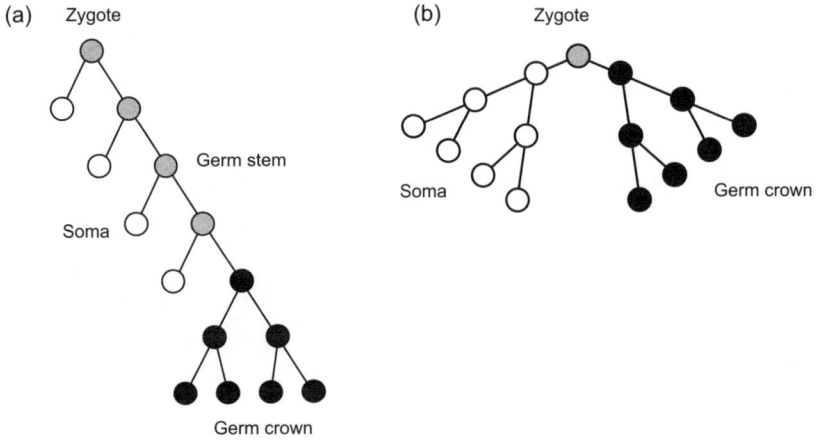

Figure 6.2. Cellular divisions during development of two hypothetical organisms, both of which contain four somatic cells (white) and four germ cells (black) in the adult body. (a) The germ stem consists of four cells. There are six cell divisions between the zygote and an adult germ cell. (b) The germ stem consists of a single cell (the zygote). There are three cell divisions between the zygote and an adult germ cell.

organisms with shorter germ stems and with delayed zygotic genome activation. Therefore, in the balance between germline and organismal fitness, greater weight will be placed on the side of germline fitness in organisms with longer germ stems or early activation of the embryonic genome. Organisms with short germ stems may be predisposed to the evolution of streamlined genomes because they are less subject to the activities of PGRs. Conversely, organisms with long, transcriptionally active germ stems will be predisposed to larger genomes with a high proportion of repetitive sequences (Haig 2024b).

Rather than attempt to review the full diversity of metazoan development, I will discuss embryonic development in a few model organisms to illustrate the variation in ontogeny.

Fragmentable Germ Stems (Planarians)

The timing of embryonic genome activation has not been determined in planarian development, but I expect it occurs early because of the unusual ectolecithal mode of development in which multiple embryos compete for an external supply of yolk deposited within an egg capsule (Laumer and Giribet 2014). The early development of planarian embryos has been described as "blastomere anarchy." Cleavage cells detach from each other, disperse through the yolk, before reaggregating to form an embryo that encloses a large yolk mass (Thomas 1986; Martín-Dúran and Eggers 2012). This is a form of matrotrophy in which there would have been strong selection for early activation of the embryonic genome. After the anarchic early development of embryos, pluripotent stem cells generate all embryonic tissues as well as the neoblasts that become the multipotent stem cells of adult bodies (Davies et al. 2017). The blastomeres that aggregate to form an embryo have been implictly assumed to be derived from a single zygote although the possibility that embryos sometimes contain cells derived from multiple zygotes should also be considered.

Adult planarians (freshwater platyhelminths) are capable of whole-body regeneration because they possess a persistent germ stem comprised of totipotent neoblasts. Somatic cells never divide but are continuously generated as daughter cells of neoblasts, which are also the progenitors of the oogonia and spermatogonia of the germ crown (Sato et al. 2006; Wang et al. 2007). After bodily fission, neoblasts in both fragments regenerate the missing bodily parts. A single neoblast, injected into a body composed of non-mitotic somatic cells, is capable of regenerating an entire adult planarian and a lineage of its clonal descendants (Wagner et al. 2011). Planarians possess germ stems of indefinite length (Figure 6.3).

Plant demographers often distinguish genets (genetic individuals) from ramets (functionally independent units) (White 1979). The growing tips (meristems) of a plant divide to produce a branching plant-body. In a lawn of a tillering grass, distant parts detach as intervening parts decay making ramets easier to count than genets. Neoblasts are the stem cells of planarian bodies, and in many ways resemble meristematic cells of plants. An asexually reproducing population of planarians is a mobile lawn, in which the number of worms (ramets) is much easier to count than the number of zygotic genotypes (genets). Neoblasts have been considered to be the fundamental individuals of planarian biology because a single neoblast could be the ancestor of a large clone of multicellular adult planarians (Fields and Levin 2018). Thus three levels of planarian individuality can be recognized: zygotic individuals (genets), individual worms (ramets), and individual neoblasts. Genets are derived from a zygote and reproduce new genets by the production of gametes. Worms reproduce asexually by fission. Neoblasts reproduce by symmetric mitotic divisions.

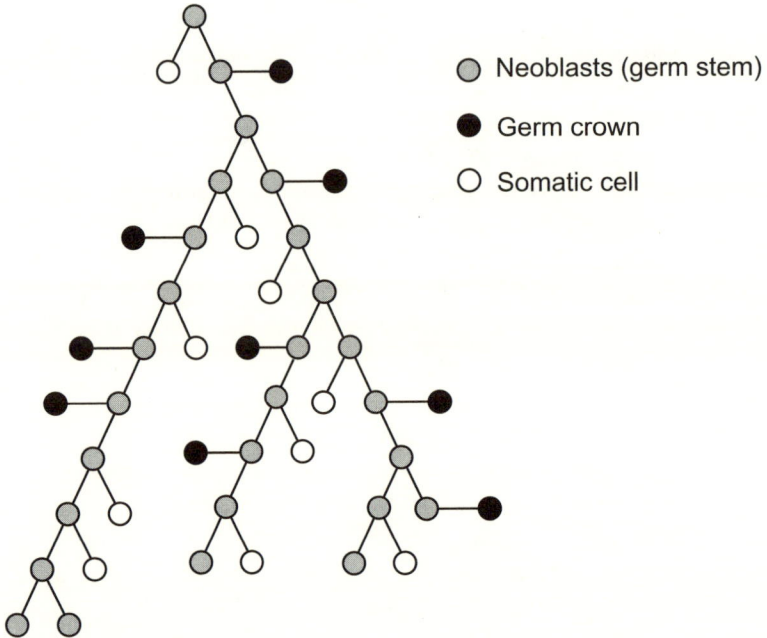

Figure 6.3. Cellular divisions in a planarian. The germ stem, comprised of neoblasts (gray), is of indefinite length and gives rise to both somatic cells (white) and germ cells (black). If the body is fragmented, each of the fragments contains a population of neoblasts.

A 4-mm-long ramet of *Schmidtea mediterranea* contained 110,000 cells including 35,200 neoblasts (32% neoblasts) whereas an 11-mm-long ramet contained 324,000 neoblasts among 1.8 million total cells (18% neoblasts). In general, neoblasts comprised 18%–35% of cells in worms of different sizes from two species of asexual planarians. The proportion of neoblasts underwent only minor fluctuations during proliferation (growth) and shrinkage (degrowth) of the total number of cells. This provides compelling evidence that the proportion of neoblasts is subject to some form of bodily regulation (Baguñá and Romero 1981). The high proportion of neoblasts has been considered surprising compared with the rarity of somatic stem cells in other organisms (Baguña and Romero 1981; Morrison and Spradling 2008). Because neoblasts are the only mitotic cells of the planarian body, a mutant neoblast can directly generate its own somatic supporting cells. Thus, the abundance of neoblasts may be explained by consideration of the interplay of cellular and organismal selection in asexual reproduction by bodily fission.

Neoblasts, like other stem cells, proliferate by symmetric divisions and generate differentiated cells by asymmetric divisions. Suppose that a ready-to-fission worm contained a million cells of which 200,000 were neoblasts and that a small proportion of these neoblasts possessed a mutation that caused a higher proportion of symmetric divisions and lower proportion of asymmetric divisions. The mutant neoblasts could be considered relatively selfish and the non-mutant neoblasts relatively altruistic. When the worm divided, its two fragments would each receive 100,000 neoblasts as a partition of the neoblasts present in the parental worm. If the genetic diversity of neoblasts in the parental worm were spatially well-mixed then the two offspring would receive very similar proportions of altruistic and selfish neoblasts. If the genetic diversity were spatially structured, then one offspring might receive a substantially higher proportion of altruistic neoblasts than the other.

Differences in neoblast proportions between daughter worms at the time of division will increase during growth of the sisters by stochastic processes that cause variation among neoblasts in the number of symmetric divisions, by cellular selection favoring selfish neoblasts that undergo more symmetric divisions, and by new mutations in individual neoblasts. Intraworm selection favors an increase in the proportion of selfish neoblasts within every worm during its development, but worms that receive a smaller proportion of selfish neoblasts after reproductive fission will be favored by interworm selection. When worms reproduce sexually, the proportion of zygotes that receive the selfish mutation, as a zygotic innovation, will be determined by the balance between these two levels of selection. All things considered, a selfish mutation in a population of altruistic neoblasts is more likely to become an innovation, transmitted by a gamete, than an altruistic mutation in a population of selfish neoblasts. A long evolutionary history of this bias favoring selfish innovations may help to explain why neoblasts comprise such a large proportion of cells in planarian bodies.

Neoblasts are immobile during tissue homeostasis (Guedelhoefer and Sánchez Alvarado 2012) and are therefore surrounded by somatic descendent cells with which they share mutations. Such immobility means that there will be greater differences between sister worms in the proportions of selfish versus altruistic neoblasts when a parental worm fissions. Stationarity thus increases the strength of interworm selection against selfish neoblasts and favors greater altruism within planarian bodies. However, neoblasts can also migrate to sites of tissue damage or regions without resident neoblasts (Guedelhoefer and Sánchez Alvarado 2012; Abnave et al. 2017; Sahu et al. 2021). The greater the mobility of neoblasts during planarian development, the greater the problem of intraworm selection favoring selfish neoblasts. Intraworm migration of neoblasts is associated with DNA breakage and repair (Sahu et al. 2021). This perhaps reflects activation of selfish genetic neoblasts within the genomes of stem cells about to proliferate.

Mutations with somatic effects will periodically occur that enhance the organismal fitness of planarian bodies. Does multilevel selection within *genetic individuals* (genets) promote or impede the likelihood that organismically favorable mutations will become zygotic innovations (Reusch et al. 2021)? For simplicity, I will consider a beneficial mutation with somatic effects that does not change the dynamics of neoblast proliferation. The effect on absolute fitness of ramets (hard selection) will presumably be related to the proportion of neoblasts within a body carrying the mutation. The extent to which competition among ramets (soft selection) would favor the mutation is presumably related to the variance among worms in the proportion of neoblasts carrying the mutation. Positive selection will be weak when the variance in proportion of mutant cells is small. Therefore, reduced mobility of neoblasts will tend to increase the variability among worms in the proportion of neoblasts with the favorable mutation and thus increase the strength of selection for the beneficial mutation.

Germline stem cells of planarians are descended from—and closely resemble—neoblasts, except for the expression of *nanos* that is used as a marker to distinguish germline stem cells from neoblasts (Newmark et al. 2008; Issigonis and Newmark 2019). Germline stem cells and the somatic supporting cells of ovaries and testes differentiate from neoblasts. I do not know the degree of ontogenetic separation between germline stem cells and the somatic supporting cells within a gonad, but it is possible that germ cells and their supporting cells are locally generated from closely related neoblasts. In planarian testes, spermatogonia undergo three rounds of mitosis to form syncytial cysts of eight spermatocytes that then undergo meiosis to form cysts of thirty-two spermatids (Iyer et al. 2016). If a planarian is capable of both asexual and sexual reproduction, PGRs and PSDs gain a benefit from sexual reproduction that is not shared by the rest of the genome, namely the transmission of their copies to more than their "fair share" of gametes. Therefore, selfish elements will evolve to promote greater investment in sexual reproduction, and less investment in asexual reproduction, than is optimal for the rest of the genome.

Planarians possess transcriptionally active germ stems of indefinite length; therefore, planarian genomes are predicted to be particularly vulnerable to the accumulation of retroelements. This may help to explain the high proportion of repetitive sequences in planarian genomes compared to the much smaller genomes of nematodes with short, transcriptionally inactive germ stems (Haig 2024b).

Unbranched Germ Stems (Nematodes)

Unlike planarians, nematodes are individuals in the strict etymological sense. They are indivisible. The germ stem of *Caenorhabditis elegans* consists of only four cells: P_0 (the zygote), P_1 (a daughter of P_0), P_2 (a daughter of P_1), and P_3 (a daughter of P_2). Each of these cells divides to produce a daughter cell with solely somatic descendants and a daughter cell that perpetuates the germ line. The fifth cell of the germ line, P_4 (a daughter of P_3), has no somatic descendants and is thus the first cell of the germ crown (Sulston et al. 1983). Note that P_0–P_4 are drone cells under remote maternal control. Transcription is activated in the somatic sister cells P_1–P_4 as soon as they diverge from the germ line (see Figure 6.4) (Robertson and Lin 2015). As such, P_4 remains mitotically and transcriptionally quiescent during active proliferation of the somatic cells that form the larval body.

The pattern of somatic cell divisions is identical among worms. Symmetric and asymmetric divisions proceed in predetermined sequence without undetermined choices of whether a cell divides symmetrically or asymmetrically. Open-ended lineages of somatic stem cells are absent in *Caenorhabditis* development. All somatic cells of adults are fully differentiated and incapable of further division (Sulston and Horvitz 1977; Sulston et al. 1983). Tumors in *Caenorhabditis* are restricted to dysregulated proliferation of stem cells of the germ crown (Kyriakis et al. 2015).

Once embryos contain about a hundred somatic cells, P_4 divides once to produce the primordial germ cells Z_2 and Z_3, which then remain in mitotic and transcriptional arrest for the remainder of embryogenesis (Schaner et al. 2003). Two somatic cells, Z_1 and Z_4, migrate to associate with Z_2 and Z_3 to form a four-celled gonadal primordium. Thus Z_1 and Z_4 are the progenitors of two distal tip cells that will form the somatic niches of the germline stem cells descended from Z_2 and Z_3 (Sulston et al.

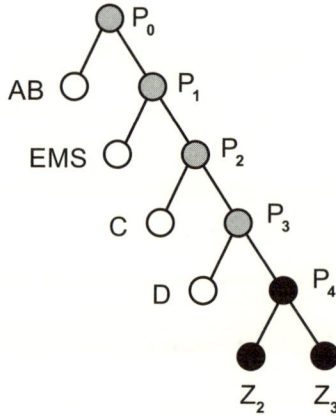

Figure 6.4. Early embryonic divisions of *Caenorhabditis elegans*. The germ stem (gray) consists of four cells (P_0–P_3). Each of these cells divides to produce a somatic daughter cell (white) in which transcription is activated, and a daughter cell that perpetuates the germ line in which transcription is inactive. The four somatic daughter cells (AB, EMS, C, D) give rise to all cells of the adult body, except the germ crown. The P_4 is the first cell of the germ crown (black). After a long delay, transcription is activated in the primordial germ cells (Z_2 and Z_3).

1983; Kipreos 2005; Kimble and Crittenden 2007). All germline stem cells of *Caenorhabditis* are descended from a single drone cell (P_4) and are more closely related to each other ontogenetically than to any somatic cell. No mutation present in a somatic cell is shared with some germ cells but not others. Therefore, every somatic cell is equally related to every cell of the germ crown and all somatic cells have a common interest in the effective management of germline stem cells. The most recent cellular ancestor of germline stem cells and gonadal supporting cells is P_1 (a daughter cell of the zygote P_0).

Robust transcription is activated in Z_2 and Z_3 only after the onset of larval feeding and is followed by the commencement of germline proliferation (Butuci et al. 2015). Cell divisions in the descendants of Z_2 and Z_3 are non-stereotypic and variable among worms, with a mixture of symmetric and asymmetric divisions (Hubbard 2007). Proliferation of germ cells depends on their physical association with a distal tip cell. Each arm of the hermaphrodite gonad is associated with a single distal tip cell (a granddaughter cell of Z_1 for one gonadal arm and of Z_4 for the other arm). The distal tip cells have processes that extend into the proliferative zone of the gonad and make contact with individual germ cells. As germ cell proliferation pushes germ cells out of the proliferative zone, they lose contact with the distal tip cell and enter meiosis (Champilas and Taver-narakis 2020; Gordon et al. 2020).

A germline mutation with somatic effects on the regulation of germ-line stem cells is subject exclusively to organismal selection for its costs and benefits for the germ line as a whole. If a mutation occurs in the germ stem, then all cells of the germ crown will possess the mutation in that generation and somatic cells can express no favoritism among germline stem cells (except for those that acquire new mutations in cells of the germ crown). If a mutation occurs in the germ crown, then it is absent from all somatic cells of that generation and will be subject to selection on its somatic effects only after it has been zygotically inherited as an innovation that is transmitted to all somatic and germline cells of the next generation. Checkmate!

Tight controls on transcription in the nematode germ stem may have limited the activities of the retroelement. However, the early differentiation of transcriptionally active somatic cells creates opportunities for embryonic PSDs that cause early embryonic death of embryos that do not inherit their copies. Indeed, a number of such systems have been described in nematodes (Ben-David et al. 2007; Seidel et al. 2011; Noble et al. 2021).

Branched Germ Stems with Somatic Stem Cells

Fundamental similarities exist in early embryonic development of three oviparous model organisms: *Drosophila melanogaster* (a maggot that needs no introduction), *Danio rerio* (zebrafish), and *Xenopus laevis* (a frog). These species have much larger eggs than *Caenorhabditis* with less favorable nucleocytoplasmic ratios. A characteristic feature of their early development is a series of rapid cleavage divisions that partition the egg's cytoplasm before the onset of embryonic transcription. Hence, all cells of blastulas are drone cells dividing under remote control of the maternal genome. In all three species, the activation of transcription in the germ line is further delayed relative to somatic cells.

Drosophila embryogenesis begins with thirteen syncytial divisions followed by cellularization of somatic cells in the fourteenth mitotic cycle. Transcription by polymerase II is suppressed until after cellularization of somatic cells (Williamson and Lehmann 1996; Blythe and Wieschaus 2015). About ten "pole cells" cellularize precociously during the tenth mitotic cycle. These are the progenitors of the germ crown (Williamson and Lehmann 1996). Thus, on a cellular path from zygote to meiocyte, roughly eight transcriptionally inactive nuclei belong to the germ stem. Activation of transcription in PGCs of the germ crown is delayed until after gastrulation (van Doren et al. 1998; Siddiqui et al. 2012).

Embryonic genome activation in zebrafish occurs after ten division cycles in both somatic and germline cells (Kane and Kimmel 1993; Knaut et al. 2000; D'Orazio et al. 2021), with four to five progenitors of the germ crown set aside after twelve to thirteen division cycles (Knaut et al. 2000). *Xenopus* embryos undergo twelve division cycles before the somatic genome is activated at the mid-blastula transition (Newport and Kirschner 1981a, b; Venkatarama et al. 2010; Sheets 2015; Blitz and Cho 2021). Five progenitors cells of the germ crown are set aside after thirteen to fourteen division cycles (Whitington and Dixon 1975), with activation of transcription delayed until PGCs separate from endoderm in late gastrulas (Zalokar 1976; Yang et al. 2015). In summary, embryonic transcription is suppressed until after somatic cells diverge from the germ stem in *Drosophila, Danio,* and *Xenopus* with activation of transcription further delayed in the germ lines of *Drosophila* and *Xenopus*. Therefore, all cells of the germ stem are embryonic drone cells. Most conflicts of the *arena germinalis* are postponed until after somatic cells are sequestered from the germ line.

The germ crowns of *Drosophila, Danio,* and *Xenopus* are descended from a small number of PGCs (rather than from a single cell as occurs in *Caenorhabditis*). In other words, their germ stems are sparsely branched with the possibility that some germ cells and somatic cells will share mutations with each other that are not shared with other germ cells and somatic cells. However, the migration of germ cells to gonads ontogenetically randomizes associations between germ cells and somatic supporting cells. Like *Caenorhabditis* these organisms possess indivisible bodies, but unlike *Caenorhabditis* these bodies contain somatic stem cells.

Somatic selection favoring excess symmetric divisions of stem cells is limited to the lifetime of the discardable body. Somatic evolution must start from scratch in each new body because its prodigies die without heirs. Germline mutations, by contrast, can survive the death of the bodies in which they reside. The resulting innovations have been selected to produce new bodies each generation that effectively manage the proliferation of somatic cells. Premature death of bodies that were less effective managers of somatic selection has resulted in preferential survival of germline innovations that postpone tumorigenesis until later in life (Haig 2015).

Prezygotic and Postzygotic Provisioning

The early cleavage divisions of *Drosophila, Danio,* and *Xenopus* proceed without transcription from the embryonic genome. Why should embryogenesis be organized in this manner? The large size of their eggs is undoubtedly a factor. An egg contains more cytoplasm per gene-copy than any other cell. Early development would proceed very slowly if it depended on transcription and translation of embryonic gene products—from a single nucleus in an enormous cell, then from two nuclei in two very large cells, then from four nuclei in somewhat smaller cells, and so on—but yolk-laden embryos are vulnerable to predation and life-histories have been subject to intense selection to pass through this defenseless stage as quickly as possible (Williams 1966, 89; Shine 1978; Strathmann et al. 2002; Farrell and O'Farrell 2014). One solution has been to supply eggs with abundant maternal proteins and RNA along with the nutritive yolk and delay embryonic transcription until after the proliferation of embryonic nuclei has created a more favorable nucleocytoplasmic ratio (Woodland 1982; O'Farrell 2015). Transcription interferes with rapid DNA replication (Saxena and Zou 2022). Therefore, a selective premium on rapid early development could help to explain both a reliance on maternal gene products and the repression of embryonic transcription.

A clutch of recently spawned eggs develops under the unified genetic control of their mother's genome. Sibling rivalry and paternal interests are suppressed because development is controlled by maternal gene products deposited in oocytes behind the meiotic veil of ignorance. Furthermore, many of the conflicts of the *arena germinalis* are postponed until after activation of the embryonic genome. In particular, activities of retroelements and embryonic PSDs are suppressed during the early cleavage divisions. Retroelements inherited via sperm must wait until after the resumption of transcription to colonize chromosomes inherited via eggs.

When mothers invest in oocytes before ovulation, the amount of yolk received by an embryo is an unconditional gift given without regard to the embryo's personal qualities. Because of this maternal largesse, resources committed to inviable embryos are sunk costs. Furthermore, when yolk is fully vested in oocytes before fertilization, embryos are unable to influence how much they receive from their mother. Because embryonic and maternal genomes have nothing to argue about, embryonic genomes should acquiesce in maternal control of early embryogenesis for its many benefits.

An alternative solution to the problem of large maternal investment per offspring has been to produce small eggs and provision embryos as they grow. Once mothers invest postzygotically in embryos that express their own genomes, resources can be withheld from embryos until after they have proved their worth. Embryos of lower expected fitness can be aborted with reallocation of resources to embryos of higher expected fitness (Stearns 1987; Haig 1990, 2010, 2019). An unconditional gift to all embryos has been replaced by conditional compensation for performance for some embryos. In this evolutionary shift to postzygotic provisioning, mothers were selected to reduce prezygotic provisioning to reduce the cost of aborted embryos. Embryos have been strongly selected to perform well at their job interviews, including providing inaccurate information about their own quality (McCoy and Haig 2020).

Roughly speaking, prezygotic provisioning is the better option for species with high fecundity and small investment per offspring (because the potential savings from reallocations among offspring are concomitantly small), whereas postzygotic provisioning is the better option for species with low fecundity and large investment per offspring (because more can be saved by early termination of investment with reallocation of resources to offspring of better quality) (Haig 1990). Postzygotic provisioning also means that all resources needed for the development of a clutch (or litter) of offspring do not need to be accumulated before ovulation.

The supply of maternal resources to offspring after ovulation is a distinctive feature of mammalian development. Although monotreme eggs contain substantial yolk at the time of ovulation, most of the mass of an oviposited egg is transferred during an extended uterine gestation (Hughes 1993). This can be considered a form of matrotrophic oviparity with a partial shift from prezygotic to postzygotic provisioning. After hatching, the bulk of maternal investment is provided to young monotremes via lactation. The minuscule eggs of marsupials and eutherians, by contrast, lack the nutritive yellow yolk and embryos hatch in the uterus (Brawand et al. 2008; Menkhorst et al. 2009). In marsupials, most maternal investment is provided during lactation after parturition. Therefore, investment can be terminated postnatally at comparatively little cost (Hayssen et al. 1985). Several marsupials give birth to more offspring than there are teats in the pouch with the siblings involved in a life-or-death scramble for a teat (Hartman 1920; Flynn 1922; Dollman 1938; Nelson and Gemmell 2003). Postnatal selection of offspring is less efficient in eutherian mammals because much greater amounts of maternal resources are transferred prenatally across the placenta. Thus eutherian reproduction, compared to marsupial reproduction, is associated with a greater emphasis on prenatal selection of embryos than postnatal selection of neonates (McCoy and Haig 2020).

Postzygotic provisioning profoundly changes the evolutionary forces acting on maternal and embryonic genomes. With postzygotic provisioning, how much an embryo receives is "up for grabs" and embryos have reasons not to acquiesce in maternal genetic control. Embryos express their genomes early to gain control of their fate. They evolve to actively solicit or seize maternal resources and to compete with genes in other (contemporary or future) embryos for maternal investment (Trivers 1974; Haig 1993). Sibling rivalry and paternal interests are no longer suppressed. As a further consequence of early activation of the embryonic genome, PGRs and PSDs gain new opportunities for proliferation in early embryos. The next section will consider the profound consequences of these changes for the intrauterine development of eutherian mammals.

Eutherian Development

Eutherian embryos develop sedately within the safe harbor of a well-defended uterus—slowly, but not steadily. Their early embryogenesis has been described as stochastic (Dietrich and Hiiragi 2007; Wennekamp et al. 2013). It has also been described as highly regulative because it is highly irregular (Dard et al. 2009). Segregation errors are common (Vázquez-Diez and FitzHarris 2018; Daughtry et al. 2019; Allais and FitzHarris 2022), with extraordinarily high rates of chromosomal missegregation in the early mitotic divisions of human embryos (Vanneste et al. 2009; McCoy 2017). The precise pattern and timing of early divisions is variable among embryos, even embryos of the same litter, in contrast to the stereotypic divisions of *Caenorhabditis, Drosophila, Danio,* or *Xenopus* embryos (Watanabe et al. 2014; White et al. 2018). This difference can be ascribed to the different nature of genetic control in early mammalian embryos. In *Drosophila, Danio,* and *Xenopus* embryos, early cleavage takes place in embryonic drone cells under remote control of the maternal genome. The conflicts of the *arena germinalis* are postponed until after sequestration of somatic cells. By contrast, transcription is active in nuclei of early mammalian embryos. As a consequence, the opening salvos of the *arena germinalis* are fired soon after syngamy with control contested between persistent maternal gene products and newly synthesized embryonic gene products, including the products of PGRs and PSDs.

The relative timing of embryonic genome activation, as measured by number of cell divisions, should be distinguished from absolute timing, as measured by hours since fertilization. Major embryonic transcription is delayed until the thirteenth cell cycle (8,000 cells) in *Drosophila,* the twelfth cell cycle (4,000 cells) in *Xenopus,* and the tenth cell cycle (1,000 cells) in zebrafish (Lee et al. 2014), whereas transcription has already begun in two-celled murine and human embryos (Vassena et al. 2011; Jukam et al. 2017). By this measure, embryonic genomes are activated early in mammals. However, embryonic transcription is initiated after 2.5 hours in *Drosophila,* 4 hours in zebrafish, 6 hours in *Xenopus,* but more than 24 hours in mice (Lee et al. 2014). By this measure, embryonic genomes are activated late in mammals. Mammalian embryos take days to produce the same number of cells that other embryos achieve within a few hours.

Chromosomal instability in mammalian embryos is already apparent before the first cellular division. In human one-cell embryos, DNA double-strand breaks are absent at the pronuclear stage but present after karyogamy and the onset of DNA replication (Palmerola et al. 2022). The first cell cycle takes a day to complete, associated with slow progression of replication forks and many double-strand breaks, with DNA synthesis extending into the G2 phase. Although this late DNA synthesis has been described as involving sequences that are "unreplicated" or "incompletely replicated" in the S phase (Palmerola et al. 2022), I conjecture that DNA synthesis in G2 phase represents excess replication of repeats that have already undergone a first round of replication in S phase but now induce double-strand breaks to recruit the machinery of homologous recombination to use the sister chromatid, or other repeats within the same array, as templates for repair (Haig 2021, 2022). Double-strand breaks in mouse zygotes are frequently repaired by insertion of reverse-transcribed sequences derived from retroelements (Ono et al. 2015).

Multiple factors contribute to more intense intra-genomic conflicts in early mammalian embryos. One factor is the replacement of transcriptionally inactive embryonic drone cells by transcriptionally active cells. As a consequence, transcriptional restraints on the activities of retrotransposons are relaxed. Another factor is a shift from hard selection (embryonic deaths are uncompensated) toward soft selection (embryonic deaths are partially compensated). With postzygotic provisioning, resources withheld from an inviable embryo become available for other embryos which may carry some of the same genes as the inviable embryo. Therefore, self-serving genetic behaviors that risk embryonic inviability for preferential transmission to the next generation carry less risk when there is postzygotic provisioning and soft selection of embryos than when there is prezygotic provisioning and hard selection of embryos.

The disorderly development of eutherian embryos resembles the blastomere anarchy of planarian embryos. In both groups, embryos compete postzygotically for maternal resources.

Embryonic Postsegregational Distorters in Early Mammalian Development

Embryonic PSDs, in which a toxin is deposited in oocytes before meiosis and an antitoxin is expressed in early embryos after fertilization, are predicted to have been a major problem during mammalian evolution. Three significant constraints on embryonic PSDs are relaxed. First, most resources are invested postzygotically. Therefore, early elimination of embryos without a PSD redistributes maternal investment to concurrent or future siblings who inherit a copy of the PSD. Second, maternally inherited toxins can kill embryos early in development because the embryonic genome is transcribed in few-celled embryos, allowing the early expression of an antitoxin that rescues embryos who inherit a copy of the PSD. Third, a "toxic" maternal effect can come not only from inside the embryo, in the form of a toxin deposited in oocytes, but also from outside the embryo from maternal cells interacting with the embryo, especially cells of the maternal immune system that are specialized for detecting the presence versus absence of antigens on target cells. Such embryonic PSDs would be examples of "gestational drive" (Haig 1993, 1996). If an embryonic PSD goes to fixation, it will rarely cause segregation distortion among embryos and may appear as a "maternal effect gene" that causes embryo inviability when its antitoxin activity is lost by mutation.

Most maternal genes segregate to offspring independently of any particular PSD. Therefore, the maternal "parliament of genes" is selected to suppress postsegregational "cabals of the few." The maternal genome has a collective interest in provisioning embryos of higher expected fitness and eliminating those of lower expected fitness (Stearns 1987; Haig 1990). Early losses of mammalian embryos are predicted to be a combination of deaths due to maladaptive PSDs (Haig 1996), and deaths due to the adaptive elimination of low quality embryos (Haig 1990, 2010; McCoy and Haig 2020).

Embryonic PSDs may contribute to high rates of embryo loss and female infertility in human couples. If PSDs have evolved because they preferentially transmit their copies to future generations, why should they be associated with infertility? First, if a PSD kills embryos that do not inherit its copies then a heterozygous mother has a probability 0.5^n of n successive pregnancy losses due to the presence of the PSD. In a large population of heterozygous women, one woman in eight would experience three successive losses, one woman in sixteen would experience four successive losses, and so on. This problem would be exacerbated if multiple unlinked PSDs segregate in the human population. Second, a PSD that has a transmission advantage to the offspring of heterozygous mothers could cause infertility of homozygous mothers (by analogy to spermatogenic PSDs that cause infertility in homozygous males) (Hartl 1969; Lyon 1986). Third, all complex genetic systems, including embryonic PSDs, are subject to mutational malfunction.

Rates of embryo loss are predicted to be higher in species that produce one offspring at a time (most primates, cattle, horses) than in species that produce litters (most rodents, pigs, insectivores). The reason for this prediction is that production of sequential singletons resembles meiotic sampling with replacement whereas production of litters resembles meiotic sampling without replacement. If embryos vary in their propensity for risky behaviors and offspring are produced sequentially, then the next embryo, who benefits from a failed gamble by the current embryo, has meiotic odds of possessing the gene responsible for the risky behavior. Conversely, if the redistribution of resources from failed gambles benefits other members of the same litter, then the beneficiaries have less than meiotic odds of carrying the gene for the risky behavior (Haig 1992).

An interesting form of postsegregation distortion may have contributed to the transfer of genetic control from maternal genomes to embryonic genomes in the context of early activation of embryonic transcription. Suppose that a dosage-dependent factor essential for early embryonic development was deposited in oocytes behind the meiotic veil of ignorance. An allelic innovation that switched expression of this factor from oocytes to early embryos would benefit from postsegregation distortion in its own favor in mothers heterozygous for the ancestral allele because all embryos of heterozygous mothers would receive half the required dose from oocytes, expressed from the ancestral allele, but this deficit would be compensated by embryonic expression only in those embryos that inherited the allelic innovation (cf. Keegan and Patten 2022).

The Minor Wave of Embryonic Genome Activation

The major wave of zygotic genome activation in two-celled mouse embryos and eight-celled human embryos is preceded by a minor wave of embryonic transcription of a smaller number of genes in one-celled mouse embryos and four-celled human embryos (Svoboda et al. 2015; Schulz and Harrison 2019). Prominent among these precociously active genes are several rapidly evolving homeobox genes that exhibit dynamic gene losses and duplications across mammalian evolution, including dispersal to new sites by retroposition. Some of these genes are associated with tandem arrays of macrosatellite repeats (Töhönen et al. 2015; Maeso et al. 2016; Lewin et al. 2021; Zou et al. 2022). They include *DUX4* and *Dux*, which are among the first genes expressed in human and mouse embryos (De Iaco et al. 2017; Hendrickson et al. 2017).

Most eutherian mammals possess a macrosatellite containing multiple copies of an intron-containing *DUXC* gene. *DUX4* and *Dux* are intronless retroposed genes derived from *DUXC* or one of its descendants (Booth and Holland 2007; Leidenroth et al. 2012). Human *DUX4* genes are associated with a 3.3-kb repeat (D4Z4) that is found on multiple human chromosomes (Clark et al. 1996; Winokur et al. 1996; Beckers et al. 2001; Ballarati et al. 2002), with occasional recombination between repeats on non-homologous chromosomes (Cacurri et al. 1998; Lemmers et al. 2010b). About twenty-two copies of *Dux* occur in a single tandem array in the mouse genome (Sugie et al. 2020; Bosnakovski et al. 2021). In addition, mouse genomes also contain a cluster of about sixty *Obox* genes, descended from an ancestral homeobox gene present as a single copy, *TPRX2,* in the human genome (Maeso et al. 2016; Royall et al. 2018). Some of the *Obox* genes are expressed in oocytes and others during the minor wave of embryonic genome activation (Ji et al. 2023).

Neither *DUX4* nor *Dux* appears essential for early development (De Iaco et al. 2020; Bosnakovski et al. 2021; Vuoristo et al. 2022).

What are their functions? Who do they serve? Are they selfless pro-
moters of embryonic development or selfish featherers of their own nest?
Their presence in tandem arrays that undergo contractions and expan-
sions in the germ stem suggests that these genes may possess features of
PGRs. *DUX*-related proteins are cytotoxic (Corona et al. 2013; Bosnakovski et al. 2023). The survival of early embryos, despite the expression
of a cytotoxin, suggests the presence of an antitoxin. This raises the
question whether *DUX*-related proteins could function as the toxic
component of a PSD that includes a nonsegregating antidote in close
linkage to the toxic protein.

The evolutionary history of *DUX4* will be considered in detail. Chromosomal adjacency between *FRG1* and *ASAH1* is conserved between
pufferfish and mice, but was disrupted in an ancestral primate with a
chromosomal fission (Clapp et al. 2003; Leidenroth et al. 2012), separating *FRG1* and *ASAH1* on separate chromosomes, and formation of a
new telomere distal to *FRG1*, corresponding to human 4qter (Leidenroth
and Hewitt 2010). Also in an ancestral primate, perhaps the same ancestor, a retroposed (intronless) copy of a *DUXC* gene was inserted between *FRG1* and the new telomere to become the ancestor of primate
DUX4 genes (Clapp et al. 2007; Leidenroth et al. 2012). Next to FRG1
at human 4q35 is *DUX4L9*, a gene that encodes a truncated *DUX4* protein (*DUX4c*), and that is not part of the D4Z4 array (Bosnakovski
et al. 2008; Claus et al. 2023). The original *DUX4* gene has been ampli-
fied as part of the D4Z4 macrosatellite of 3.3-kb repeats both at 4qter
and on other human chromosomes (Lyle et al. 1995; Gabriëls et al. 1999).
The copy number of *FRG1* has also undergone amplification from a single
ancestral copy to twenty-three paralogs at multiple sites in the human
genome (Grewal et al. 1999; Nurk et al. 2022). At some of these locations, an *FRG1*-related sequence is found in close proximity to D4Z4 repeats, suggesting they moved to the new chromosomal location together
(Grewal et al. 1999).

Human chromosome 4q35 typically contains many D4Z4 repeats,
each containing a *DUX4* coding sequence. Arrays with ten or fewer repeats are associated with facioscapulohumeral muscular dystrophy
(FSHD) (Gabriëls et al. 1999; Lemmers et al. 2022). Several cases of germ-
line/somatic mosaicism in FSHD pedigrees show that variation in D4Z4
copy number can arise in early embryonic divisions before sequestration
of somatic cells from the germ line (van der Maarel et al. 2000). Although
DUX4 is a cytotoxic protein encoded by many gene copies, translation of

the protein is highly inefficient because all *DUX4* genes (exceept the terminal copy on disease-causing haplotypes) lack polyadenylation signals with their transcripts subject to nonsense-mediated decay (Lemmers et al. 2010a; Feng et al. 2015).

FRG1 and *DUX4L9* are candidates to be the antitoxin to *DUX4* in a system of postsegregation distortion. *DUX4* activates its own promoter via its C-terminal domain, which is also responsible for the protein's cytotoxicity and increases expression of human *FRG1* (Ferri et al. 2015; Mitsuhashi et al. 2021). *FRG1* promotes nonsense-mediated decay (Palo et al. 2023), and would therefore destabilize *DUX4* transcripts. *DUX4c*, the truncated protein encoded by *DUX4L9*, is not cytotoxic (Bosnakovski et al. 2008), and it has antagonistic effects toward *DUX4* (Ganassi et al. 2022).

Other PGRs would be expected to evolve to cue their own transcriptional activation to gene products reliably present in early embryos. Many retroelements active in early human embryos possess binding sites for *DUX4* that are associated with transcriptional activation of the retroelements (Geng et al. 2012; Young et al. 2014; Mitsuhashi et al. 2021; Vuorista et al. 2022). Similarly, *Dux* is associated with activation of retroelements in early mouse embryos (De Iaco et al. 2017; Sakashita et al. 2023). *DUX4* activates human-specific elements but not mouse-specific elements; *Dux* activates mouse-specific elements but not human-specific elements (Whiddon et al. 2017). I conjecture that this species specificity reflects adaptations of retroelements to the shifting landscape of early embryonic gene expression.

Transcription during the minor wave of zygotic genome activation in mice occurs predominantly from the paternally inherited pronucleus (Aoki et al. 1997). A one-celled embryo is the first opportunity for a paternally inherited preferential replicator to colonize maternally inherited chromosomes. Therefore, it is not surprising that some retroelements have evolved to be active in very early embryos.

Morulas and Blastocysts

Eight-celled human and mouse embryos undergo compaction to form a tight ball of cells known as a morula. Such compaction is absent in marsupial (Kress and Selwood 2004) and afrotherian mammals (Goetz 1937; van der Horst 1942). It thus appears to be a derived feature of boreoeutherian mammals (Pfeffer 2018). (Note that the relevant stages have not been described in xenarthran mammals [Frankenberg et al. 2016]). In these taxa, as in non-mammals, a hollow blastocyst expands without prior compaction. In mouse embryos, compaction of the morula can proceed without synthesis of new proteins mediated by maternally encoded E-cadherin (Levy et al. 1986; Sefton et al. 1992; de Vries et al. 2004).

Some cells of eight-celled morulas are completely surrounded by outer cells (White et al. 2018). In the next round of mitotic divisions, some internal cells divide symmetrically to produce two internal cells, while some outer cells divide asymmetrically to produce an outer cell and an inner cell, and others divide symmetrically to produce two outer cells (Dard et al. 2009). Outer cells of sixteen-cell embryos appear to jostle for position (Watanabe et al. 2014) in what has been described as a tug-of-war (Niwayama et al. 2019). A cell that exerts stronger tension on neighboring cells is able to pull its neighbors over itself like a blanket and thus occupy an internal position (Samarage et al. 2015). Stable assignment of inner and outer cells is established by the end of the sixteen-cell stage.

Outer cells of thirty-two-cell embryos give rise to the trophectodermal epithelium. Pumping of fluid across the trophectodermal epithelium results in the expansion of a blastocyst with inner cells aggregated as an inner cell mass (ICM) (Zhang and Hiiragi 2018). Maintenance of the integrity of the trophectodermal epithelium requires expression of embryonically encoded E-cadherin (Stephenson et al. 2010). After a further round of cell divisions, five day-old human blastocysts contain an average of fifty-eight cells, of which thirty-eight are trophectoderm and twenty belong to the ICM (Hardy et al. 1989). The shortfall from the expected number of sixty-four cells reflects asynchronous divisions or apoptosis of some blastomeres. In what appears to be a protracted stochastic process, cells of the ICM differentiate as either primitive endoderm or epiblast (Chazaud et al. 2006; Saiz et al. 2016; Allègre et al. 2022). Because epiblast gives rise to all cells of the fetal body, including PGCs, I predict that PGRs that have undergone preferential replication in the ICM will bias cells with extra copies of the PGR toward differentiation as epiblast rather than primitive endoderm.

The arrangement in which epiblast forms inside of the blastocyst contrasts with the unilaminar blastocysts of monotremes and marsupials (Selwood and Johnson 2006). In these blastocysts, epiblast occupies a peripheral center, surrounded by concentric circles of hypoblast (the precursor of primitive endoderm), and then trophoblast (Selwood and

Johnson 2006). Morulas can be considered to have evolved by the internalization of the precursors of hypoblast and epiblast.

Why did morulas evolve? A definitive answer is lost in the fog of evolutionary time but I surmise compaction evolved because early embryonic cells competed for occupation of the epiblastic center because this was where germ cells developed. In the process, the precursors of hypoblast and epiblast became internalized within the trophoblastic epithelium. Competition for representation in the germ line implies that genetic differences among cells can arise during the first few cellular divisions. The activities of multicopy sequences including retrotransposons were likely contributors to genetic variation in early embryos. Significantly, stored RNAs are preferentially inherited by inner cells after asymmetric divisions in the morula (Hawdon et al. 2023). If a PGR replicates via an RNA product, then the PGR benefits if its RNAs preferentially segregate to inner cells of morulas.

Pilgrim's Progress

PGCs are first detected in murine epiblast as 4–8 *Blimp1*-positive cells prior to gastrulation (Ohinata et al. 2005), specifically when epiblast contains in the order of 500 cells (Snow 1977). Human PGCs differentiate from epiblast about ten divisions after fertilization (Rahbari et al. 2016). Not all PGCs may be recruited from epiblast at the same time. One would like to know the number of independent founding germ cells (FGCs) defined as the founders of cellular clones without somatic descendants. Estimates of the number of FGCs are indirect and based on various modeling assumptions. The number of FGCs in mice is estimated to be small, but at least three cells (Soriano and Jaenisch 1986; Ueno et al. 2009). Detailed analysis of mutations that are mosaic in both somatic cells and germ cells show that the human germ crown is descended from more than one FGC (Samuels and Friedman 2015; Rahbari et al. 2016). Estimates of the number of human FGCs range from two to three cells (Zheng et al. 2005) to an unspecified several cells (Jónsson et al. 2018).

FGCs differentiate from epiblast but it is not yet understood how particular cells are picked out from the epiblastic multitude to continue the germ line (Cooke and Moris 2021). This is a consequential decision with potentially many candidates for the job. Mammals specify PGCs inductively by signals from other cells. The precise location and detailed mechanisms of primordial germ-cell specification differ between humans and mice (Hancock et al. 2021), perhaps reflecting an evolutionary history of germline hide-and-seek involving PGRs and the conjugal genome.

From the small number of FGCs specified in early epiblast, PGCs undergo clonal expansion as they migrate through multiple tissues before arriving at their final destination in the developing gonad. In the process, many PGCs fall by the wayside and undergo apoptotic cell death. This arduous process creates possibilities of germline selection on the performance of many basic cellular functions and in interactions with other cells (Nguyen et al. 2015). An important feature of this migratory route is that it involves interactions with multiple distinct populations of somatic cells that are necessary for continued progress and survival of PGCs (Cooke and Moris 2021). The location where PGCs need to be to survive is continually shifting in what has been described as a "traveling" or "motility" niche (Gu et al. 2009, 2011b).

An unusual feature of mammalian PGCs is that they undergo proliferation during migration. Time spent dividing is time spent falling behind. PGCs must balance the costs and benefits of proliferation as stragglers lose the somatic support of their dynamically shifting somatic niche (Cantú et al. 2016, 2017). From the small number of PGCs that depart from murine epiblast, their numbers swell to about 22,000 cells on arrival at the gonad where they establish stable associations with gonadal supporting cells and undergo further proliferation (Tam and Snow 1981).

Retrotransposons, Placental Retroviruses, and Other Preferential Replicators

Many retroelements are expressed in early mammalian embryos (Zamudio and Bourc'his 2010; Fadloun et al. 2013; Göke et al. 2015; Grow et al. 2015; Thompson et al. 2016; Senft and Macfarlan 2021; Vuoristo et al. 2022; Sakashita et al. 2023). Retroelements are said to perform essential functions during early embryogenesis, often with the implication that these functions are the *raison d'être* for the expression of the retroelements at this time (Peaston et al. 2004; Beraldi et al. 2006; Kunarso et al. 2010; Jachowicz et al. 2017; Guo et al. 2024). Maybe so, but I would urge caution in interpreting such observations until the objectives of the different players are better understood. There is a marked lack of evolutionary conservation in the kinds of retroelements that perform these essential functions in different species.

Mutations that inactivate retroelements are subject to positive selection because inactivation reduces the cellular and organismal costs of active transposition but only transposition-competent elements undergo preferential replication. For a lineage of retroelements to remain active, new copies must move to new loci faster than old copies are inactivated by mutations at existing loci (Haig 2016b). Because active elements have evolved regulatory motifs that cue their activity to cellular markers of germline status, the genome will be littered with remnants of retroelements that possessed regulatory motifs promoting germline transcription. Some of these motifs will have been co-opted to drive germline expression of other genes. Because gene products of retrotransposons are a dependable feature of early embryonic cells, other genes with embryonic functions may cue their own expression to these gene products, including genes whose function is to target retroelements.

The first somatic cells to diverge from the mammalian germ stem comprise the trophectodermal epithelium of the blastocyst. These extraembryonic cells give rise to the trophoblastic cells of the placenta that interact directly with maternal cells and secrete substances into, and absorb substances from, the mother's body. The need for the rapid deployment of trophoblast was probably one reason for early activation of the mammalian embryonic genome. A distinctive feature of trophoblasts is that they are permissive environments for expression of endogenous retroviruses (Haig 2012a, 2013; Chuong 2013; Chuong et al. 2013). A probable reason is that, with the origin of placentas, trophoblast became an important way station for the infectious transmission of retroviruses, either from mothers to embryos, or from embryos to mothers and other embryos (Panem 1979; Haig 2012a). A hint that paternally inherited endogenous retroviruses may have infected mothers via expression in trophoblast is provided by the observation that all imprinted genes derived from retroviruses are paternally expressed and none are maternally expressed (Walter and Paulsen 2003).

Many retroviruses infect new cells by membrane fusions involving retroviral envelope proteins. The genes for retroviral envelope proteins have been repeatedly co-opted as a source of syncytins that mediate membrane fusion in the placenta (Dupressoir et al. 2012). Syncytins have also been implicated in the fusion of placentally derived exosomes with maternal cells (Vargas et al. 2014). A fascinating, retrovirally derived gene is *Paternally expressed gene 10,* which packages its own mRNA as a cargo that can be transmitted from cell to cell (Segel et al. 2021).

The origin of early genome activation in mammalian embryos was probably accompanied by a massive surge in the activities of retrotransposons and retroviruses before newly evolved defenses of the conjugal genome brought these elements under partial control. One component of these defenses was a massive expansion of genes encoding KRAB-ZFP proteins that block the transcription of retroelements in early embryonic cells (Thomas and Schneider 2011; Wolf et al. 2015; Bruno et al. 2019). Many KRAB-ZFP genes are clustered on human chromosome 19 (Lukic et al. 2014).

The long arm of human chromosome chromosome (19q) is a conserved eutherian linkage group notable for many other genes involved in maternal–embryonic relations including clusters of chorionic gonadotropin, pregnancy-specific b_1-glycoprotein, and killer cell inhibitory receptor genes (Haig 1999, 1465). Since I compiled that list many years ago the number of placentally related gene clusters mapping to this linkage group has continued to grown to include, but is not limited to, the primate-specific C19MC cluster of imprinted micro-RNAs (Bortolin-Cavaillé et al. 2010; Noguer-Dance et al. 2010), a cluster of placenta-specific galectins (Ely et al. 2019), a cluster of sialic acid-binding Ig-like lectins (Cao et al. 2009), and the *TPRX1, TPRX2, LEUTX,* and *Obox* homeodomain genes expressed in early embryos (Maeso et al. 2016; Royall et al. 2018; Ji et al. 2023). I suspect that this linkage group was a hot-bed of PSDs during the early evolution of eutherian placentation and has continued to play an important role in maternal–embryonic interactions.[2]

Genomic Imprinting

No organism is an island, entire unto itself. The social contract of the *pax somatica* is guaranteed by the shared interest of all somatic genes in the survival and reproduction of their associated germ line. But the contract has a major loophole that can destabilize the intra-genomic harmony of somatic cells. Conflicts may arise over the replication of genes present in the germ lines of other bodies, because then some, but not all, genes of somatic cells are present in the germ lines of related individuals (Haig 2011).

---

[2] A historical note: Matt Ridley graciously attributed his inspiration for writing his best-selling "autobiography" of the human genome to an off-hand remark of mine in the early 1990s that chromosome 19 was my favorite chromosome because it contained "all sorts of mischievous genes" (Ridley 1999, 3).

If some genes of individual X are present in the germ line of individual Y and somatic expression of a gene in X has fitness consequences for both X and Y, then genes that are shared with Y will have a different optimal outcome of trade-offs between the fitnesses of X and Y than have genes that are not shared. Such conflicts can be considered problems that arise from the divided loyalties of dual citizenship: Some genes may be citizens of X and Y whereas other genes are citizens of X but not Y (Haig 1993, 497). These conflicts can also be viewed as indirect forms of postsegregation distortion: A gene in X confers benefits on Y contingent on the probability that the gene also has a copy in Y.

Conflicts between shared and unshared genes will be particularly pronounced in the interactions of mammalian embryos with maternal bodies because the simple cue of parental origin reveals which genes are shared and which are unshared. Given information about their parental origin, embryonic genes of maternal and paternal origin will disagree about how much to demand from mothers (Haig 2000b, 2004b). Such considerations seemingly divide the embryonic genome into two evenly matched parties with divergent interests—namely, the maternally derived haploid genome and the paternally derived haploid genome (Haig 2006). But the unity of the genome has not fractured in this way. Only a small minority of genes are differentially expressed when maternally derived and paternally derived—these genes are said to be imprinted—but most genes in mammalian genomes are unimprinted (Edwards et al. 2023). Thus, the caucus of unimprinted genes constitutes a genomic majority opposed to the minority interests of smaller factions of maternally expressed and paternally expressed imprinted genes. This parliamentary majority is made possible by what I will call an *epigenetic veil of ignorance* that erases information about their parental origin for most genes of the genome.

Epigenetic Veils of Ignorance

At the time of fertilization, the genomes of maternal and paternal pronuclei are differentially methylated. In one-celled embryos, there is active erasure of DNA methylation from paternal pronuclei by the action of maternal gene products deposited in oocytes (Reik and Walter 2001). Natural selection favors trans-acting maternal factors that erase cis-acting paternal imprints (Wilkins and Haig 2002; Wilkins 2005). Most methylation marks in the sperm-derived genome are erased in early mouse embryos (Gu et al. 2011a; Iqbal et al. 2011; Wossidlo et al. 2011) and human embryos (Okae et al. 2014; Smith et al. 2014). There is also extensive loss of methylation from the egg-derived genome (Shen et al. 2014; Wang et al. 2014). Despite the general pattern of imprint erasure by demethylation there is remarkable diversity among mammals in early epigenetic mechanisms (Lu et al. 2021).

The extensive epigenetic reprogramming that is a distinctive feature of early mammalian embryos may have evolved as a mechanism of minimizing conflicts between genes of maternal and paternal origin, especially in early embryos. This wipes the slate clean of most information about a gene's recent past that could be used to favor nepotistic maldistribution of resources. Epigenetic reprogramming in the early mammalian embryos can be contrasted with a very different pattern in zebrafish. The paternal genome of early zebrafish embryos is more heavily methylated than the maternal genome but does not undergo embryonic demethylation; rather, the maternal genome gradually acquires methylation similar to the paternal genome (Jiang et al. 2013; Skvortsova et al. 2019). Despite differential methylation of parental genomes in zebrafish embryos, these differences have not been used as cues for imprinted gene expression (Martin and McGowan 1995). An absence of functionally significant imprinting in zebrafish embryos is not unexpected because genes of paternal origin cannot influence the amount of yolk received by their egg. There is no need for an epigenetic veil of ignorance.

A small subset of loci maintain differential DNA methylation of maternal and paternal alleles despite the erasure of most methylation differences in early embryos. These loci become the nuclei of imprinted clusters of genes expressed in somatic cells. Demethylation of DNA in early embryos is followed by extensive remethylation in epiblast that is nonspecific with respect to parental origin. After PGCs differentiate from epiblast, their genomes undergo a second wave of genome-wide demethylation before the establishment of sex-specific imprinted marks for transmission to the next generation (Seki et al. 2005; Seisenberger et al. 2013; Reik and Surani 2015). One consequence of the second wave of DNA demethylation would be to erase higher-order imprints of sex of grandparental origin (second-order imprints), or sex of great-grandparental origin (third-order imprints) (Haig 2000a, 2014).

Because information about whether a gene was inherited from a male or female body can be transmitted between the generations, and can be the basis of conditional gene expression, the question arises whether other adaptive information could be transmitted via this epigenetic route (Haig 2000b, 28). I conjecture that most of this potentially useful information is erased by the epigenetic veils of ignorance that evolved to erase information about sex of parental origin.

The mammalian maternal–embryonic unit can be thought of as a triploid entity comprised of a haploid genome shared by mother and embryo (the inherited maternal genome), a haploid genome present in the mother but not in the embryo (the non-inherited maternal genome), and a haploid genome present in the embryo but not in the mother (the inherited paternal genome). The death of an embryo has different consequences for the fitnesses of these three sets of genes (Haig 2004a, 2019). Potential conflicts within maternal and embryonic genomes are reduced by two veils of ignorance. The meiotic veil of ignorance reduces conflicts within maternal genomes: most genes in mothers have no way of knowing whether they are inherited or non-inherited by any particular embryo. The epigenetic veil of ignorance reduces conflicts within embryonic genomes because most genes have no way of knowing whether they were maternally inherited or paternally inherited.

Synopsis

Many biologists have considered the disorderliness of early mammalian development to be counterintuitive and have been puzzled by why early cleavage should be so error-prone (Akera and Lampson 2019; Schneider and Ellenberg 2019; Duro and Nilsson 2021; Regin et al. 2022). However, the relative disorder of early mammalian development becomes readily explicable when considered as a downstream consequence of the early activation of the embryonic genome and elimination of drone cells from the germ stem. Early cleavage comes under contested control for two major reasons. First, the early activation of the embryonic genome has created new opportunities for the self-aggrandizement of PGRs and PSDs. Second, the transition to postzygotic provisioning of early embryos enables the expression of strong conflicts between maternal and embryonic genes, and between maternally derived and paternally derived genes of embryos. These conflicts are concentrated within nuclei of the early germ stem and in interactions between maternal tissues and trophoblast that forms the feeding structures of the developing embryo, but may also be expressed in somatic cells when gene expression affects the fitness of non-descendant kin. The developmental melee is brought under more unified

control with the differentiation of somatic cells because PGRs have
evolved to be silent in somatic cells and because the conflicts of the germ
line come under the supervisory oversight of somatic supporting cells.

Organismal Republics

Perhaps the oldest metaphor of political science is of society as a body,
with its members allotted designated roles in serving the well-being of
the commonwealth. In this metaphor, *members* are *organs* of the *body
politic* or *corporate body* (an etymological tautology). A "member" was
originally a part of a body (hence dismemberment); an organ was a machine or tool used in work. This chapter reverses the metaphor by viewing
the body as a society or, more precisely, by viewing the genome as a society of genes for which the body is the corporate infrastructure. These
organized bodies are what we call organisms. One "paradox of the organism" is that genes that engage in intense conflicts among organisms
coexist peaceably within organisms. Another paradox is that ephemeral
organisms exert more complex agency than their immortal genes.

A central question in political theory concerns how societal goods
should be distributed among citizens and how competing interests can
be reconciled. John Rawls (1971) introduced the *veil of ignorance* as a
theoretical device for reaching reflective equilibrium about the nature of
a just or fair society. Societal goods would be justly distributed if the distribution would be agreed upon by all members of society in the same
"original position" in which they did not know their place in society (and
hence did not know which goods they would personally receive). This
chapter has argued that the theoretical device of a *veil of ignorance* society has been practically implemented in organismal development as a
means of resolving internal conflicts. Meiotic, ontogenetic, and epigenetic
erasures of information about genetic identities place each gene in the
same "original position" with respect to the distribution of bodily goods.

Organismal bodies approximate the utopian ideal of holding all
property in common. Executive power is broadly distributed with the
absence of a ruling class that expropriates a disproportionate share of
resources. The somatic parliament of genes rules by consensus with their
collective actions judged by collective performance. Germline profits are
shared equally. Each gene contributes according to its abilities; each receives according to its needs. Under this egalitarian arrangement, "com-

pensation" (representation in progeny) is decoupled from each partner's contribution to societal success: All genes replicate equally, all are represented equally in gametes of the next generation, all receive an equal share of bodily resources. Some might call this replicative communism. Whatever label is used, the distribution of goods among the members of the genome has no resemblance to a market mechanism. The *pax somatica* could also be presented as a dystopian totalitarian state in which there is little freedom for individual action.

The peace of the body coexists with the tribulations of the *arena germinalis*. Germline genetic assemblies are fractious communities of competing interests. Replicative equality is subverted by multicopy sequences that are able to replicate faster than the rest of the genome. Other genes, or coalitions of genes, exploit meiotic segregation to promote their own progeny by sabotaging progeny that do not inherit their copies. But the organismal republic continues to function because the somatic parliament of genes implements policies for collective benefit from behind an ontogenetic veil of ignorance. This is perhaps the most remarkable paradox of the organism, that genes can work selflessly in somatic cells despite their selfish interests in the germ line.

Appendix: August Weismann and Continuity of the Germ Plasm

> As the thousands of cells which constitute an organism possess very different properties, *the chromatin* which controls them *cannot be uniform; it must be different in each kind of cell. . . . Ontogeny, or the development of the individual, depends therefore on a series of gradual qualitative changes in the nuclear substance of the egg-cell.* (Weismann 1893, 32; emphases in original)

The purpose of this appendix is to clear up some terminological tangles about what Weismann actually wrote about continuity of the germ plasm, rather than what he is said to have written. Between 1883 and 1885, Weismann became convinced that the hereditary determinants resided in nuclear chromatin (see Weismann 1890). Before this shift in his thinking, Weismann's concept of continuity of the germ plasm was expressed in terms of a distinction between somatic and germ cells. After this shift, he recognized the fundamental distinction as existing between nucleus and cytoplasm. His categorical rejection of the inheritance of acquired characters was not based on an early separation of germ cells from somatic cells but on the distinction between determinants of form in the nuclear germ plasm and the effects of these determinants in the cytoplasmic body. This can be considered an early formulation of Johannsen's (1911) later separation of genotype from phenotype.

Weismann defined the germ line (*Keimbahn*) as the cellular path by which germ plasm (*Keimplasma*) was transmitted from one generation to the next. This path included all cells that connected a fertilized egg to a reproductive cell (Weismann 1892, 242). By this definition, all organisms possess continuous germ lines that extend back to the origins of life. Weismann (1904) was well-informed about plant development and phenomena of regeneration in planarians. He recognized that organisms varied in the degree to which the *Keimbahn* passed through somatic cells. However, Wilson (1925, 311, 313) defined germ lines as restricted to cells without somatic descendants in the current generation. His definition excluded cells with both somatic and gametic descendants, which he called stem cells and which I call the germ stem. By Wilson's definition, the germ lines of multicellular organisms are discontinuous. Weismann's and Wilson's definitions differ as to how they classify cells with both somatic and gametic progeny.

Different conceptions of the germ line are entangled with different conceptions of somatic cells. This relates to the classification of cells of the germ stem such as the neoblasts of planarians. Many embryologists classify neoblasts as adult somatic stem cells. Weismann might have agreed with this description because he recognized that somatic cells could belong to the germ line. However, in this chapter, I defined a somatic cell as a cell that belongs to neither the germ stem nor the germ crown. By this definition, neoblasts are not somatic cells. My definition is not without problems when cells of the germ stem perform somatic functions or when somatic cells can be reprogrammed to pluripotency. Because of these problems, I did not follow Weismann and accept that "somatic cells" can also belong to the germ line. My reason was that I wanted a term that distinguished cells that could not leave descendants in future generations (corresponding to the *pax somatica*) from cells that could (corresponding to the *arena germinalis*).

Weismann's determinants (we would now call them genes) resided in the germ plasm of nuclear chromatin and "determined" the form of the body. However, many of his contemporaries identified germ plasm with cytoplasmic granules that determined whether a cell would develop as a germ cell or somatic cell. The latter definition has stuck. In current usage, germ plasm usually refers to microscopically visible particles in the cytoplasm of eggs that are inherited by germ cells. The nineteenth-century disagreement over the location of determinants was actually a semantic disagreement over different senses of the word "determine." Weismann was interested in how *information* was transmitted between generations, whereas most proponents of cytoplasmic germ plasm were interested in physical *causes* of development. Weismann's use of informational metaphors come to the fore in the following passage:

> If we take our stand upon the theory of determinants, it would be necessary to a transmission of acquired strength of memory that the states of these brain-cells should be communicated by the telegraphic path of the nerve-cells to the germ-cells, and should there modify only the determinants of the brain-cells, and should do so in such a way that, in the subsequent development of an embryo from the germ-cell, the corresponding brain-cells should turn out to be capable of increased functional activity. But as the determinants are not miniature brain-cells . . . I can only compare the assumption of the transmission of the results of memory-exercise to the telegraphing of a poem, which is handed in in German, but at the place of arrival appears on the paper translated into Chinese. (Weismann 1904, 107–108)

Weismann has also been accused of defending the wrong theory of development in which determinants (now understood as genes) segregate during cellular differentiation rather than the correct theory that all cells receive a complete set of determinants. I believe this criticism is misdirected for three reasons. First, Weismann did not believe that the segregation of determinants was a necessary feature of differentiation. Cells could possess inactive determinants. When differentiated cells gave rise to reproductive cells, these cells possessed "unalterable and inactive germ-plasm in addition to their own active idioplasm" (Weismann 1893, 211). Second, the phenomenon of chromatin-diminution during somatic development of *Ascaris* provided a clear-cut example in which somatic cells did not inherit all the determinants of the germ line (Boveri 1887). Third, it is only a minor modification to go from "differentiated cells inherit a subset of determinants" to "differentiated cells inherit an epigenetic state in which a subset of determinants are active."

Weismann (along with Wilhelm Roux [1881, 2024]) incorporated ideas of competition within or among cells into his evolutionary theories. He recognized both *intra-selection* and *germinal selection* (Haig 2024a). Intra-selection occurred among somatic cells and its effects were not bequeathed to offspring. Germinal selection involved differential growth and reproduction of determinants in cells of the germ line and its effects could be bequeathed.

The fact that "germ line" has been used to refer to two distinct concepts has been a continuing source of confusion. How should we disentangle this semantic knot? A writer should make clear which definition they are using, but it would also be good to have some simple terms to distinguish between the two concepts. Haig (2016b) translated Weismann's *Keimbahn* as "germ track" to distinguish it from Wilson's "germ line." I now favor using adjectives to distinguish an "inclusive germ line" (which includes the germ stem) from an "exclusive germ line" (which excludes the germ stem). When employing the inclusive concept, "germ crown" is a synonym of the "exclusive germ line."

References

Abnave, P., Aboukhatwa, E., Kosaka, N., Thompson, J., Hill, M. A., and Aboobaker, A. A. (2017). Epithelial-mesenchymal transition transcription factors control pluripotent adult stem cell migration *in vivo* in planarians. *Development, 144,* 3440–3453.

Akera, T., and Lampson, M. A. (2019). Chromosome segregation: Poor supervision in the early stage of life. *Current Biology, 29,* R150–R172.

Alié, A., Hayashi, T., Sugimura, I., et al. (2015). The ancestral gene repertoire of animal stem cells. *Proceedings of the National Academy of Sciences of the United States of America, 112,* E7093–E7100.

Allais, A., and FitzHarris, G. (2022). Absence of a robust mitotic timer mechanism in early preimplantation mouse embryos leads to chromosome instability. *Development, 149,* dev200391.

Allègre, N., Chazaud C., Dennis C., et al. (2022). NANOG initiates epiblast fate through the coordination of pluripotency genes expression. *Nature Communications, 13,* 3550.

Aoki, F., Worrad, D. M., and Schultz, R. M. (1997). Regulation of transcriptional activity during the first and second cell cycles in the preimplantation mouse embryo. *Developmental Biology, 181,* 296–307.

Aoki, S. T., Kershner, A. M., Bingman, C. A., Wickens, M., and Kimble, J. (2016). PGLgerm granule assembly protein is a base-specific, single-stranded RNase.

Proceedings of the National Academy of Sciences of the United States of America, 113, 1279–1284.

Aravin, A. A. (2020). Pachytene piRNAs as beneficial regulators or a defense system gone rogue. *Nature Genetics, 52*, 644–645.

Aravin, A. A., Hannon, G. J., and Brennecke, J. (2007). The Piwi-piRNA provides an adaptive defense in the transposon arms race. *Science, 318*, 761–764.

Baguña, J., and Romero, R. (1981). Quantitative analysis of cell types during growth, degrowth and regeneration in the planarians *Dugesia mediterranea* and *Dugesia tigrina*. *Hydrobiologia, 84*, 181–194.

Ballarati, L., Piccini, I., Carbone, L., et al. (2002). Human genome dispersal and evolution of 4q35 duplications and interspersed LSau repeats. *Gene, 296*, 21–27.

Becker, J. S., Nicetto, D., and Zaret, K. S. (2016). H3K9me3-dependent heterochromatin: Barrier to cell fate changes. *Trends in Genetics, 32*, 29–41.

Beckers, M. C., Gabriëls J, van de Maarel, S., et al. (2001). Active genes in junk DNA? Characterization of DUX genes embedded within 3.3 kb repeated elements. *Gene, 264*, 51–57.

Beeman, R. W., Friesen, K. S., and Denell, R. E. (1992). Maternal effect selfish genes in flour beetles. *Science, 256*, 89–92.

Bely, A. E., and Nyberg, K. G. (2010). Evolution of animal regeneration: From germ cell specification to gonadal colonization in mammals. *Trends in Ecology and Evolution, 25*, 161–170.

Ben-David, E., Burga, A., and Kruglyak, L. (2017). A maternal-effect selfish genetic element in *Caenorhabditis elegans*. *Science, 356*, 1051–1055.

Bengtsson, B. O. (1990). The effect of biased conversion on the mutation load. *Genetical Research, 55*, 183–187.

Beraldi, R., Pittoggi, C., Sciamanna, I., Mattei, E., and Spadafora, C. (2006). Expression of LINE-1 retroposons is essential for murine preimplantation development. *Molecular Reproduction and Development, 73*, 279–287.

Bhutani, K., Stansifer, K., Ticau, S., et al. (2021). Widespread haploid-biased gene expression enables sperm-level natural selection. *Science, 371*(6533), 1008.

Blondel, L., Besse, S., Rivard, E. L., Ylla, G., and Extavour, C. G. (2021). Evolution of a cytoplasmic determinant: Evidence for the biochemical basis of functional evolution of the novel germ line regulator oskar. *Molecular Biology and Evolution, 38*, 5491–5513.

Blitz, I. L., and Cho, K. W. Y. (2021). Control of zygotic genome activation in *Xenopus*. *Current Topics in Developmental Biology, 145*, 167–204.

Blythe, S. A., and Wieschaus, E. F. (2015). Coordinating cell cycle remodeling with transcriptional activation at the *Drosophila* MBT. *Current Topics in Developmental Biology, 113*, 113–148.

Bontems, F., Stein, A., Marlow, F., et al. (2009). Bucky ball organizes germ plasm assembly in zebrafish. *Current Biology, 19*, 414–422.

Booth, H. A. F., and Holland, P. W. H. (2007). Annotation, nomenclature and evolution of four novel homeobox genes expressed in the human germ line. *Gene, 387,* 7–14.

Bortolin-Cavaillé, M. L., Dance, M., Weber, M., and Cavaillé, J. (2010). C19MC microRNAs are processed from introns of large Pol-II, non-protein-coding transcripts. *Nucleic Acids Research, 37,* 3464–3473.

Bosnakovski, D., Gearheart, M. D., Choi, S. H., and Kyba, M. (2021). Dux facilitates post-implantation development, but is not essential for zygotic genome activation. *Biology of Reproduction, 104,* 83–93.

Bosnakovski, D., Lamb, S., Simsek, T., et al. (2008). DUX4c, an FSHD candidate gene, interferes with myogenic regulators and abolishes myoblast differentiation. *Experimental Neurology, 214,* 87–96.

Bosnakovski, D., Toso, E. A., Ener, E. T., et al. (2023). Antagonism among DUX family members evolved from an ancestral toxic single homeodomain protein. *iScience, 26,* 107823.

Boveri, T. (1887). Über Differenzierung der Zellkerne während der Furchung des Eies von *Ascaris megalocephala. Anatomischer Anzeiger, 2,* 688–693.

Brangwynne, C. P., Eckmann, C. R., Courson, D. S., et al. (2009). Germline P granules are liquid droplets that localize by controlled dissolution/condensation. *Science, 324,* 1729–1732.

Bravo Núñez, M. A., Nuckolls, N. L., and Zanders, S. E. (2018). Genetic villains: Killer meiotic drivers. *Trends in Genetics, 34,* 424–433.

Brawand, D., Wahli, W., and Kaessmann, H. (2008). Loss of egg yolk genes in mammals and the origin of lactation and placentation. *PLoS Biology, 6,* e63.

Bruno, M., Mahgoub, M., and Macfarlan, T. S. (2019). The arms race between KRAB-zinc finger proteins and endogenous retroelements and its impact on mammals. *Annual Review of Genetics, 53,* 393–416.

Butuci, M., Williams, A. B., Wong, M. M., Kramer, B., and Michael, W. M. (2015). Zygotic genome activation triggers chromosome damage and checkpoint signaling in *C. elegans* primordial germ cells. *Developmental Cell, 34,* 85–95.

Cacurri, S., Piazzo, N., Deidda, G., et al. (1998). Sequence homology between 4qter and 10qter loci facilitates the instability of subtelomeric KpnI repeat units implicated in facioscapulohumeral muscular dystrophy. *American Journal of Human Genetics, 63,* 181–190.

Cantú, A. V., Altshuler-Keylin, S., and Laird, D. J. (2016). Discrete somatic niches coordinate proliferation and migration of primordial germ cells via Wnt signaling. *Journal of Cell Biology, 214,* 215–229.

Cantú, A. V., and Laird, D. J. (2017). A pilgrim's progress: seeking meaning in primordial germ cell migration. *Stem Cell Research, 24,* 181–187.

Cao, H., de Bono, B., Belov, K., Wong, E. S., Trowsdale, J., and Barrow, A. D. (2009). Comparative genomics indicates the mammalian CD33rSiglec locus evolved by an ancient large-scale inverse duplication and suggests all Siglecs share a common ancestral region. *Immunogenetics, 61,* 401–417.

Chaigne, A., and Brunet, T. (2022). Incomplete abscission and cytoplasmic bridges in the evolution of eukaryotic multicellularity. *Current Biology, 32,* R385–R397.

Chalopin, D., Naville, M., Plard, F., Galiana, D., and Volff, J. N. (2015). Comparative analysis of transposable elements highlights mobilome diversity and evolution in vertebrates. *Genome Biology and Evolution, 7,* 567–580.

Chatfield, J., O'Reilly, M. A., Bachvarova, R. F., et al. (2014). Stochastic specification of primordial germ cells from mesoderm precursors in axolotl embryos. *Development, 141,* 2429–2440.

Chazaud, C., Yamanaka, Y., Pawson, T., and Rossant, J. (2006). Early lineage segregation between epiblast and primitive endoderm in mouse blastocysts through the Grb2–MAPK pathway. *Developmental Cell, 10,* 615–624.

Chen, P., Kotov, A. A., Godneeva, B. K., Bazylev, S. S., Olenina, L. V., and Aravin, A. A. (2020). piRNA-mediated gene regulation and adaptation to sex-specific transposon expression in *D. melanogaster* male germline. *Genes and Development, 35,* 914–935.

Chen, P., Luo, Y., and Aravin, A. A. (2021). RDC complex executes a dynamic piRNA program during spermatogenesis to safeguard male fertility. *PLoS Genetics, 17,* e1009591.

Chiappetta, A., Liao, J., Tian, S., and Trcek, T. (2022). Structural and functional organization of germ plasm condensates. *Biochemical Journal, 479,* 2477–2495.

Choi, H., Wang, Z., and Dean, J. (2021). Sperm acrosome overgrowth and infertility in mice lacking chromosome 18 pachytene piRNA. *PLoS Genetics, 17,* e1009485.

Choi, S. K., Yoon, S. R., Calabrese, P., and Arnheim, N. (2012). Positive selection for new disease mutations in the human germline: Evidence from the heritable cancer syndrome multiple endocrine neoplasia type 2B. *PLoS Genetics, 8,* e1002420.

Chuong, E. B. (2013). Retroviruses facilitate the rapid evolution of the mammalian placenta. *BioEssays, 35,* 853–861.

Chuong, E. B., Rumi, M. A. K., Soares, M. J., and Baker, J. C. (2013). Endogenous retroviruses function as species-specific enhancer elements in the placenta. *Nature Genetics, 45,* 325–329.

Cinalli, R. M., Rangan, P., and Lehmann, R. (2008). Germ cells are forever. *Cell, 132,* 559–562.

Clapp, J., Bolland, D. J., and Hewitt, J. E. (2003). Genomic analysis of facioscapulohumeral muscular dystrophy. *Briefings in Functional Genomics and Proteomics, 2,* 213–223.

Clapp, J., Mitchell, L. M., Bolland, D. J., et al. (2007). Evolutionary conservation of a coding function for D4Z4, the tandem DNA repeat mutated in facioscapulohumeral muscular dystrophy. *American Journal of Human Genetics, 81*, 264–279.

Clark, F. E., and Akera, T. (2021). Unravelling the mystery of female meiotic drive: Where we are. *Open Biology, 11*, 210074.

Clark, L. N., Hoehler, U., Ward, D. C., Wienberg, J., and Hewitt, J. E. (1996). Analysis of the organisation and localisation of the FSHD-associated tandem array in primates: Implications for the origin and evolution of the 3.3 kb repeat family. *Chromosoma, 105*, 180–189.

Claus, C., Slavin, M., Ansseau, E., et al. (2023). The double homeodomain protein DUX4c is associated with regenerating muscle fibers and RNA-binding proteins. *Skeletal Muscle, 13*, 5.

Cooke, C. B., and Moris, N. (2021). Tissue and cell interactions in mammalian PGC development. *Development, 148*, dev200093.

Corona, E. D., Jacquelin, D., Gatica, L., and Rosa, A. L. (2013). Multiple protein domains contribute to nuclear import and cell toxicity of DUX4, a candidate pathogenic protein for facioscapulohumeral muscular dystrophy. *PLoS One, 8*, e75614.

Cosmides, L. M., and Tooby, J. (1981). Cytoplasmic inheritance and intragenomic conflict. *Journal of Theoretical Biology, 89*, 83–129.

Curnutte, H. A., Lan, X., Sargen, M., et al. (2023). Proteins rather than mRNAs regulate nucleation and persistence of Oskar germ granules in *Drosophila. Cell Reports, 42*, 112723.

Dard, N., Louvet-Vallée, S., and Maro, B. (2009). Orientation of mitotic spindles during the 8- to 16-cell stage transition in mouse embryos. *PLoS One, 12*, e8171.

Daughtry, B. L., Rosenkrantz, J. L., Lazar, N. H., et al. (2019). Single-cell sequencing of primate preimplantation embryos reveals chromosome elimination via cellular fragmentation and blastomere exclusion. *Genome Research, 29*, 367–382.

Davies, E. L., Lei, K., Seidel, C. W., et al. (2017). Embryonic origin of adult stem cells required for tissue homeostasis and regeneration. *eLife, 6*, e21052.

Dawkins, R. (1976). *The Selfish Gene.* Oxford University Press.

Dawkins, R. (1982). *The Extended Phenotype.* Oxford University Press.

De Iaco, A., Planet, E., Coluccio, A., Verp, S., Duc, J., and Trono, D. (2017). DUX-family transcription factors regulate zygotic genome activation in placental mammals. *Nature Genetics, 49*, 941–945.

De Iaco, A., Verp, S., Offner, S., Grun, D., and Trono, D. (2020). DUX is a nonessential synchronizer of zygotic genome activation. *Development, 147*, dev177725.

de Vries, W. N., Eviskov, A. V., Haac, B. E., et al. (2004). Maternal b-catenin and E-cadherin in mouse development. *Development, 131*, 4435–4445.

Dietrich, J. E., and Hiiragi, T. (2007). Stochastic patterning in the mouse preimplantation embryo. *Development, 134*, 4219–4231.

Dollman, G. (1938). Post-natal development of kangaroos. *Proceedings of the Linnean Society of London, 151,* 19–23.

D'Orazio, F. M., Balwierz, P. J., González, A. J., et al. (2021). Germ cell differentiation requires Tdrd7-dependent chromatin and transcriptome reprogramming marked by germ plasm relocalization. *Developmental Cell, 56,* 641–656.

Dupressoir, A., Lavialle, C., and Heidmann, T. (2012). From ancestral infectious retroviruses to bona fide cellular genes: Role of the captured syncytins in placentation. *Placenta, 33,* 663–671.

Duret, L., and Galtier, N. (2009). Biased gene conversion and the evolution of mammalian genomic landscapes. *Annual Review of Genomics and Human Genetics, 10,* 285–311.

Duro, J., and Nilsson, J. (2021). SAC during early cell divisions: Sacrificing fidelity over timely division, regulated differently across organisms. *BioEssays, 43,* 2000174.

Eddy, E. M. (1975). Germ plasm and the differentiation of the germ cell line. *International Review of Cytology, 43,* 229–280.

Edwards, C. A., Watkinson, W. M. D., Telerman, S. B., Hulsmann, L. C., Hamilton, R. S., and Ferguson-Smith, A. C. (2023). Reassessment of weak parent-of-origin expression bias shows it rarely exists outside of known imprinted regions. *eLife, 12,* e83364.

Ely, Z. A., Moon, J. M., Sliwoski, G. R., et al. (2019). The impact of natural selection on the evolution and function of placentally expressed galectins. *Genome Biology and Evolution, 11,* 2574–2592.

Ephrussi, A., and Lehmann, R. (1992). Induction of germ cell formation by oskar. *Nature, 358,* 387–392.

Extavour, C. G., and Akam, M. (2003). Mechanisms of germ cell specification across the metazoans: Epigenesis and preformation. *Development, 130,* 5869–5884.

Fadloun, A., Le Gras, S., Jost, B., et al. (2013). Chromatin signatures and retrotransposon profiling in mouse embryos reveal regulation of LINE-1 by RNA. *Nature Structural and Molecular Biology, 20,* 332–338.

Farrell, J. A., and O'Farrell, P. H. (2014). From egg to gastrula: How the cell cycle is remodeled during the *Drosophila* mid-blastula transition. *Annual Review of Genetics, 48,* 269–294.

Fayomi, A. P., and Orwig, K. E. (2018). Spermatogonial stem cells and spermatogenesis in mice, monkeys and men. *Stem Cell Research, 29,* 207–214.

Feng, Q., Snider, L., Jagannathan, S., et al. (2015). A feedback loop between nonsense-mediated decay and the retrogene DUX4 in facioscapulohumeral muscular dystrophy. *eLife, 4,* e04996.

Ferri, G., Huichalaf, C. H., Caccia, R., and Gabellini, D. (2015). Direct interplay between two candidate genes in FSHD muscular dystrophy. *Human Molecular Genetics, 24,* 1256–1266.

Fields, C., and Levin, M. (2018). Are planaria individuals? What regenerative biology is telling us about the nature of multicellularity. *Evolutionary Biology, 45,* 237–247.

Fierro-Constaín, L., Schenkelaars, Q., Gazave, E., et al. (2017). The conservation of the germline multipotency program, from sponges to vertebrates: A stepping stone to understanding the somatic and germline origins. *Genome Biology and Evolution, 9,* 474–488.

Fishman, L., and McIntosh, M. (2019). Standard deviations: The biological bases of transmission ratio distortion. *Annual Review of Genetics, 53,* 347–372.

Flynn, T. T. (1922). Notes on certain reproductive phenomena in some Tasmanian marsupials. *Annals and Magazine of Natural History, 10*(56), 225–231.

Frankenberg, S. R., de Barros, F. R. O., Rossant, J., and Renfree, M. B. (2016). The mammalian blastocyts. *WIREs Developmental Biology, 5,* 210–232.

Funayama, N. (2018). The cellular and molecular bases of the sponge stem cell systems underlying reproduction, homeostasis and regeneration. *International Journal of Developmental Biology, 62,* 513–525.

Gabriëls, J., Beckers, M. C., Ding, H., et al. (1999). Nucleotide sequence of the partially deleted D4Z4 locus in a patient with FSHD identifies a putative gene within each 3.3 kb element. *Gene, 236,* 25–32.

Gallo, C. M., Wang, J. T., Motegi, F., and Seydoux, G. (2010). Cytoplasmic partitioning of P granule components is not required to specify the germline in C. *elegans. Science, 330,* 1685–1689.

Ganassi, M., Figeac, N., Reynaud, M., Quiroga, H. P. O., and Zammit, P. S. (2022). Antagonism between DUX4 and DUX4c highlights a pathomechanism operating through b-catenin in facioscapulohumeral muscular dystrophy. *Frontiers in Cell and Developmental Biology, 10,* 802573.

Ganot, P., Bouquet, J. M., Kallesøe, T., and Thompson, E. M. (2007). The *Oikopleura* coenocyst, a unique chordate germ cell permitting rapid, extensive modulation of oocyte production. *Developmental Biology, 302,* 591–600.

Gardner, A. (2019). The greenbeard effect. *Current Biology, 29,* R425–R473.

Geng, L. N., Yao, Z., Snider, L., et al. (2012). DUX4 activates germline genes, retroelements, and immune mediators: Implications for facioscapulohumeral muscular dystrophy. *Developmental Cell, 22,* 38–51.

Gerhold, A. R., Labbé, J. C., and Singh, R. (2022). Uncoupling cell division and cytokinesis during germline development in metazoans. *Frontiers in Cell and Developmental Biology, 10,* 1001689.

Giannoulatou, E., McVean, G., Taylor, I. B., et al. (2013). Contributions of intrinsic mutation rate and selfish selection to levels of de novo *HRAS* mutations in the paternal germline. *Proceedings of the National Academy of Sciences of the United States of America, 110,* 20152–20157.

Goetz, R. H. (1937). Studien zur Placentation der Centetiden. II. Die Implantation und Frühentwicklung von *Hemicentetes semispinosus* (Cuvier). *Zeitschrift für Anatomie und Entwicklungsgeschichte, 107,* 274–318.

Göke, J., Lu, X., Chan, Y. S., et al. (2015). Dynamic transcription of distinct classes of endogenous retroviral elements marks specific populations of early human embryonic cells. *Cell Stem Cell, 16*, 135–141.

Gordon, K. L., Zussman, J. W., Li, X., Miller, C., and Sherwood, D. R. (2020). Stem cell niche exit in *C. elegans* via orientation and segregation of daughter cells by a cryptic cell outside the niche. *eLife, 9*, e56383.

Gou, L. T., Dai, P., Yang, J. H., et al. (2014). Pachytene piRNAs instruct massive mRNA elimination during late spermiogenesis. *Cell Research, 24*, 680–700.

Greenbaum, M. P., Iwamori, T., Buchold, G. M., and Matzuk, M. (2011). Germ cell intercellular bridges. *Cold Spring Harbor Perspectives in Biology, 3*, a005850.

Grewal, P. K., van Geel, M., Frants, R. R., de Jong, P., and Hewitt, J. E. (1999). Recent amplification of the human *FRG1* gene during primate evolution. *Gene, 227*, 79–88.

Grosberg, R. K., and Strathmann, R. R. (2007). The evolution of multicellularity: A minor major transition? *Annual Review of Ecology, Evolution, and Systematics, 38*, 621–654.

Grow, E. J., Flynn, R. A., Chavez, S. L., et al. (2015). Intrinsic retrovirus reactivation in human preimplantation embryos and pluripotent cells. *Nature, 522*, 221–225.

Gu, T. P., Guo, F., Yang, H., et al. (2011a). The role of Tet3 DNA dioxygenase in epigenetic reprogramming by oocytes. *Nature, 477*, 606–610.

Gu, Y., Runyan, C., Shoemaker, A., Surani, A., and Wylie, C. (2009). Steel factor controls primordial germ cell survival and motility from the time of their specification in the allantois, and provides a continuous niche throughout their migration. *Development, 136*, 1295–1303.

Gu, Y., Runyan, C., Shoemaker, A., Surani, A., and Wylie, C. (2011b). Membrane-bound steel factor maintains a high local concentration for mouse primordial germ cell motility, and defines the region of their migration. *PLoS One, 6*, e25984.

Guedelhoefer, O. C., and Sánchez Alvarado, A. (2012). Amputation induces stem cell mobilization to sites of injury during planarian regeneration. *Development, 139*, 3510–3520.

Gumienny, T. L., Lambie, E., Hartwieg, E., Horvitz, H. R., and Hengartner, M. O. (1999). Genetic control of programmed cell death in the *Caenorhabditis elegans* hermaphrodite germline. *Development, 126*, 1011–1022.

Guo, Y., Li, T. D., Modzelewski, A. J., and Siomi, H. (2024). Retrotransposon renaissance in early embryos. *Trends in Genetics, 40*, 39–51.

Haig, D. (1990). Brood reduction and optimal parental investment when offspring differ in quality. *The American Naturalist, 136*, 550–556.

Haig, D. (1992). Genomic imprinting and the theory of parent-offspring conflict. *Seminars in Developmental Biology, 3*, 153–160.

Haig, D. (1993). Genetic conflicts in human pregnancy. *Quarterly Review of Biology, 68*, 495–532.

Haig, D. (1996). Gestational drive and the green-bearded placenta. *Proceedings of the National Academy of Sciences of the United States of America,* 6547–6551.

Haig, D. (1997). The social gene. In *Behavioural Ecology,* 4th ed., edited by J. R. Krebs and N. B. Davies, 284–304. Blackwell Scientific.

Haig, D. (1999). A brief history of human autosomes. *Philosophical Transactions of the Royal Society B, 354,* 1447–1470.

Haig, D. (2000a). Genomic imprinting, sex-biased dispersal, and social behavior. *Annals of the New York Academy of Sciences, 907,* 149–163.

Haig, D. (2000b). The kinship theory of genomic imprinting. *Annual Review of Ecology and Systematics, 31,* 9–32.

Haig, D. (2004a). Evolutionary conflicts in pregnancy and calcium metabolism—A review. *Placenta, 25*(suppl. A), S10–S15.

Haig, D. (2004b). Genomic imprinting and kinship: How good is the evidence? *Annual Review of Genetics, 38,* 553–585.

Haig, D. (2006). Intragenomic politics. *Cytogenetic and Genome Research, 113,* 68–74.

Haig, D. (2010). Fertile soil or no man's land: Cooperation and conflict in the placental bed. In *Placental Bed Disorders,* edited by R. Pijnenborg, I. Brosens, and R. Romero, 165–173. Cambridge University Press.

Haig, D. (2011). Genomic imprinting and the evolutionary psychology of human kinship. *Proceedings of the National Academy of Sciences of the United States of America, 108,* 10878–10885.

Haig, D. (2012a). Retroviruses and the placenta. *Current Biology, 22,* R609–R613.

Haig, D. (2012b). The strategic gene. *Biology and Philosophy, 27,* 461–479.

Haig, D. (2013). Genomic vagabonds: Endogenous retroviruses and placental evolution. *Bioessays, 35,* 845–846.

Haig, D. (2014). Genetic dissent and individual compromise. *Biology and Philosophy, 29,* 233–239.

Haig, D. (2015). Maternal-fetal conflict, genomic imprinting, and mammalian vulnerabilities to cancer. *Philosophical Transactions of the Royal Society B, 370,* 20140178.

Haig, D. (2016a). Intracellular evolution of mitochondrial DNA (mtDNA) and the tragedy of the cytoplasmic commons. *Bioessays, 38,* 549–555.

Haig, D. (2016b). Transposable elements: Self-seekers of the germline, team players of the soma. *Bioessays, 38,* 1158–1166.

Haig, D. (2019). Cooperation and conflict in human pregnancy. *Current Biology, 29,* R455–R458.

Haig, D. (2020). *From Darwin to Derrida. Selfish Genes, Social Selves and the Meanings of Life.* MIT Press.

Haig, D. (2021). Concerted evolution of ribosomal DNA: Somatic peace amid germinal strife. *BioEssays, 43,* 2100179.

Haig, D. (2022). Paradox lost: Concerted evolution and centromeric instability. *BioEssays, 44,* 2200023.

Haig, D. (2024a). The afterlife of *Der Kampf der Theile im Organismus.* In *The Struggle of Parts,* edited by W. Roux, 237–254. Harvard University Press.

Haig, D. (2024b). Germline ecology: Managed herds, tolerated flocks, and pest control. *Journal of Heredity, 115*(6), 643–659.

Haig, D., and Bergstrom, C. T. (1995). Multiple mating, sperm competition, and meiotic drive. *Journal of Evolutionary Biology, 8,* 265–282.

Haig, D., and Grafen A. (1991). Genetic scrambling as a defence against meiotic drive. *Journal of Theoretical Biology, 153,* 531–558.

Hanazawa, M., Yonetani, M., and Sugimoto, A. (2011). PGL proteins self associate and bind RNPs to mediate germ granule assembly in *C. elegans. Journal of Cell Biology, 192,* 929–937.

Hancock, G. V., Wamaitha, S. E., Peretz, L., and Clark, A. T. (2021). Mammalian primordial germ cell specification. *Development, 148,* dev189217.

Hardy, K., Handyside, A. H., and Winston, R. M. L. (1989). The human blastocyst: Cell number, death and allocation during late preimplantation development in vitro. *Development, 107,* 597–604.

Hartl, D. L. (1969). Dysfunctional sperm production in *Drosophila melanogaster* males homozygous for the segregation distorter elements. *Proceedings of the National Academy of Sciences of the United States of America, 63,* 782–789.

Hartman, C. G. (1920). Studies in the development of the opossum *Didelphys virginiana* L. V. The phenomena of parturition. *Anatomical Record, 19,* 251–261.

Hastings, I. M. (1989). Potential germline competition in animals and its evolutionary implications. *Genetics, 123,* 191–197.

Hastings, I. M. (1991). Germline selection: population genetic aspects of the sexual/asexual life cycle. *Genetics, 129,* 1167–1176.

Hawdon, A., Geoghegan, N. D., Mohenska, M., et al. (2023). Apicobasal RNA asymmetries regulate cell fate in the early mouse embryo. *Nature Communications, 14,* 2909.

Hayssen, V., Lacy, R. C., and Parker, P. J. (1985). Metatherian reproduction: Transitional or transcending? *The American Naturalist, 126,* 617–632.

Hegner, R. W. (1911). Germ cell determinants and their significance. *The American Naturalist, 45,* 385–397.

Hendrickson, P. G., Doráis, J. A., Grow, E. J., et al. (2017). Conserved roles of mouse DUX and human DUX4 in activating cleavage-stage genes and MERVL/HERVL retrotransposons. *Nature Genetics, 49,* 925–934.

Hinnant, T. D., Merkle, J. A., and Ables, E. T. (2020). Coordinating proliferation, polarity, and cell fate in the *Drosophila* female germline. *Frontiers in Cell and Developmental Biology, 8,* 19.

Hubbard, E. J. A. (2007). *Caenorhabditis elegans* germ line: A model for stem cell biology. *Developmental Dynamics, 236,* 3343–3357.

Hubbard, E. J. A., and Schedl, T. (2019). Biology of the *Caenorhabditis elegans* germline stem system. *Genetics, 213,* 1145–1188.

Hughes, R. L. (1993). Monotreme development with particular reference to the extraembryonic membranes. *Journal of Experimental Zoology, 266,* 480–494.

Iqbal, K., Jin, S. G., Pfeifer, G. P., and Szabó, P. E. (2011). Reprogramming of the paternal genome upon fertilization involves genome-wide oxidation of 5-methylcytosine. *Proceedings of the National Academy of Sciences of the United States of America, 108,* 3642–3647.

Jachowicz, J. W., Bing, X., Pontabry, J., Boskovic, A., Rando, O. J., and Torres-Padilla, M. E. (2017). LINE-1 activation after fertilization regulates global chromatin accessability in the early mouse embryo. *Nature Genetics, 49,* 1502–1510.

Janssen, A., Colmenares, S. U., and Karpen, G. H. (2018). Heterochromatin: Guardian of the genome. *Annual Review of Cell and Developmental Biology, 7,* 311–336.

Ji, S., Chen, F., Stein, P., et al. (2023). *OBOX* regulates mouse zygotic genome activation and early development. *Nature, 620,* 1047–1053.

Jiang, L., Zhang, J., Wang, J. J., et al. (2013). Sperm, but not oocyte, DNA methylome is inherited by zebrafish early embryos. *Cell, 153,* 773–784.

Johannsen, W. (1911). The genotype conception of heredity. *The American Naturalist, 45,* 129–159.

Johnson, A. D., and Alberio, R. (2015). Primordial germ cells: The first cell lineage or the last cells standing? *Development, 142,* 2730–2739.

Johnson, A. D., Crother, B., White, M. E., et al. (2003). Regulative germ cell specification in axolotl embryos: A primitive trait conserved in the mammalian lineage. *Philosophical Transactions of the Royal Society B, 358,* 1371–1379.

Johnson, A. D., Richardson, E., Bachvarova, R. F., and Crother, B. I. (2011). Evolution of the germ line–soma relationship in vertebrate embryos. *Reproduction, 141,* 291–300.

Jónsson, H., Sulem, P., Arnadottir, G. A., et al. (2018). Multiple transmissions of de novo mutations in families. *Nature Genetics, 50,* 1674–1680.

Jukam, D., Shariati, S. A. M., and Skotheim, J. M. (2017). Zygotic genome activation in vertebrates. *Developmental Cell, 42,* 316–332.

Juliano, C., Swartz, S. Z., and Wessell, G. (2010). A conserved germline multipotency program. *Development, 137,* 4113–4126.

Kane, D. A., and Kimmel, C. B. (1993). The zebrafish midblastula transition. *Development, 119,* 447–456.

Keegan, G., and Patten, M. M. (2022). Selfish evolution of placental hormones. *Evolution, Medicine, and Public Health, 10*(1), 391–397.

Keinath, M. C., Timoshevskiy, V. A., Timoshevskaya, N. Y., Tsonis, P. A., Voss, S. R., and Smith, J. J. (2015). Initial characterization of the large genome of the salamander *Ambystoma mexicanum* using shotgun and laser capture chromosome sequencing. *Scientific Reports, 5,* 16413.

Kimble, J., and Crittenden, S. L. (2007). Control of germline stem cells, entry into meiosis, and the sperm/oocyte decision in *Caenorhabditis elegans*. *Annual Review of Cell and Developmental Biology, 23,* 405–433.

Kimura, J. O., Bolaños, D. M., Ricci, L., and Srivastava, M. (2022). Embryonic origins of adult pluripotent stem cells. *Cell, 185,* 4756–4769.

Kipreos, E. T. (2005). *C. elegans* cell cycles: Invariance and stem cell divisions. *Nature Reviews Molecular Cell Biology, 6,* 766–776.

Kloc, M., Bilinski, S., Dougherty, M. T., Brey, E. M., and Etkin, L. D. (2004). Formation, architecture and polarity of female germline cyst in *Xenopus*. *Developmental Biology, 266,* 43–61.

Knaut, H., Pelegri, F., Bohmann, K., Schwarz, H., and Nüsslein-Volhard, C. (2000). Zebrafish *vasa* RNA but not its protein is a component of the germ plasm and segregates asymmetrically before germline specification. *Journal of Cell Biology, 149,* 875–888.

Kress, A., and Selwood, L. (2004). Precedence of cell-zona adhesion over cell-cell adhesion during marsupial blastocyst formation prohibits morula formation and ensures that both the pluriblast and trophoblast are superficial. *Cells Tissues Organs, 177,* 87–103.

Kruger, A. N., and Mueller, J. L. (2021). Mechanisms of meiotic drive in symmetric and asymmetric meiosis. *Cellular and Molecular Life Sciences, 78,* 3205–3218.

Kulkarni, A., and Extavour, C. G. (2015). Convergent evolution of germ granule nucleators: A hypothesis. *Stem Cell Research, 24,* 188–194.

Kumano, G. (2015). Evolution of germline segregation processes in animal development. *Development, Growth and Differentiation, 57,* 324–332.

Kunarso, G., Chia, N. Y., Jeyakani, J., et al. (2010). Transposable elements have rewired the core regulatory network of human embryonic stem cells. *Nature Genetics, 42,* 631–634.

Kyriakis, E., Markaki, M., and Tavernarakis, N. (2015). *Caenorhabditis elegans* as a model for cancer research. *Molecular and Cellular Oncology, 2,* e975027.

Lai, F., and King, M. L. (2013). Repressive translational control in germ cells. *Molecular Reproduction and Development, 80,* 665–676.

Laumer, C. E., and Giribet, G. (2014). Inclusive taxon sampling suggests a single, stepwise origin of ectolecithality in Platyhelminthes. *Biological Journal of the Linnean Society, 111,* 570–588.

Leatherman, J. L., and Jongens, T. A. (2003). Transcriptional silencing and translational control: Key features of early germline development. *BioEssays, 25,* 326–335.

Lebedeva, L. A., Yakovlev, K. V., Kozlov, E. N., et al. (2018). Transcriptional quiescence in primordial germ cells. *Critical Reviews in Biochemistry and Molecular Biology, 53,* 579–595.

Leclère, L., Jager, M., Barreau, C.,. et al. (2012). Maternally localized germ plasm mRNAs and germ cell/stem cell formation in the cnidarian *Clytia*. *Developmental Biology, 364,* 236–248.

Lee, M. T., Bonneau, A. R., and Giraldez, A. J. (2014). Zygotic genome activation during the maternal-to-zygotic transition. *Annual Review of Cell and Developmental Biology, 30,* 581–613.

Lehmann, R. (2015). Germ plasm biogenesis—An oskar-centric perspective. *Current Topics in Developmental Biology, 116,* 679–707.

Lei, L., and Spradling, A. C. (2013). Mouse primordial germ cells produce cysts that partially fragment prior to meiosis. *Development, 140,* 2075–2081.

Lei, L., and Spradling, A. C. (2016). Mouse oocytes differentiate through organelle enrichment from sister cyst germ cells. *Science, 352,* 95–99.

Leidenroth, A., Clapp, J., Mitchell, L. M., et al. (2012). Evolution of DUX gene macrosatellites in placental mammals. *Chromosoma, 121,* 489–497.

Leidenroth, A., and Hewitt, J. E. (2010). A family history of DUX4: Phylogenetic analysis of DUXA, B, C and Duxbl reveals the ancestral DUX gene. *BMC Evolutionary Biology, 10,* 364.

Leigh, E. G. (1977). How does selection reconcile individual advantage with the good of the group? *Proceedings of the National Academy of Sciences of the United States of America, 74,* 4542–4546.

Lemmers, R. J. L. F., van der Vliet, P. J., Blatnik, A., et al. (2022). Chromosome 10q–linked FSHD identifies *DUX4* as principal disease gene. *Journal of Medical Genetics, 59,* 180–188.

Lemmers, R. J. L. F., van der Vliet, P. J., Klooster, R., et al. (2010a). A unifying genetic model for facioscapulohumeral muscular dystrophy. *Science, 329,* 1650–1653.

Lemmers, R. J. L. F., van der Vliet, P. J., van der Gaag, K. J., et al. (2010b). Worldwide population analysis of the 4q and 10q subtelomeres identifies only four discrete interchromosomal transfers in human evolution. *American Journal of Human Genetics, 86,* 364–377.

Levy, J. B., Johnson, M. H., Goodall, H., and Maro, B. (1986). The timing of compaction: Control of a major developmental transition in mouse early embryogenesis. *Journal of Embryology and Experimental Morphology, 95,* 213–237.

Lewin, T. D., Royall, A. H., and Holland, P. W. H. (2021). Dynamic molecular evolution of mammalian homeobox genes: Duplication, loss, divergence and gene conversion sculpt PRD class repertoires. *Journal of Molecular Evolution, 89,* 396–414.

Li, L., and Xie, T. (2015). Stem cell niche: Structure and function. *Annual Review of Cell and Developmental Biology, 21,* 605–631.

Longo, F. J. (1973). Fertilization: A comparative ultrastructural review. *Biology of Reproduction, 9,* 149–215.

Lu, K., Jensen, L., Lei, L., and Yamashita, Y. M. (2017). Stay connected: A germ cell strategy. *Trends in Genetics, 33,* 971–978.

Lu, X., Zhang, Y., Wang, L., et al. (2021). Evolutionary epigenomic analyses in mammalian early embryos reveal species-specific innovations and conserved principles of imprinting. *Science Advances, 7,* eabi6178.

Lukic, S., Nicolas, J. C., and Levine, A. J. (2014). The diversity of zinc-finger genes on human chromosome 19 provides an evolutionary mechanism of defense against inherited endogenous retroviruses. *Cell Death and Differentiation, 21,* 381–387.

Lyle, R., Wright, T. J., Clark, L. N., and Hewitt, J. E. (1995). The FSHD-associated repeat, D4Z4, is a member of a dispersed family of homeobox-containing repeats, subsets of which are clustered on the short arms of the acrocentric chromosomes. *Genomics, 28,* 389–397.

Lynch, J. A., Özüak, O., Khila, A., Abouheif, E., Desplan, C., and Roth, S. (2011). The phylogenetic origin of *oskar* coincided with the origin of maternally provisioned germ plasm and pole cells at the base of the Holometabola. *PLoS Genetics, 7,* e1002029.

Lyon, M. F. (1986). Male sterility of the mouse t-complex is due to homozygosity for the distorter genes. *Cell, 44,* 357–363.

Lyttle, T. W. (1991). Segregation distorters. *Annual Review of Genetics, 25,* 511–557.

Madison, J. [Publius]. (1787). *The Same Subject Continued. The Union as a Safeguard Against Domestic Faction and Insurrection.* Federalist Papers No. 10.

Maeso, I., Dunwell, T. L., Wyatt, C. D. R., et al. (2016). Evolutionary origin and functional divergence of totipotent cell homeobox genes in eutherian mammals. *BMC Biology, 14,* 45.

Magor, B. G., de Tomaso, A., Rinkevich, B., and Weissman, I. L. (1999). Allorecognition in colonial tunicates: Protection against predatory cell lineages? *Immunological Reviews, 167,* 69–79.

Martin, C. C., and McGowan, R. (1995). Parent-of-origin specific effects on the methylation of a transgene in the zebrafish, *Danio rerio. Developmental Genetics, 17,* 233–239.

Martín-Durán, J. M., and Eggers, B. (2012). Developmental diversity in free-living flatworms. *EvoDevo, 3,* 7.

Matova, N., and Cooley, L. (2001). Comparative aspects of animal oogenesis. *Developmental Biology, 231,* 291–320.

McCoy, D. E., and Haig, D. (2020). Embryo selection and mate choice: Can "honest signals" be trusted? *Trends in Ecology and Evolution, 35,* 308–318.

McCoy, R. C. (2017). Mosaicism in preimplantation human embryos: When chromosomal abnormalities are the norm. *Trends in Genetics, 33,* 448–463.

Menkhorst, E., Nation, A., Cui, S., and Selwood, L. (2009). Evolution of the shell coat and yolk in amniotes: A marsupial perspective. *Journal of Experimental Zoology, 312B,* 625–638.

Meshorer, E., and Misteli, T. (2006). Chromatin in pluripotent embryonic stem cells and differentiation. *Nature Reviews Molecular Cell Biology, 7,* 540–546.

Mitsuhashi, H., Ishimaru, S., Homma, S., et al. (2021). Functional domains of the FSHD-associated DUX4 protein. *Biology Open, 7,* bio033977.

Morrison, S. J., and Kimble, J. (2006). Asymmetric and symmetric stem cell divisions in development and cancer. *Nature, 441,* 1068–1074.

Morrison, S. J., and Spradling, A. C. J. (2008). Stem cells and niches: Mechanisms that promote stem cell maintenance throughout life. *Cell, 132,* 598–611.

Mukherjee, N., and Mukherjee, C. (2021). Germ cell ribonucleoprotein granules in different clades of life: From insects to mammals. *WIREs RNA, 12,* e1642.

Nakamura, A., and Seydoux, G. (2008). Less is more: Specification of the germline by transcriptional repression. *Development, 135,* 3817–3827.

Nelson, J. E., and Gemmell, R. T. (2003). Birth in the northern quoll, *Dasyurus hallucatus* (Marsupialia: Dasyuridae). *Australian Journal of Zoology, 51,* 187–198.

Newport, J., and Kirschner, M. (1981a). A major developmental transition in early *Xenopus* embryos: I. Characterization and timing of cellular changes at the midblastula stage. *Cell, 30,* 675–686.

Newport, J., and Kirschner, M. (1981b). A major developmental transition in early *Xenopus* embryos: II. Control of the onset of transcription. *Cell, 30,* 687–696.

Nguyen, D. H., Jaszczak, R. G., and Laird, D. J. (2015). Heterogeneity of primordial germ cells. *Current Topics in Developmental Biology, 135,* 155–201.

Niklas, K. J., and Kutschera, U. (2014). Amphimixis and the individual in evolving populations: Does Weismann's Doctrine apply to all, most or a few organisms? *Naturwissenschaften, 101,* 357–372.

Niwayama, R., Mogha, P., Liu, Y. J., et al. (2019). A tug-of-war between cell shape and polarity controls division orientation to ensure robust patterning in the mouse blastocyst. *Developmental Cell, 51,* 564–574.

Noble, L. M., Yuen, J., Stevens, L., et al. (2021). Selfing is the safest sex for *Caenorhabditis tropicalis*. *eLife, 10,* e62587.

Noguer-Dance, M., Abu-Amero, S., Al-Khatib, M., et al. (2010). The primate-specific microRNA gene cluster (C19MC) is imprinted in the placenta. *Human Molecular Genetics, 19,* 3566–3582.

Nousch, M., and Eckmann, C. R. (2013). Translational control in the *Caenorhabditis elegans* germ line. In *Germ Cell Development in C. elegans,* edited by T. Schedl, 205–247. Springer.

Nurk, S., Koren, S., Rhie, A., et al. (2022). The complete sequence of a human genome. *Science, 376,* 44–53.

O'Farrell, P. H. (2015). Growing an embryo from a single cell: A hurdle in animal life. *Cold Spring Harbor Perspectives in Biology, 7,* a019042.

Ohinata, Y., Payer, B., O'Carroll, D., et al. (2005). *Blimp1* is a critical determinant of the germ cell lineage in mice. *Nature, 436,* 207–213.

Okae, H., Chiba, H., Hiura, H., et al. (2014). Genome-wide analysis of DNA methylation dynamics during early human development. *PLoS Genetics, 10,* e1004868.

Okasha, S. (2012). Social justice, genomic justice and the veil of ignorance: Harsanyi meets Mendel. *Economics and Philosophy, 28,* 43–71.

Ono, R., Ishii, M., Fujihara, Y., et al. (2015). Double strand break repair by capture of retrotransposon sequences and reverse-transcribed spliced mRNA sequences in mouse zygotes. *Scientific Reports, 5,* 12281.

Otto, S. P. (2009). The evolutionary enigma of sex. *The American Naturalist, 174*(S1), S1–S14.

Otto, S. P., and Hastings, I. M. (1998). Mutation and selection within the individual. *Genetica, 102*(103), 507–524.

Palmerola, K. L., Amrane, S., De Los Angeles, A., et al. (2022). Replication stress impairs chromosome segregation and preimplantation development in human embryos. *Cell, 185,* 2988–3007.

Palo, A., Patel, S. A., Sahoo, B., Chowdary, T. K., and Dixit, M. (2023). FRG1 is a direct transcriptional regulator of nonsense-mediated mRNA decay genes. *Genomics, 115,* 110539.

Panem, S. (1979). C-type virus expression in the placenta. *Current Topics in Pathology, 66,* 175–189.

Patten, M. M., Schenkel, M. A., and Ågren, J. A. (2023). Adaptation in the face of internal conflict: The paradox of the organism revisited. *Biological Reviews, 98,* 1796–1811.

Peaston, A. E., Evsikov, A. V., Graber, J. H., et al. (2004). Retrotransposons regulate host genes in mouse oocytes and preimplantation embryos. *Developmental Cell, 7,* 597–606.

Pepling, M. E., de Cuevas M., and Spradling A. C. (1999). Germline cysts: A conserved phase of germ cell development? *Trends in Cell Biology, 9,* 257–262.

Pepling, M. E., and Spradling, A. C. (2001). Mouse ovarian germ cell cysts undergo programmed breakdown to form primordial follicles. *Developmental Biology, 234,* 339–351.

Pfeffer, P. L. (2018). Building principles for constructing a mammalian blastocyst embryo. *Biology, 7,* 41.

Piasecka, B., Lichocki, P., Moretti, S., Bergmann, S., and Robinson-Rechavi, M. (2013). The hourglass and early conservation models—Co-existing patterns of developmental constraints in vertebrates. *PLoS Genetics, 9,* e1003476.

Piccinini, G., and Milani, L. (2023). Germline-related molecular phenotype in Metazoa: Conservation and innovation highlighted by comparative transcriptomics. *EvoDevo, 14,* 2.

Pontecorvo, G. (1944). Synchronous mitoses and differentiation, sheltering the germ track. *Drosophila Information Service, 18,* 54–55.

Quan, H., and Lynch, J. A. (2014). The evolution of insect germline specification strategies. *Current Opinion in Insect Science, 13,* 99–105.

Queller, D. C. (2000). Relatedness and the fraternal major transitions. *Philosophical Transactions of the Royal Society B, 355,* 1647–1655.

Raff, R. A. (1996). *The Shape of Life: Genes, Development, and the Evolution of Animal Form.* University of Chicago Press.

Rahbari, R., Wuster, A., Lindsay S. J., et al. (2016). Timing, rates and spectra of human germline mutation. *Nature Genetics, 48,* 126–133.

Rawls, J. (1971). *A Theory of Justice.* Harvard University Press.

Regin, M., Spits, C., and Sermon K. (2022). On the origins and fate of chromosomal abnormalities in human preimplantation embryos: An unsolved riddle. *Molecular Human Reproduction, 28,* gaaco11.

Reik, W., and Surani, M. A. (2015). Germline and pluripotent stem cells. *Cold Spring Harbor Perspectives in Biology, 7,* a019422.

Reik, W., and Walter, J. (2001). Evolution of imprinting mechanisms: The battle of the sexes begins in the zygote. *Nature Genetics, 27,* 255–256.

Reusch, T. B. H., Baums, I. B., and Werner, B. (2021). Evolution via somatic genetic variation in modular species. *Trends in Ecology and Evolution, 36,* 1083–1092.

Richardson, B. E., and Lehmann, R. (2010). Mechanisms guiding primordial germ cell migration: Strategies from different organisms. *Nature Reviews Molecular and Cellular Biology, 11,* 37–49.

Ridley, M. (1999). *Genome: The Autobiography of a Species in 23 Chapters.* HarperCollins.

Rinkevich, B., and Weissman, I. L. (1987). Chimeras in colonial invertebrates: A synergistic symbiosis or somatic- and germ-cell parasitism? *Symbiosis, 4,* 117–134.

Robertson, S., and Lin, R. (2015). The maternal-to-zygotic transition in *C. elegans. Current Topics in Developmental Biology, 113,* 1–42.

Rouhana, L., Shibata, N., Nishimura, O., and Agata, K. (2010). Different requirements for conserved post-transcriptional regulators in planarian regeneration and stem cell maintenance. *Developmental Biology, 341,* 429–443.

Roux, W. (1881). *Der Kampf der Theile im Organismus.* Engelmann.

Roux, W. (2024). *The Struggle of Parts.* Translated by D. Haig and R. Bondi. Harvard University Press.

Royall, A. H., Maeso, I., Dunwell, T. L., and Holland, P. W. H. (2018). Mouse *Obox* and *Crxos* modulate preimplantation transcriptional profiles revealing similarity between paralogous mouse and human homeobox genes. *EvoDevo, 9,* 2.

Sahu, S., Sridhar, D., Abnave, P., et al. (2021). Ongoing repair of migration-coupled DNA damage allows planarian adult stem cells to reach wound sites. *eLife, 10,* e63779.

Saiz, N., Williams, K. M., Seshan, V. E., and Hadjantonakis, A. K. (2016). Asynchronous fate decisions by single cells collectively ensure consistent lineage composition in the mouse blastocyst. *Nature Communications, 7,* 13463.

Sakashita, A., Kitano T., Ishizu H., et al. (2023). Transcription of MERVL retrotransposons is required for preimplantation embryo development. *Nature Genetics, 55,* 484–495.

Samarage, C. R., White, M. D., Álvarez, Y. D., et al. (2015). Cortical tension allocates the first inner cells of the mammalian embryo. *Developmental Cell, 34,* 435–447.

Samuels, M. E., and Friedman, J. M. (2015). Genetic mosaics and the germ cell lineage. *Genes, 6,* 216–237.

Sato, K., Shibata, N., Orii, H., et al. (2006). Identification and origin of the germline stem cells as revealed by the expression of *nanos*-related gene in planarians. *Development, Growth, and Differentiation, 48,* 615–628.

Saxena, S., and Zou, L. (2022). Hallmarks of DNA replication stress. *Molecular Cell, 82,* 2298–2314.

Schaner, C. E., Deshpande, G., Schedl, P. D., and Kelly, W. G. (2003). A conserved chromatin architecture marks and maintains the restricted germ cell lineage in worms and flies. *Developmental Cell, 5,* 747–757.

Schneider, I., and Ellenberg, J. (2019). Mysteries in embryonic development: How can errors arise so frequently at the beginning of mammalian life? *PLoS Biology, 17,* e3000173.

Schulz, K. N., and Harrison, M. M. (2019). Mechanisms regulating zygotic genome activation. *Nature Reviews Genetics, 20,* 221–332.

Sefton, M., Johnson, M. H., and Clayton, M. H. (1992). Synthesis and phosphorylation of uvomorulin during mouse early development. *Development, 115,* 313–318.

Segel, M., Lash, B., Song, J., et al. (2021). Mammalian retrovirus-like protein PEG10 packages its own mRNA and can be pseudotyped for mRNA delivery. *Science, 373,* 882–889.

Seidel, H. S., Ailion, M., Li, J., van Oudenaarden, A., Rockman, M. V., and Kruglyak, L. (2011). A novel sperm-delivered toxin causes late-stage embryo lethality and transmission ratio distortion in C. *elegans. PLoS Biology, 9,* e1001115.

Seisenberger, S., Peat, J. R., Hore, T. A., Santos, F., Dean, W., and Reik, W. (2013). Reprogramming DNA methylation in the mammalian life cycle: Building and breaking epigenetic barriers. *Philosophical Transactions of the Royal Society B, 368,* 20110330.

Seki, Y., Hayashi, K., Itoh, K., Mizugaki, M., Saitou, M., and Matsui, Y. (2005). Extensive and orderly reprogramming of genome-wide chromatin modifications associated with specification and early development of germ cells in mice. *Developmental Biology, 278,* 440–458.

Selwood, L., and Johnson, M. H. (2006). Trophoblast and hypoblast in the mono-treme, marsupial and eutherian mammal: Evolution and origins. *BioEssays, 28*, 128–145.

Senft, A. D., and Macfarlan, T. S. (2021). Transposable elements shape the evolution of mammalian development. *Nature Reviews Genetics, 22*, 691–711.

Seydoux, G., and Braun, R. E. (2006). Pathway to totipotency: Lessons from germ cells. *Cell, 127*, 891–903.

Sheets, M. D. (2015). Building the future: Post-transcriptional regulation of cell fate decisions prior to the *Xenopus* midblastula transition. *Current Topics in Developmental Biology, 113*, 233–270.

Shen, L., Inoue, A., He, J., Liu, Y., Lu, F., and Zhang, Y. (2014). Tet3 and DNA replication mediate demethylation of both the maternal and paternal genomes in mouse zygotes. *Cell Stem Cell, 15*, 459–470.

Shinde, D. N., Elmer, D. P., Calabrese, P., Boulanger, J., Arnheim, N., and Tiemann-Boege, I. (2013). New evidence for positive selection helps explain the paternal age effect observed in achondroplasia. *Human Molecular Genetics, 22*, 4117–4126.

Shine, R. (1978). Propagule size and parental care: The "safe harbor" hypothesis. *Journal of Theoretical Biology, 75*, 417–424.

Siddiqui, N. U., Li, X., Luo, H., et al. (2012). Genome-wide analysis of the maternal-to-zygotic transition in *Drosophila* primordial germ cells. *Genome Biology, 13*, R11.

Silva, D. M. Z. A., and Akera, T. (2023). Meiotic drive of noncentromeric loci in mammalian meiosis, II Eggs. *Current Opinion in Genetics and Development, 81*, 102082.

Simons, B. D., and Clevers, H. (2011). Strategies of homeostatic stem cell self-renewal in adult tissues. *Cell, 145*, 851–862.

Skvortsova, K., Tarbashevich, K., Stehling, M., et al. (2019). Retention of paternal DNA methylome in the developing zebrafish germline. *Nature Communications, 10*, 3054.

Slaidina, M., and Lehmann, R. (2014). Translational control in germline stem cell development. *Journal of Cell Biology, 207*, 13–21.

Smith, Z. D., Chan, M. M., Humm, K. C., et al. (2014). DNA methylation dynamics of the human preimplantation embryo. *Nature, 511*, 611–615.

Snee, M. J., and Macdonald, P. M. (2003). Live imaging of nuage and polar granules: Evidence against a precursor–product relationship and a novel role for *Oskar* in stabilization of polar granule components. *Journal of Cell Science, 117*, 2109–2120.

Snow, M. H. L. (1977). Gastrulation in the mouse: Growth and regionalization of the epiblast. *Journal of Embryology and Experimental Morphology, 42*, 293–303.

Solana, J. (2013). Closing the circle of germline and stem cells: The primordial stem cell hypothesis. *EvoDevo, 4*, 2.

Soriano, P., and Jaenisch, R. (1986). Retroviruses as probes for mammalian development: Allocation of cells to the somatic and germ cell lineages. *Cell, 46,* 19–29.

Sotero-Caio, C. G., Platt, R. N., Suh, A., and Ray, D. A. (2017). Evolution and diversity of transposable elements in vertebrate genomes. *Genome Biology and Evolution, 9,* 161–177.

Spichal, M., Heestand, B., Billmyre, K. K., Frenk, S., Mello, C. C., and Ahmed, S. (2021). Germ granule dysfunction is a hallmark and mirror of Piwi mutant sterility. *Nature Communications, 12,* 1420.

Srivastava, M. (2022). Studying development, regeneration, stems cells, and more in the acoel *Hostenia miamia. Current Topics in Developmental Biology, 147,* 153–172.

Stearns, S. C. (1987). The selection-arena hypothesis. In *The Evolution of Sex and its Consequences,* edited by S. C. Stearns, 337–380. Birkhäuser Verlag.

Steger, K. (1999). Transcriptional and translational regulation of gene expression in haploid spermatids. *Anatomy and Embryology, 199,* 471–487.

Stephenson, R. O., Yamanaka, Y., and Rossant, J. (2010). Disorganized epithelial polarity and excess trophectoderm cell fate in preimplantation embryos lacking E-cadherin. *Development, 137,* 3383–3391.

Strathmann, R. R., Staver, J. M., and Hoffman, J. R. (2002). Risk and the evolution of cell-cycle durations of embryos. *Evolution, 56,* 708–720.

Strome, S., and Lehmann, R. (2007). Germ versus soma decisions: Lessons from flies and worms. *Science, 316,* 392–393.

Strome, S., and Updike, D. (2015). Specifying and protecting germ cell fate. *Nature Reviews Molecular Cell Biology, 16,* 406–416.

Sugie, K., Funaya, S., Kawamura, M., Nakamura, T., Suzuki, M. G., and Aoki, F. (2020). Expression of *Dux* family genes in early preimplantation embryos. *Scientific Reports, 10,* 19396.

Sulston, J. E., and Horvitz, H. R. (1977). Post-embryonic cell lineages of the nematode, *Caenorhabditis elegans. Developmental Biology, 56,* 110–156.

Sulston, J. E., Schierenberg, E., White, J. G., and Thomson, J. N. (1983). The embryonic cell lineage of the nematode *Caenorhabditis elegans. Developmental Biology, 100,* 64–119.

Sun, C., Shepard, D. B., Chong, R. A., et al. (2012). LTR retrotransposons contribute to genomic gigantism in plethodontid salamanders. *Genome Biology and Evolution, 4,* 168–183.

Svoboda, P., Franke, V., and Schultz, R. M. (2015). Sculpting the transcriptome during the oocyte-to-embryo transition in mouse. *Current Topics in Developmental Biology, 113,* 305–349.

Tam, P. P. L., and Snow, M. H. L. (1981). Proliferation and migration of primordial germ cells during compensatory growth in mouse embryos. *Journal of Embryology and Experimental Morphology, 64,* 133–147.

Thomas, J. H., and Schneider, S. (2011). Coevolution of retroelements and tandem zinc finger genes. *Genome Research, 21,* 1800–1812.

Thomas, M. B. (1986). Embryology of the Turbellaria and its phylogenetic significance. *Hydrobiologia, 132,* 105–115.

Thompson, P. J., Macfarlan, T. S., and Lorincz, M. C. (2016). Long terminal repeats: From parasitic elements to building blocks of the transcriptional regulatory repertoire. *Molecular Cell, 62,* 766–776.

Tiwari, B., Kurtz, P., Jones, A. E., et al. (2017). Retrotransposons mimic germ plasm determinants to promote transgenerational inheritance. *Current Biology, 27,* 3010–3016.

Töhönen, V., Katayama, S., Vesterlund, L., et al. (2015). Novel PRD-like homeodomain transcription factors and retrotransposon elements in early human development. *Nature Communications, 6,* 8207.

Trivers, R. L. (1974). Parent-offspring conflict. *American Zoologist, 14,* 249–264.

Ueno, H., Turnbull, B. B., and Weissman, I. L. (2009). Two-step oligoclonal development of male germ cells. *Proceedings of the National Academy of Sciences of the United States of America, 106,* 175–180.

van der Horst, C. J. (1942). Early stages in the embryonic development of *Elephantulus*. *South African Journal of Medical Sciences, 7*(suppl.), 55–64.

van der Maarel, S. M., Deidda, G., Lemmers, R. J. F. L., et al. (2000). De novo facioscapulohumeral muscular dystrophy: Frequent somatic mosaicism, sex-dependent phenotype, and the role of mitotic transchromosomal repeat interaction between chromosomes 4 and 10. *American Journal of Human Genetics, 66,* 26–35.

van Doren, M., Williamson, A. L., and Lehmann, R. (1998). Regulation of zygotic gene expression in *Drosophila* primordial germ cells. *Current Biology, 8,* 243–246.

Vanneste, E., Voet, T., Le Ciagnec, C., et al. (2009). Chromosome instability is common in human cleavage-stage embryos. *Nature Medicine, 15,* 577–583.

Vargas, A., Zhou, S., Éthier-Chiasson, M., et al. (2014). Syncytin proteins incorporated in placenta exosomes are important for cell uptake and show variation in abundance in serum exosomes from patients with preeclampsia. *FASEB Journal, 28,* 3703–3719.

Vasquez Kuntz, K. L., Kitchen, S. A., Conn, T. L., et al. (2022). Inheritance of somatic mutations by animal offspring. *Science Advances, 8,* eabn0707.

Vassena, R., Boué, S., González-Roca, E., et al. (2011). Waves of transcriptional activation and pluripotency program initiation during human preimplantation development. *Development, 138,* 3699–3709.

Vázquez-Diez, C., and FitzHarris, G. (2018). Causes and consequences of chromosome segregation error in preimplantation embryos. *Reproduction, 155,* R63–R76.

Venkatarama, T., Lai, F., Luo, X., Zhou, Y., Newman, K., and King, M. L. (2010). Repression of zygotic gene expression in *Xenopus* germline. *Development, 137,* 651–660.

Voronina, E., Seydoux, G., Sassone-Corsi, P., and Nagamori, I. (2011). RNA granules in germ cells. *Cold Spring Harbor Perspectives in Biology, 3,* a002744.

Vuoristo, S., Bhagat, S., Hydén-Granskog, C., et al. (2022). *DUX4* is a multifunctional factor priming human embryonic genome activation. *iScience, 25,* 104137.

Wade, M. J., and Beeman, R. W. (1994). The population dynamics of maternal-effect selfish genes. *Genetics, 138,* 1309–1314.

Wagner, D. E., Wang, I. E., and Reddien, P. W. (2011). Clonogenic neoblasts are pluripotent adult stem cells that underlie planarian regeneration. *Science, 332,* 811–816.

Walter, J., and Paulsen, M. (2003). The potential role of gene duplications in the evolution of imprinting mechanisms. *Human Molecular Genetics, 12,* R215–R220.

Wang, L., Dou, K., Moon, S., Tan, F. J., and Zhang, Z. Z. Z. (2018). Hijacking oogenesis enables massive propagation of LINE and retroviral transposons. *Cell, 174,* 1082–1094.

Wang, L., Zhang, J., Duan, J., et al. (2014). Programming and inheritance of parental DNA methylomes in mammals. *Cell, 157,* 979–991.

Wang, X., Gou, L. T., and Liu, M. F. (2022). Noncanonical functions of PIWIL1/piRNAs in animal male germ cells and human diseases. *Biology of Reproduction, 107,* 101–108.

Wang, X., Ramat, A., Simonelig, M., and Liu, M. F. (2023). Emerging roles and functional mechanisms of PIWI-interacting RNAs. *Nature Reviews Molecular Cell Biology, 24,* 123–141.

Wang, Y., Zayas, R. M., Guo, T., and Newmark, P. A. (2007). *nanos* function is essential for development and regeneration of planarian germ cells. *Proceedings of the National Academy of Sciences of the United States of America, 104,* 5901–5906.

Watanabe, T., Biggins, J. S., Tannan, N. B., and Srinivas, S. (2014). Limited predictive value of blastomere angle of division in trophectoderm and inner cell mass specification. *Development, 141,* 2279–2288.

Weismann, A. (1890). Prof. Weismann's theory of heredity. *Nature, 41,* 317–323.

Weismann, A. (1892). *Das Keimplasma: Eine Theorie der Vererbung.* Gustav Fischer.

Weismann, A. (1893). *The Germ-Plasm: A Theory of Heredity.* Translated by W. N. Parker and H. Rönnfeldt. Charles Scribner.

Weismann, A. (1904). *The Evolution Theory.* 2 vols. Translated by J. A. Thomson and M. R. Thomson. Edward Arnold.

Wells, J. N., and Feschotte, C. (2020). A field guide to eukaryotic transposable elements. *Annual Review of Genetics, 54,* 539–561.

Wennekamp, S., Mesecke, S., Nédélec, F., and Hiiragi, T. (2013). A self-organization framework for symmetry breaking in the mammalian embryo. *Nature Reviews Molecular Cell Biology, 14,* 452–459.

Westerich, K. J., Tarbashevich, K., Schick, J., et al. (2023). Spatial organization and function of RNA molecules within phase-separated condensates in zebrafish are controlled by Dnd1. *Developmental Cell, 58*, 1578–1592.

Whiddon, J. L., Langford, A. T., Wong, C. J., Zhong, J. W., and Tapscott, S. J. (2017). Conservation and innovation in the DUX4-family gene network. *Nature Genetics, 49*, 935–940.

White, J. (1979). The plant as a metapopulation. *Annual Review of Ecology and Systematics, 10*, 109–145.

White, M. D., Zenker, J., Bissiere, S., and Plachta, N. (2018). Instructions for assembling the early mammalian embryo. *Developmental Cell, 45*, 667–679.

Whitington, P. M., and Dixon, K. E. (1975). Quantitative studies of germ plasm and germ cells during early embryogenesis of Xenopus laevis. *Journal of Embryology and Experimental Morphology, 33*, 57–74.

Wilkins, J. F. (2005). Genomic imprinting and methylation: Epigenetic canalization and conflict. *Trends in Genetics, 21*, 356–365.

Wilkins, J. F., and Haig, D. (2002). Parental modifiers, antisense transcripts and loss of imprinting. *Proceedings of the Royal Society B, 269*, 1841–1846.

Williams, G. C. (1966). *Adaptation and Natural Selection.* Princeton University Press.

Williamson, A., and Lehmann, R. (1996). Germ cell development in *Drosophila*. *Annual Review of Cell and Developmental Biology, 12*, 365–391.

Wilson, E. B. (1925). *The Cell in Development and Heredity,* 3rd ed. MacMillan.

Winokur, S. T., Bengtsson, U., Vargas, J. C., Wasmuth, J. J., and Altherr, M. R. (1996). The evolutionary distribution and structural organization of the homeobox-containing repeat D4Z4 indicates a functional role for the ancestral copy in the FSHD region. *Human Molecular Genetics, 5*, 1567–1575.

Wolf, G., Greenberg, D., and Macfarlan, T. S. (2015). Spotting the enemy within: Targeted silencing of foreign DNA in mammalian genomes by the Krüppel-associated box zinc finger protein family. *Mobile DNA, 6*, 17.

Woodland, H. (1982). The translational control phase of early development. *Bioscience Reports, 2*, 471–491.

Wossidlo, M., Nakamura, T., Lepikhov, K., et al. (2011). 5-hydroxymethylcytosine in the mammalian zygote is linked with epigenetic reprogramming. *Nature Communications, 2*, 241.

Wu, P. H., Yu, T., Cecchini, K., et al. (2020). The evolutionarily conserved piRNA-producing locus pi6 is required for male mouse fertility. *Nature Genetics, 52*, 728–739.

Yang, J., Aguero, T., and King, M. L. (2015). The *Xenopus* maternal-to-zygotic transition from the perspective of the germline. *Current Topics in Developmental Biology, 113*, 271–303.

Yoon, S. R., Choi, S. K., Eboreime, J., Gelb, B. D., Calabrese, P., and Arnheim, N. (2013). Age-dependent germline mosaicism of the most common Noonan syndrome mutation shows the signature of germline selection. *American Journal of Human Genetics, 92,* 917–926.

Yoshida, S. (2019). Heterogeneous, dynamic, and stochastic nature of mammalian spermatogenic stem cells. *Current Topics in Developmental Biology, 135,* 245–285.

Zalokar, M. (1976). Autoradiographic study of protein and RNA formation during early development of *Drosophila* eggs. *Developmental Biology, 49,* 425–437.

Zamudio, N., and Bourc'his, D. (2010). Transposable elements in the mammalian germline: A comfortable niche or a deadly trap? *Heredity, 105,* 92–104.

Zhang, H. T., and Hiiragi, T. (2018). Symmetry breaking in the mammalian embryo. *Annual Review of Cell and Developmental Biology, 34,* 405–426.

Zheng, C. J., Luebeck, E. G., Byers, B., and Moolgavkar, S. H. (2005). On the number of founding germ cells in humans. *Theoretical Biology and Medical Modelling, 2,* 32.

Zheng, K., and Wang, P. J. (2012). Blockade of pachytene piRNA biogenesis reveals a novel requirement for maintaining post-meiotic germline genome integrity. *PLoS Genetics, 8,* e1003038.

Zou, Z., Zhang, C., Wang, Q., et al. (2022). Translatome and transcriptome co-profiling reveals a role for TPRXs in human zygotic genome activation. *Science, 378,* 41.

CHAPTER SEVEN

Selfish Genetic Elements and the Evolution of Sex Determination and Inheritance Systems

MARTIJN A. SCHENKEL, LAURA ROSS, AND NINA WEDELL

ABSTRACT A major strength of the paradox-of-the-organism concept is that it makes us look at old facts in new ways. By its textbook definition, evolution is the change in the frequencies of genetic variants in a population over time. Individuals struggle to survive and compete to reproduce, and whatever genes help improve the odds of success should become more common. In this description, the transmission of genes from one generation to the next is the central component and is usually assumed to be a conflict-free process as genes share a common goal. This chapter argues that internal conflict between genes with different agendas and transmission patterns impinges on reproduction and heredity in numerous ways, shaping the mechanisms that underlie these two essential processes. By surveying the mechanisms that distort inheritance and sex, we determine how the changes these conflicts bring about within individuals link up to evolutionary change. We then apply the paradox-of-the-organism framework to one of the most profound internal conflicts—the conflict between the cytoplasm and the nucleus over the individual's sex—as a case study to demonstrate the explanatory power of intra-genomic conflict. Taken together, we conclude that organismal ontogeny, and the selfish elements present within the resulting individual, are best viewed as two interacting and coevolving components of organismal biology. Consequently, the origin and development of biological individuals are ever-changing owing to internal conflicts.

Introduction

Evolution occurs through changes in the frequencies of different genetic variants in a population over time, with adaptations arising through the spread of beneficial variants and the loss of deleterious ones. Organisms have traditionally been ascribed a pivotal role in this process, and what constitutes a "beneficial" or "deleterious" genetic variant has often been cast in terms of its effect at an organismal level. This view maintains that adaptive evolution leads to the accumulation of genes that enhance organismal (inclusive) fitness, for example by increasing survival or fecundity. By extension, organismal traits are then expected to reflect the organism's optimum. As the field of evolutionary biology underwent major advances during the twentieth century, it became apparent that this traditional view does not hold up. Instead, organisms themselves are composed of suborganismal components such as genes (but also cells) that may have markedly different interests. This difference can lead to intra-genomic conflict, which occurs when different genes in an individual optimize their fitness under mutually exclusive conditions (Gardner and Úbeda 2017).

Under intra-genomic conflict, some genes may enhance their own evolutionary success, but not by the conventional means of enhancing organismal fitness. Instead, these genes may distort the biology of their organismal host by interfering with organismal traits or the mechanisms of inheritance (Patten et al. 2023). While advantageous to the gene, such tampering generally comes at a cost to organismal fitness, which has led them to be called "selfish genetic elements." We define a selfish genetic element as a genetic, or in some cases genomic, element that exhibits some selfish effect often associated with increased transmission. This effect acts in such a way as to enhance the fitness of the selfish element but may simultaneously impose a cost on the organism, thus interfering with the fitness optimization of the host. Following their discovery, selfish genetic elements were initially regarded as evolutionary curiosities, exceptions that proved the rules of genetics and evolutionary biology (Werren et al. 1988). However, their existence also helped spark the levels-of-selection debate, as it cast further doubts on the organismal monopoly on adaptation. Moreover, it became clear that the fundamentals of Mendelian genetics are not impervious to criticism. Together with a growing appreciation of inheritance systems that deviate from the norm, this discovery helped show that the mechanisms by which genetic material is transmitted are malleable, both within living organisms and across evolutionary time.

The different ways in which selfish genetic elements tamper with the principles of heredity are reflected in their diverse associations with sexual reproduction. First, sexual reproduction has been touted as the key requisite for selfish genetic elements to invade because it enables them to switch lineages (Partridge and Hurst 1998). By not being tied down to a single host lineage, the costs selfish genetic elements impose on their hosts need not apply to the selfish genetic elements themselves. Without this dissociation, fitness of the host and any would-be selfish genetic element would be completely aligned—and selfishness would by default be detrimental to both. Second, the sexual identity of an organism is the cause

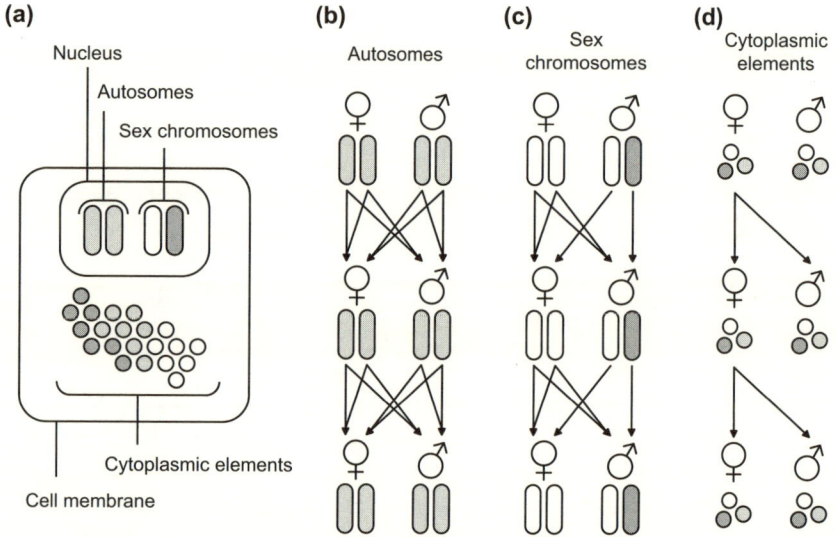

Figure 7.1. Inheritance patterns of different genetic elements across females and males. (a) Eukaryotic cells contain nuclear elements (autosomes and sex chromosomes) as well as cytoplasmic elements (mitochondria, chloroplasts, and endosymbionts) that segregate differently to and from females and males. (b) Autosomes segregate freely to and from females and males. (c) Sex chromosomes (X chromosomes in white, Y chromosome in dark gray) are transmitted in biased ways between males and females. X chromosomes in females are transmitted equally to daughters and sons, but in males are transmitted only to daughters; Y chromosomes are inherited strictly from fathers to sons and thus are male-limited. (d) Cytoplasmic elements are inherited only maternally, but are transmitted to both sexes alike; however, only daughters continue to transmit these elements, resulting in a quasi-female-restricted inheritance pattern.

of much intra-genomic conflict, because different genomic elements are transmitted to and from males and females in different ways (Cosmides and Tooby 1981; Gardner and Úbeda 2017). Each of these elements is selected to bias the host's sexual identity in such a way that its own fitness will be maximized, generally by somehow favoring the sex in which the element is transmitted. For example, cytoplasmic elements are often only maternally transmitted, and hence only beneficial when they end up in females (Figure 7.1). Control over sex determination can evolve rapidly as antagonistic genomic entities evolve to one-up each other, much like other antagonistic parties (e.g., hosts and parasites) that engage in continuous coevolution (Brockhurst et al. 2014). Third, the peculiarities of female and male biology lend themselves to different types of selfish genetic elements, as these use different mechanisms of exploiting their hosts. The effect of sex differences on the capacity of selfish genetic elements to get their way has even led some selfish genetic elements to evolve ways of distorting sex determination, reminiscent of the genetic conflict over host sex determination that comes from the differences in segregation rates of different genomic elements (Wedell 2020).

Intra-genomic conflicts can take on many different forms, resulting in an astonishingly diverse array of selfish genetic elements that distort organismal biology to further their own agenda. Each of these elements may exploit different components, or weaknesses, of the biology of the organism. Understanding how this exploitation impacts individuals, and how this links up with evolutionary changes in heredity and sex is the key focus of this chapter. Adopting a paradox-of-the-organism viewpoint on the evolution of these fundamental processes in organismal biology is essential to understand their evolutionary malleability. To this end, we have structured this chapter into three parts. First, we discuss the ways in which intra-genomic conflict has been theorized to lead to evolutionary changes in the mechanisms and modes of inheritance. Second, we discuss past theories on the role of intra-genomic conflict in driving the evolution of sex-determination mechanisms. These two sections lay the foundation for the third section, where we consider one of the most profound and best-studied intra-genomic conflicts—namely, the conflict between an organism's nuclear genome and the cytoplasmic factors. In many species, cytoplasmic material is strictly maternally inherited, whereas the nuclear genome segregates freely to and from males and females (with the exception of sex chromosomes, whose transmission dynamics deviate in different ways). These different transmission patterns cause cytoplasmic material to favor females over males, whereas the nuclear genome may value them equally. Consequently, nuclear and cytoplasmic genes engage in a perennial conflict over the individual's sexual identity, through which different cytoplasmic agents have evolved myriad methods of interfering with the process of sex determination. In this final section, we explore how different agents impinge on sex determination with the aim of inferring general principles that underlie this fundamental internal conflict, and how these relate to past predictions from evolutionary theory.

Genetic Conflict and the Evolution of Mechanisms of Genetic Inheritance

In textbook genetics, Mendelian inheritance yields an organism with a diploid genome that is effectively a mash-up of chromosomes selected randomly from its diploid parents. Here, each parent generates a gamete that contains one of each chromosome pair, and fusion of the maternal and paternal gametes (egg and sperm) results in the diploid zygote. This single-celled zygote then develops into a multicellular adult through clonal division. The adult may similarly transmit random subsets of its genetic material to its offspring. While this framework is a useful heuristic that applies broadly to many organisms, it fails to capture the many cases in which parents do not contribute equally to an organismal genome. In this section, we will break down this organismal blueprint and determine how deviations from the heuristic of Mendelian inheritance may both originate in specific genetic conflicts and beget other genetic conflicts in turn.

Separation of Soma and Germline

Multicellular organisms typically exhibit a fundamental division of labor, in which many cells perform somatic functions but only a small minority contribute to the production of gametes: the germline. In the context of the major transitions, multicellularity may arise when the division of labor enhances the reproductive output of a (clonal) group of cells beyond the cumulative output of a similarly sized collection of unicellular copies (Michod 2006, 2007). Division of labor allows certain cells, effectively precursors to the would-be germline, to commit to gametogenesis. Other cells, representing the soma, differentiate and focus on other tasks, for example enhancing the survivability of the group. As group relatedness is high (and possibly even complete), the sacrifice made by somatic copies to forgo reproduction themselves is offset by the benefit through enhanced fitness to the germline, so that indirectly somatic cells (or more specifically the genes within them) are represented in future generations as well.

However, other selective processes may also contribute to the evolution of a distinct germline, with genetic conflict potentially playing two different roles. First, the separation of soma and germline establishes two interacting suborganismal ecosystems that shape the action of selfish genetic elements. Selfish genetic elements are favored to exert their effects in germline cells, through which they become represented in the next generation. Yet somatic activity confers no evolutionary benefit to selfishness. A selfish genetic element may increase in frequency among the genome or among somatic cells, but these benefits are transient because these cells, and the genes within them, are lost upon the death of the organism (Haig 2016, 2021). A second explanation for soma-germline sequestration was proposed by Johnson (2008). Here, the detrimental effects of somatic replication are assumed to affect the whole organism, whereas germline replication is thought to only affect gamete viability. This dichotomy can promote soma-germline sequestration even when it is deleterious to the host but advantageous to the selfish genetic element. Selfish variants that enhance this sequestration thus are expected to invade, followed by variants that exhibit higher differentiation of germline versus somatic activity (the former increasing, the latter decreasing). This establishes a positive feedback loop where both the level of soma-germline sequestration and the initial selfishness increase. Recent evidence supporting such a positive feedback loop is found in ciliates, which despite being unicellular exhibit soma-germline sequestration at the nuclear level, with one nucleus carrying the mostly transcriptionally silent "germline" copy of the genome, while the other "somatic" nucleus is transcribed. The formation of the somatic nucleus involves the elimination of germline-specific DNA sequences, mostly repetitive DNA, which appears to have evolved as a way for transposable elements to cut themselves out of the somatic nucleus to reduce their deleterious effects. However, the result is the proliferation of these elements in the germline and an ever-increasing genetic differentiation between the somatic and germline genomes (Seah and Swart 2023).

Evolution of Maternal Inheritance of Cytoplasmic Elements

In anisogamic eukaryotes, cytoplasmic elements such as mitochondria (and in plants, chloroplasts) are almost exclusively maternally inherited. Possibly, the strongly reduced size of sperm makes paternal inheritance of large cytoplasmic elements infeasible. However, even in isogamic eukaryotes, systems that dictate the uniparental inheritance of cytoplasmic elements have evolved, such as in the ulvophycean *Monostroma angicava* (Togashi et al. 2023). In early eukaryotes, inheritance of cytoplasmic elements may have occurred through both parents, yielding heterogeneous cytoplasms. Some even suggest that biparental inheritance promotes competition between various cytoplasmic organelles and allows the spread of selfish cytoplasmic elements (Havird et al. 2019). For example, in the evening primrose *Oenothera*, fast-replicating chloroplasts (plastids) compete against each other, and their success is determined by rapidly evolving metabolic genes. These chloroplasts are often incompatible with the hybrid nuclear genome (Sobanski et al. 2019). Competition between unrelated, competing cytoplasmic elements may have generated such a cost at the organismal level that conflict suppressors were favored among the nuclear DNA (Cosmides and Tooby 1981). Here, suppression of the conflict may have been achieved by selective elimination of minor-frequency cytoplasmic variants, so that only the most-prevalent variant remained. This homogenization would abolish conflict among the cyto-plasm, as all cytoplasmic elements would then be genetically identical. With sperm being the smaller gamete, its contribution to the zygotic cytoplasm is similarly reduced, and its cytoplasmic elements would gen-erally be eliminated under this model. Alternatively, sperm may be vul-nerable to invasion by foreign elements like endosymbionts because their role in gamete fusion requires that they are comparatively naked in terms of, for example, neighboring cells and other extracellular structures that preserve their integrity. Elimination of sperm-derived cytoplasmic ele-ments may then have been favored as a means of host defense against would-be genetic parasites that enter the egg via the sperm (Coleman 1982). Consequently, it may have become favorable to eliminate cyto-plasmic elements from sperm altogether. However, the consequences of maternal transmission of cytoplasmic elements such as mitochondria means there is no selection on their function in males that can result in adverse effects on sperm function, fertility, and male fitness. This is termed the "mother's curse" (Gemmell et al. 2004). For example, in *Drosophila melanogaster* a mitochondrial hypomorph of cytochrome oxidase II causes defects in sperm development and function, thereby impairing male fertility without affecting other male or female functions (Patel et al. 2016).

Elimination of DNA

In the section above we discussed the potential evolutionary route to maternal-only transmission of cytoplasmic elements, such as mitochon-dria and chloroplasts. One step in this process is the elimination of any minor variants of mitochondria that are present in the cytoplasm (e.g., Sobanski et al. 2019). Here, competition between cytoplasmic variants are assumed to be costly to the organism as whole, so that nuclear re-pressors that eliminate the "losing" party are favored. Mendelian inheri-tance predicts that a given organism will be equally likely to transmit genetic material regardless of whether it is paternally or maternally inherited. The maternal-only inheritance of mitochondrial DNA provides an example where this prediction does not hold.

Cytoplasmic male sterility (i.e., the suppression of spermatogenesis to redirect resources to oogenesis) has evolved independently multiple times in hermaphroditic plant species, effectively converting would-be hermaphrodites into females. Such sterility may be achieved in a variety of ways, including by generating sterile or inviable pollen, or by preventing the development of male reproductive organs. Cytoplasmic male sterility frequently favors the evolution of nuclear genes that suppress male sterility, thereby restoring the sex ratio, and their presence is only evident in population crosses involving different cytoplasmic male sterility factors and suppressor genes (Schnable and Wise 1998). Until recently, there were no known cases of mitochondrial sex ratio distortion in animals (Perlman et al. 2015). However, a first case of cytoplasmic male sterility in an animal was recently reported in the freshwater snail *Physa acuta*. In this case, males are sterilized by selfish mitochondria, thereby turning hermaphrodites into functional females—a process that may increase female fitness and hence favor the selfish mitochondria (David et al. 2022).

DNA elimination is not limited to the cytoplasm, however. Nuclear DNA may also be eliminated at times. In fact, nuclear DNA elimination is an integral part of the life cycle of many organisms (Smith et al. 2021; Dedukh and Krasikova 2022). Generally, DNA elimination is specific to certain cell types and often, but not always, restricted to one sex. Here, we present two of the most widespread examples of DNA elimination in the nucleus: paternal genome elimination (PGE) and germline-specific DNA. We then discuss the possible involvement of genetic conflict in the origin of each form and explain its effects on shaping future genetic conflicts.

PGE begins with the normal formation of zygotes through gametic fusion of an egg and a sperm. But while maternally inherited DNA is maintained and transmitted as predicted by Mendelian genetics, paternally inherited DNA is treated very differently. Three systems of PGE have been described in arthropods so far, defined according to the time at which the paternal DNA is eliminated: (1) germline PGE, where paternal DNA is lost during or just prior to spermatogenesis, is found in some scale insects, beetles, and booklice as well as all parasitic lice; (2) embryonic PGE, where DNA is lost during embryonic development, is found in some scale insects and mites; and (3) X-PGE, which is a mixture of the two, where paternal X chromosomes are lost during embryonic development while paternal autosomes are lost during spermatogenesis; this last system is found in two fly families, the fungus gnats and gall midges and the globular springtails (Normark 2003; Normark and Ross 2014; Herbette and Ross 2023). PGE is thought to evolve as a result of genomic conflict between the maternal and paternal genomes over who gets access to mature gametes, with the maternal genome succeeding in being preferentially included in male gametes (Bull 1979; Gardner and Ross 2014).

A second form of DNA elimination is encountered in germline-restricted DNA. This was first discovered in the late 1800s and early 1900s (Boveri 1895; Metz 1926) but has only recently become an active area of research (Hodson and Ross 2021; Sellis et al. 2021; Borodin et al. 2022). In these cases, specific DNA sequences, which can make up from 25% to 90% of the genome, are present in germline cells but absent from somatic cells. These sequences are transmitted between generations and stably eliminated or maintained within each generation along the soma/germline divide. In some groups, such as songbirds, flies, and lampreys, the eliminated DNA comprise entire chromosomes (Smith et al. 2018; Wang et al. 2020; Smith et al. 2021), while in other groups, such as ciliates and nematodes, the genome undergoes fragmentation during early development, with only some pieces of DNA retained in the soma while others are eliminated (Sellis et al. 2021).

In a sense, germline-restricted DNA represent the inverse of PGE; rather than eliminating DNA in the germline and maintaining it in the soma, germline DNA is eliminated in the soma but maintained in the germline. However, a key difference is that it is not entire genome sets but rather specific regions or chromosomes that are eliminated or maintained. Their absence from somatic cells suggests that they, like B chromosomes, are at least partly dispensable, though they likely perform critical functions in the germline, as has been suggested for songbirds whose genome-restricted chromosomes carry a number of meiosis and oogenesis genes (Vontzou et al. 2023). Alternatively, they might be selfish genetic elements that are eliminated from the soma to reduce their negative fitness consequences. This is likely to be the case in ciliates, where many of the eliminated fragments are derived from transposable elements (Seah and Swart 2023). The result is that these elements can essentially hide from purifying selection in the germline-specific portion of the genome and therefore further accumulate, leading to a positive feedback loop with increasing expansion of the germline genome and stronger selection for somatic elimination to mitigate negative fitness consequences.

Evolution of Sex Determination Through Genetic Conflict

Sex determination is the developmental process that makes individuals into females or males (or both, in case of hermaphrodites). Its proper execution is essential to ensure fertility and thus fitness, but its regulatory mechanisms show a remarkable diversity across the tree of life. Numerous selective processes have been found to drive evolutionary diversification of these mechanisms, many of which involve genetic conflict (van Doorn 2014). Here, we discuss the role of intra-genomic conflicts in driving evolutionary change in sex-determination mechanisms.

Sex Chromosome Turnover Through Segregation Distortion

Segregation distortion occurs when specific genetic elements are transmitted at rates that deviate from equal (i.e., Mendelian) rates. Even when the result is effectively the same (one variant is overrepresented among offspring, the other underrepresented), the mechanisms for achieving these distorted rates vary substantially. Segregation distorters can act either before or after fertilization and be transmitted through males and/or females. Segregation distorters that act during meiosis are referred to as meiotic drivers and are either autosomally inherited or linked to the sex chromosomes, in which case they can cause sex ratio distortion and promote suppression and new sex-determination mechanisms (Jaenike 2001; Ågren and Clark 2019). Most known meiotic drivers act during male gametogenesis, resulting in differential sperm killing (Courret et al. 2019). Female meiotic drivers, of which there are fewer examples, often involve centromere drive in plants due to the asymmetry of female meiosis (Finseth 2023), although there is indirect evidence that centromeric drive might be much more widespread (Yoshida and Kitano 2012; Malik 2022).

When segregation distorters evolve on sex chromosomes, their ability to become overrepresented has the additional effect of sex ratio distortion. Rather than producing daughters and sons in more or less equal numbers, bearers of sex segregation distorters will produce either more daughters or more sons. The direction of sex ratio distortion hinges on the sex chromosome that exhibits segregation distortion. Assuming male heterogamety (females XX, males XY), Y-chromosomal segregation distorters will result in more sons, whereas X-chromosomal ones will result in more daughters. In the latter case, this assumes that segregation distortion occurs in males. (It may occur in females as well, but this does not affect offspring sex ratio.)

While autosomal segregation distorters may sweep to fixation, sex-chromosomal segregation distorters are generally kept in check due to sex ratio selection. A classic paper by Hamilton (1967) predicted sex-chromosomal segregation distorters would lead to population extinction as successive bouts of segregation distortion left fewer and fewer individuals of the rarer sex (e.g., females for Y-chromosomal segregation distorters). But this makes rather stringent assumptions about the level of segregation distortion and the limits of female fertility. Under natural conditions, sex ratios may become biased in proportion to the level of segregation distortion, but populations may nonetheless persist despite the rarity of one sex, as seen, for example, in the butterfly *Hypolimnas bolina* (Charlat et al. 2005).

The possibility of novel sex-determining genes spreading through linkage to a segregation distorter has been explored using a variety of models. Kozielska et al. (2010) found that segregation distortion by a variant Y chromosome leads it to replace the ancestral, Mendelian-segregating Y chromosome. At equilibrium, sex ratio distortion is equal to segregation distortion: for example, a segregation rate of 80% yields a sex ratio of 80% males. Biased sex ratios favor the invasion of novel sex-determining genes that restore the sex ratio to parity (Fisher 1930; Wilkins 1995); however, there are many known exceptions to this principle (Uller et al. 2007). In the Kozielska et al. (2010) model, this replacement occurs through the invasion of a feminizing gene that over-rules the segregation-distorting Y chromosome, yielding a polymorphic system as costs of YY homozygosity initially prevent fixation of the Y chromosome. At this point, a nonsegregation-distorting Y chromosome, though initially displaced, can invade and spread. This dynamic establishes the full turnover from an XY-male heterogamety system to a ZW-female heterogamety system. A similar model—in which a male-determining gene spreads through transposition and yields a biased sex ratio—also yields transitions between male and female heterogamety (Schenkel et al. 2023). Transposing sex determination genes have been described in a range of species, including strawberries (Tennessen et al.

2018), salmon (McKinney et al. 2021), and houseflies (Sharma et al. 2017; Li et al. 2024). For example, in the African pygmy mouse (*Mus minutoides*) a novel third feminizing X chromosome has evolved in response to male sex-chromosome drive (Saunders et al. 2022).

Yet another model for sex-determination evolution through segregation distortion explores the impact of sex differences in segregation rates (Úbeda et al. 2014). Segregation distortion in males and females can occur through markedly different pathways. In males, it typically occurs through postsegregational killers. These eliminate gametes that do not bear the segregation-distortion complex (Burga et al. 2020). In females, segregation distortion can also be achieved through meiotic drive (preferential segregation to the egg instead of the polar bodies), which leads to overrepresentation through less detrimental means. The differences in gametogenesis between the sexes provides different possibilities for segregation distorters to work their magic. Consequently, the level of segregation distortion can vary substantially depending on whether this selfish genetic element finds itself in a male or a female. When segregation distortion differs between the sexes, a distorting element might increase in frequency by becoming linked to a sex-determining gene. Through a series of steps, the establishment of a sex-determining gene near a sex-specific segregation distorter can produce a genetically distinct sex chromosome pair. This yields an XY-male heterogamety system when segregation distortion is higher in males and a ZW-female heterogamety when it is higher in females (Úbeda et al. 2014).

B Chromosomes

B chromosomes are essentially optional chromosomes present in an estimated 10% of all eukaryotic species, but not necessarily present in all individuals (Ahmad and Martins 2019). This distinguishes them from the A set of chromosomes, which represents the core genome comprising all essential chromosomes. B chromosomes were the first genetic element to be classified as a potential "parasite," merely because they have no essential function (Östergren 1945). For the same reason, they may be lost at meiosis without severe consequences to offspring fitness. Such leakage can lead to their loss through genetic drift–like effects. As a countermeasure, many B chromosomes exhibit characteristics that enhance their transmission rate in various ways, including many examples of meiotic or pre- and postmeiotic drive. Particularly striking are the paternal sex ratio (PSR) B chromosomes encountered in several species of parasitoid wasp (Werren et al. 1981; Stouthamer et al. 2001), which have been called the "ultimate selfish genetic element" (Werren and Stouthamer 2003). In haplodiploid wasps, a PSR B chromosome causes the paternally inherited genome to be lost during mitosis in fertilized eggs, so that fertilized offspring develop into haploid males (similar to PGE) (Nur et al. 1988). This loss is caused by the failure to process the paternal chromatin into individualized chromosomes. This benefits the PSR B chromosome because it is only stably transmitted by males, whereas in female meiosis it may be lost. Converting would-be females into males ensures that its transmission will continue (Werren 1991).

B chromosomes have been suggested as potential precursors to sex chromosome pairs. A B-chromosomal origin of the Y chromosome was initially proposed to explain its lack of homology to the X chromosome in *Drosophila* (Hackstein et al. 1996). Although this hypothesis has been challenged in the case of *Drosophila,* some other systems have indeed co-opted B chromosomes as their Y or W chromosomes (Johnson Pokorná and Reifová 2021). For example, in three species of butterflies and moths, the W chromosome corresponds to a B chromosome having gained a female-determining factor (Fraïsse et al. 2017). Likewise, in the butterfly *Dryas iulia,* the W chromosome appears to represent a captured B chromosome (Lewis et al. 2021). Another convincing example is found in a cavefish where the male-determining locus is located on a B chromosome (Imarazene et al. 2021). Given that B chromosomes tend to exhibit sex differences in segregation rates, these systems suggest that the model by Úbeda et al. (2014) may indeed yield novel sex chromosomes, though the evolutionary pathway is quite different from the one they proposed. There is also some suggestion that the germline-restricted chromosomes of songbirds might be derived from B chromosomes (Vontzou et al. 2023).

Cytonuclear Conflict over Sex Determination

The maternal inheritance of the cytoplasm, as discussed above, lays the foundation for one of the most potent forms of intra-genomic conflict: that between the cytoplasm and the nucleus over the sexual identity of the individual. Owing to their (almost) exclusively female-limited inheritance, cytoplasmic genetic elements such as mitochondria, chloroplasts,

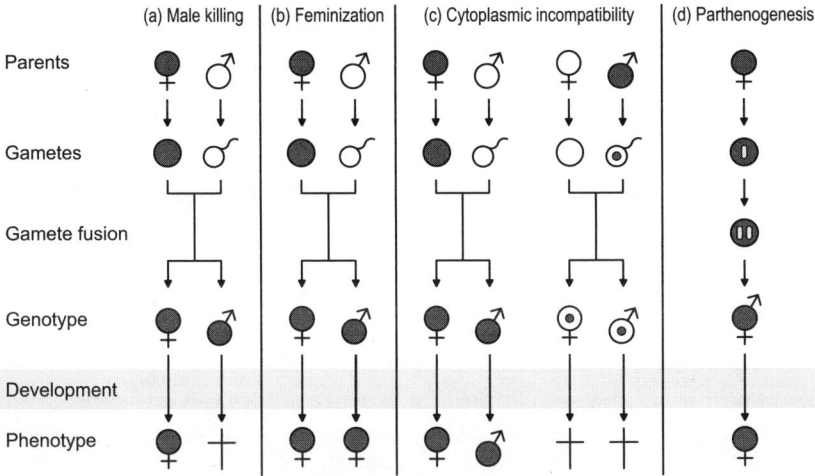

Figure 7.2. Mechanisms for manipulating sex determination. (a) Male-killing. Infected females produce female and male offspring, but male offspring die at some point during development. (b) Feminization. Genetic males are converted into phenotypic females during development. (c) Cytoplasmic incompatibility. Infected females can mate with uninfected and infected (not depicted) males to produce viable, infected offspring, but uninfected females that mate with infected males produce only inviable offspring. This is likely caused by some form of modification of sperm derived from infected males. (d) Parthenogenesis. Infected females reproduce asexually to produce infected daughters. Here, haploid eggs are diploidized, and the resulting zygote is converted into a female individual during early development.

and endosymbiotic bacteria residing in males have no means of propagating themselves to future generations. Consequently, they face strong selection pressures to bias the sex exclusively toward the transmitting female sex, whereas the nuclear DNA (particularly the autosomes) are selected to maintain a mixed-sex ratio (but see Uller et al. 2007). Cytoplasmic factors across a wide range of taxa have evolved various means to enhance their own transmission to the next generation, which fall in four broad categories: (1) parthenogenesis, (2) feminization of genetic males, (3) male killing, and (4) cytoplasmic incompatibility (Figure 7.2). This last mechanism, which causes reproductive incompatibilities between males with the specific cytoplasmic factor and females without it, generally does not result in sex ratio distortion and thus will not be discussed here. However, the first three can cause severe or even complete sex ratio distortion among their host. While the selective pressures involved have long been well understood (Cosmides and Tooby 1981), a mechanistic basis for studying the actual evolutionary changes underlying reproductive manipulations has so far been lacking. Advances in our understanding of sex-determination mechanisms and the mechanisms employed by cytoplasmic elements to manipulate sex determination allow us to determine (1) whether different host sex-determination mechanisms are differently liable to different types of reproductive manipulation; (2) whether different reproductive manipulation phenotypes use similar or distinct mechanisms of interfering with host sex determination; and (3) whether similar exploitation mechanisms by selfish cytoplasmic elements may lead to different reproductive manipulation phenotypes in different hosts. Here, we discuss recent advances that help us answer these questions. We focus on insects and other arthropods, as sex-determination mechanisms in this group are comparatively simple and well studied. They also show that high degrees of conservation and endosymbionts (and therefore endosymbiont-induced reproductive manipulation) are highly prevalent in this group (Duron et al. 2008; Zug and Hammerstein 2012).

Parthenogenesis

Parthenogenesis is a reproductive system in which females produce viable, fertile daughters without fertilization of their eggs with sperm. Effectively, parthenogenesis induction represents a reversal to asexual reproduction by females, though their oocytes are produced through meiotic division. Parthenogenesis induction by cytoplasmic bacteria (endosymbionts) has so far only been described in haplodiploid arthropods (Verhulst et al. 2023). Endosymbiont-induced parthenogenesis in insects, where demonstrated, tends to occur after meiosis (Ma and Schwander 2017). It is possible that specific components of the host sex-determination mechanisms determine the ability of endosymbionts to induce parthenogenesis (Verhulst et al. 2023). Mechanistically, parthenogenesis induction occurs through diploidization of haploid oocytes and subsequent feminization of the diploidized embryos. In some species, the act of diploidization is sufficient to induce feminization (one-step systems), whereas in others feminization is executed separately from diploidization (two-step systems). However, in both systems, the gamete needs to be diploidized by endosymbiont modification of meiosis or mitosis. To induce feminization, endosymbionts are thought to directly manipulate genes in the sex-determination cascade, possibly by inducing a female-specific splice variant of *transformer*, which is a key part of the conserved sex-determining cascade in many insect groups (Verhulst et al. 2010; Geuverink and Beukeboom 2014). Parthenogenesis-inducing endosymbionts have not been found in stinging bees, ants, or wasps, which may be because the process is incompatible with the female developmental pathways leading to either queens or workers (Verhulst et al. 2023). There may also be a link between parthenogenesis induction and endosymbionts that causes cytoplasmic incompatibility. This is because cytoplasmic incompatibility causes male-biased offspring in haplodiploids; unfertilized embryos will survive and develop into haploid males whereas only the fertilized eggs destined to become females cease to develop (Verhulst et al. 2023). Theoretical models have shown that cytoplasmic incompatibility–inducing endosymbionts are highly susceptible to invasion by sex-ratio-distorting variants, possibly indicating that cytoplasmic incompatibility might even be a forerunner of parthenogenesis induction (Hurst et al. 2002).

Feminization

Feminization is the process of converting genetic males into functional, fertile females. For cytoplasmic agents, this converts a nontransmitting individual into a transmitting individual, thus enhancing their success. Endosymbionts have evolved various mechanisms that induce feminization, and these are best characterized in butterflies and moths, isopods, and leafhoppers. In the isopod *Armadillidium vulgare,* male differentiation is induced by the insulin-like androgenic gland hormone. *Wolbachia* endosymbionts target the insulin-like androgenic gland hormone to feminize males by inducing insulin resistance, likely involving defunct insulin receptors (Herran et al. 2020). In the leafhopper *Zyginidia pullula,* the endosymbiont *Wolbachia* feminizes males, which can display some intersex traits, retaining their male genotype and some of their phenotypic male features (Negri et al. 2006). This leafhopper has an XX/XO male heterogametic sex-determining system. *Wolbachia* likely feminizes males by disrupting male imprinting, which influences the expression of genes involved in sex differentiation and development. However, this disruption only occurs when *Wolbachia* is present at a high density (Negri et al. 2009). In the butterfly *Eurema mandarina,* feminization by *Wolbachia* (*w*Fem) involve the dual process of targeting sex determination, possibly by interfering with sex-specific splicing of *doublesex* (*dsx*), and by preventing the inheritance of the female sex chromosomes. This means that only the paternally derived Z chromosome is inherited by offspring, potentially through the process of meiotic drive (Kern et al. 2015; Kageyama et al. 2017). Antibiotic treatment of wFem+ females results in male-only offspring, indicating that the female-specific W chromosome has been lost in wFem+ linages generating ZO individuals. This in turn strongly suggests that the loss of the W chromosome in wFem+ individuals (ZO males) are compensated for by *Wolbachia*-induced conversion of sex determination (Kageyama et al. 2017). Intriguingly, in female-heterospecific species, the W chromosome and cytoplasmic agents are inherited as a single linkage group, and hence it will be challenging to separate the impact of distortion of the sex chromosome from distortion caused by selfish cytoplasmic agents (Hurst et al. 2002).

Male Killing

Finally, endosymbionts may induce male killing to enhance their own evolutionary success. Here, male killing is favored when females gain from the death of their brothers, for example by reducing competition over resources or reducing the risk of inbreeding. For the causal male-killing endosymbionts, sacrificing themselves by killing their male hosts similarly benefits their clonal relatives that inhabit these female siblings, or—in cases where endosymbionts are also horizontally transmitted—their nest mates (e.g., Parratt et al. 2016). Theoretical models have shown that male-killing bacteria can have dramatic population-level impacts, in addition to distorting the sex ratio, by impeding the spread of beneficial alleles, facilitating the spread of deleterious alleles, and reducing genetic variation (Engelstädter and Hurst 2007). This is due to the low fitness of infected females, along with the severe reduction or total lack of gene flow between infected and uninfected individuals, both of which result in reduced effective population size (Engelstädter and Hurst 2007).

Male-killing endosymbionts are taxonomically diverse and include members of the bacteria *Wolbachia, Rickettsia, Spiroplasma, Arsenophonus,* and *Flavobacteria,* as well as some RNA viruses. Male killers are found in a range of insects and arthropods that lay their eggs in clutches, such as beetles, butterflies and moths, flies, hymenopteran wasps, lacewings, ticks, and a pseudoscorpion (Hornett et al. 2022). Male killing has been extensively studied, revealing a diversity of mechanisms and targeted genes, many of which are associated with sex determination. In general, it is thought that dosage-compensation failure during development in insects results in sex-specific lethality and hence male killing. Dosage-compensation mechanisms are therefore ideal targets for symbiont-induced sexual manipulation. Symbionts can directly target the sex-specific alternative splicing of *dsx*, which is widely conserved across insect taxa. Alternatively, genes that are upstream of *dsx* in the sex-determination cascade could be targeted, but these exhibit a lower level of conservation than *dsx*. For example, Lepidoptera and Diptera do not share any known sex-determining genes other than the widely conserved *dsx* gene (Beukeboom 2012). In Lepidoptera, male killing involves targeting genes involved in sex determination, such as *Masculinizer* (*Masc*). *Masc* is required for masculinization and dosage compensation, as it is needed to induce the male-specific splice variant of *dsx* (Katsuma et al. 2022). In the Asian corn borer (*Ostrinia furnicalis*) and the silkworm (*Bombyx mori*), a *Wolbachia* protein called *Oscar* directly inhibits *Masc*-induced masculinization, resulting in male killing by also preventing dosage compensation. Furthermore, in *O. furnicalis* and *O. scapulalis,* removal of *Wolbachia* from male-killing strains results in female killing, instead of restoring the sex ratio to unity (Sugimoto and Ishikawa 2012; Fukui et al. 2015). This suggests that the feminizing gene in *Ostrinia* has been degraded. This suggestion is supported by the observation that in lines that have coevolved with *Wolbachia,* all *Wolbachia*-infected individuals, regardless of genetic sex, express the female version of *dsx,* and males die (Sugimoto and Ishikawa 2012). Using trans-infected *O. scapulalis* cell lines, a recent study demonstrated that male-killing *Wolbachia* enhance the female-specific splice variant of *dsx* while suppressing the male-specific variant, indicating that *Wolbachia* affect sex determination in vivo (Herran et al. 2023).

The available data so far indicate that bacteriophages are involved in the evolution and diversity of male-killing mechanisms (Arai et al. 2023a). The male-killing factor in moths is *Oscar,* a novel protein that contains large number of ankyrin repeats (Katsuma et al. 2022). A different factor, *wmk,* is the male-killing agent found in other insects such as flies (Perlmutter et al. 2019). So far, *Oscar* homologs have only been identified in male-killing *Wolbachia* in butterflies and moths (i.e., *w*Fur, *w*Bol, and *w*Hm-t), suggesting that the mechanisms of *Wolbachia*-induced male killing likely differ between taxa (Arai et al. 2023a). *Oscar* is unique to endosymbionts infecting lepidopteran insects and does not show any sequence homology to *wmk* (Katsuma et al. 2022). In general, the prophage WO appears to be an important factor for various *Wolbachia*-induced reproductive manipulations. WO encodes CI-inducing factors (CifA and

CifB), which inhibit normal embryonic development by preventing chromosome segregation after fertilization (Perlmutter et al. 2019). Male killing in the moth *O. furnacalis* is induced by the *Oscar* protein that contains a CifB C-terminus-like domain and ankyrin repeats (Katsuma et al. 2022). The *w*Fur and *w*Sca genomes in *O. furnicalis* and *O. scapulalis* encode greater than thirty ankyrin repeat–containing proteins, respectively (Katsuma et al. 2022). In *Homona magnanima* moths a 76 kbp prophage region is associated with male-killing *Wolbachia* that is a homolog to *Oscar* in *Ostrinia* moths and the *wmk* gene in *D. melanogaster* (Arai et al. 2023a). This pattern indicates that ankyrin repeat–containing proteins responsible for male killing have evolved rapidly in *Wolbachia* (Arai et al. 2023a). Some suggest that an evolutionary arms race between *Wolbachia* and its host have resulted in the increased diversity of genes associated with male killing (i.e., *wmk* gene pairs in flies and *Oscar* in moths) in the *Wolbachia* genome (Hornett et al. 2022; Arai et al. 2023a).

Male killing in Diptera also involves targeting the sex-determination and dosage-compensation system. In general, *Spiroplasma*-induced male killing in flies involves abnormal male development, apoptosis, and disruption of the male nervous tissue. In *D. melanogaster,* male killing is caused by a strain of *Spiroplasma* that targets the dosage-compensation system (Veneti et al. 2005). Male-killing *Spiroplasma* can also target and damage the DNA on the dosage-compensated X chromosome in males that interacts with the functional dosage-compensation system (Harumoto et al. 2016). Further studies have shown that one gene is sufficient for male killing involving a plasmid gene *spaid* (*S. poulsonii* androcidin) variant that encodes a specific ankyrin-repeat protein. In transgenic systems, *spaid* mediates its male-killing effects through the dosage-compensation system (Harumoto and Lemaitre 2018). There are some similarities in *Wolbachia*-induced male killing in flies that show signs of abnormal male development, apoptosis, and damage to the dosage-compensated X chromosome in males. However, male neural development is normal (Harumoto and Lemaitre 2018). The candidate gene in *Wolbachia* for male killing in flies is *wmk* (WO-mediated killing), a likely transcriptional regulator encoded on prophage WO (Perlmutter et al. 2019), with the male-killing phenotype being determined by a single silent nucleotide (Perlmutter et al. 2021).

Viruses have also been shown to induce male killing, but it is less clear what transmission benefit biasing the sex ratio toward females would have for viruses, if it indeed benefits transmission at all. However, it is worth noting that phages (viruses) in bacteria have been shown to be the agents that bias the sex ratio. The bacteriophage, for example, is a critical driver of the evolution of male-killing *Wolbachia*. It is unclear whether the bacteria have co-opted the male-killing effects of the viruses to promote their own transmission or integrated the phages symbiotically such that their transmission success is now aligned. Some viruses are maternally inherited and hence favor male killing. For example, the maternally inherited, double-stranded RNA (dsRNA) virus belonging to the family Partitiviridae (DbMKPV1) induces male killing in *Drosophila* (Kageyama et al. 2023). Similarly, a maternally transmitted Osugoroshi virus (OGV), belonging to the family Partitiviridae, induces male killing in the moth *Homona magnanima* (Fujita et al. 2021).

Some taxa are infected by multiple male-killing agents. For example, the moth *H. magnanima* harbors two male-killing bacteria: *Wolbachia* and *Spiroplasma*, in addition to a larval male-killing OGV (Arai et al. 2023b). These agents use different mechanisms of male killing. *Wolbachia* and *Spiroplasma* disrupt the sex-determination cascade of males by inducing female-type splice variants of *dsx*. However, male-killing *Wolbachia* and *Spiroplasma* alter *dsx* splicing using different processes. *Wolbachia* impairs the dosage-compensation system, whereas *Spiroplasma* and Osugoroshi viruses do not. Moreover, *Wolbachia* and *Spiroplasma*, but not Osugoroshi viruses, trigger abnormal apoptosis in male embryos. The diversity of mechanisms used by male-killing agents in the same host illustrates the wide variety of ways selfish agents can induce the male-killing host phenotype.

General Discussion

In this chapter, we viewed the evolution of sex and heredity through a paradox-of-the-organism lens, revealing how intra-genomic conflict is a key driver of evolutionary change in mechanisms of inheritance and sex determination. Although both processes are essential to the generation of seemingly coherent organisms, this does not mean that they occur peacefully. Nor does it imply that they evolve in a manner that is wholly beneficial at an organismal level. Instead, internal conflicts promote selfish adaptations among different genetic elements that distort these processes to their benefit, resulting in mechanisms of heredity and sex determination that may be liable to evolutionary change at any time.

Sex Determination, Selfish Elements, and Their Reciprocal Evolution

The sex-specific inheritance of many genetic elements imbues them with a selfish interest to distort the sexual development of an organism. Similarly, the sex-specific action of other selfish genetic elements may lead them to be selected to interfere with sex determination. When successful, the resulting sex ratio distortion causes selection to favor mutations that restore the sex ratio. In some cases, these can be suppressors of the selfish genetic element; in others, a novel sex-determining gene that is impervious to the action of the selfish genetic element may arise. Consequently, sex-determination evolution may reflect evolutionary responses to selfish genetic elements, and internal conflict therefore may strongly contribute to the diversity and rapid evolution of these mechanisms. The interplay between how sex is currently determined, how this process can be exploited by selfish genetic elements, and what evolutionary changes can negate this exploitation is an active area of investigation. As discussed, sex ratio-distorting cytoplasmic elements target a variety of host genetic pathways and therefore present different possibilities for counteradaptations by the host to restore the sex ratio to unity. Frequently, sex-ratio-biasing genomic parasites exploit conserved genetic pathways like sex determination and target many different genes in this pathway. However, not all sex-determination mechanisms lend themselves to all forms of reproductive parasitism, and vice versa. For example, sex-chromosome meiotic drive might be harder to evolve in XO/ZO systems because X/Z-chromosomal killer loci lack a functional target locus on their O-chromosomal counterpart. Hence, only toxin-antidote systems may achieve segregation distortion in such systems, and sex ratio biases can only lead to an increase in females in XO and males in ZO systems, respectively. Similarly, as the O chromosomes are characterized by the absence of genetic material, they cannot themselves carry any segregation-distorting genes, which ought to negate the possibility that these "genetic elements" interfere with fair Mendelian segregation. A deeper understanding of the diversity of sex-determination mechanisms and their susceptibility to exploitation will be essential to more quantitative inquiries into the evolutionary impact of selfish genetic elements on host sex determination. In turn, this information will vastly expand our ability to convert this understanding into applications in pest control, insect breeding, and more.

As with all selfish genetic elements, cytoplasmic agents and segregation distorters cause intra-genomic conflict that promotes antagonistic coevolution between nuclear genes and suppressors, potentially resulting in population divergence and incompatibility and fertility reduction in crosses between populations (Werren 2011; Verspoor et al. 2018). Importantly, the intra-genomic conflict generated by selfish genetic elements has the potential to promote new sex-determination systems, as discussed above (Meiklejohn and Tao 2009; Hornett et al. 2022). While the role of intra-genomic conflict stemming from sexually antagonistic alleles in shaping sex-determination system have been extensively studied (Pomiankowski et al. 2004; Muralidhar and Veller 2018; Hitchcock and Gardner 2020), the role of selfish genetic elements has garnered less attention—except, perhaps, for sex-linked segregation distorters (see Cocquet et al. 2012). Yet, there are similarities between the intra-genomic conflict generated by sexually antagonistic alleles and conflicts caused by selfish genetic elements. For example, models of sexual antagonism reveal that the masculinizing and feminizing effects of specific sexually antagonistic alleles promote rapid antagonistic coevolution favoring the emergence of novel sex-determination systems—a feature they share with sex-ratio-distorting selfish genetic elements (Wedell 2020). In addition, sexually antagonistic alleles can result in reduced recombination between the sex chromosomes (Rice 1987), something that is also predicted to occur when involving selfish genetic elements that affect sex determination (see below).

Inheritance Under Internal Conflict

As we have discussed, there is ample evidence of genetic conflict driving evolutionary changes in the mechanisms and modes of inheritance patterns in diploid organisms, including major evolutionary transitions like the separation of germline and soma. Intra-genomic conflicts have also played a key role in the near-universal pattern of uniparental inheritance of cytoplasmic elements, such as mitochondria and chloroplasts in eukaryotes, which has profound consequences for ongoing conflict between nuclear and cytoplasmic genes, as well as the individuals that inherit them. Elimination of nuclear DNA is an integral part of the life cycle of many organisms. Often involving only one sex, such elimination has its roots in genetic conflict and promotes ongoing internal conflicts through a variety of mechanisms. These mechanisms extend the impact of selfish genetic elements on sex determination and exert a powerful influence on inheritance. A similar pattern emerges here of continuous coevolution between selfish genetic elements and their suppressors, so that the transmission of genetic material may at any point be subject to evolutionary change. For example, in *D. simulans,* which harbors an X-linked meiotic driver, there is a network of X-linked genes of the *Distorter on the X (Dox)* that is suppressed by multiple small interfering RNA (siRNA) genes (Vedanayagam et al. 2023a). Moreover, recent findings show that many previously unknown genes that are silenced by the RNA interference (RNAi) pathway in *D. simulans* have recently evolved and have a potential role in the conflict between the sex chromosomes. This suggests that an ongoing coevolutionary cycle between distorters and suppressors shapes sex chromosome evolution (Vedanayagam et al. 2023b). Similarly, in *D. melanogaster,* sperm nuclear basic proteins (protamines) are involved with packaging sperm and also appear to suppress selfish DNA. Comparative analyses have revealed that young protamines on the sex chromosomes enhance their own success through meiotic drive, whereas autosomal protamines silence the selfishly driving ones. Such conflicts between the sex chromosomes and the autosomes cause protamine function during spermatogenesis to evolve rapidly across *Drosophila* (Chang et al. 2023). By unraveling the coevolutionary molecular basis of selfish genes' capacity for manipulating sex and reproduction, these newly emerging data illustrate just how potent selfish genes are in shaping sex determination in general.

Sex chromosome differentiation relies on the process of suppressing the recombination of emerging sex chromosomes. Several models have explored how the accumulation of sexually antagonistic alleles near the sex-determining region can promote suppression of recombination (Rice 1987). Reduced recombination can lead to degeneration in nonrecombining regions and to the emergence of heterochromatic sex chromosomes (reviewed in Bachtrog 2013). However, the possibility that selfish genetic elements, such as meiotic drivers that promote reduced recombination (often present in non-recombining inversions), could serve as an analogous agent to sexual antagonism in favoring sex chromosome divergence has yet to be explored. Recent models have shown that segregation distorters might facilitate the spread of novel sex-determining genes through linkage (Kozielska et al. 2010; Úbeda et al. 2014). But the extent to which selfish genetic elements specifically target regions of reduced recombination, such as sex chromosomes, as opposed to being an attractor for novel sex-determining genes due to their association with reduced recombination, has yet to be determined. In both cases the outcome is an association between heterochromatic sex chromosomes and suppressed recombination. However, not all sex chromosomes show extensive differentiation, possibly due to occasional recombination (Perrin 2009), or due to rapid turnover of sex determination and hence emergence of new sex chromosomes (Smith et al. 2023). If selfish genetic elements are more often associated with non-differentiated (young) sex chromosomes, this would support the idea that selfish genetic elements play a role in promoting novel sex-determination systems. Complex sex-determination systems (e.g., X_1X_2Y, XY_1Y_2, or Z_1Z_2W) can arise because of fission, fusion, and translocation between sex chromosomes (Roco et al. 2015). It is an outstanding question whether selfish genetic elements play a role in this process, but transposable elements and other selfish genetic elements are known to be a major contributor to novel chromosome rearrangement (Werren 2011; Klein and O'Neill 2018).

Organismal Blueprints in Light of Intra-genomic Conflict

In examining internal conflicts and selfish genetic elements, we find that core components of the organismal blueprint—inheritance of genetic material and development into reproductively functional males and females—are distorted in a bewildering variety of ways. By compromising the organism in this manner, selfish genetic elements drive evolutionary changes in how organisms are generated. With different blueprints providing different opportunities for selfish elements to exploit, organismal ontogenies and the selfish genetic elements that they engender engage in continuous coevolution, much like the various elements within organisms themselves. To conclude, we can view organisms simultaneously as coherent entities and as aggregates of warring factions. Similarly, we can see the emergence of organisms as a process aimed at producing a functionally cohesive whole, but also one that is riven by conflict—much in the spirit of the paradox of the organism.

References

Ågren, J. A., and Clark, A. G. (2019). Selfish genetic elements. *PLoS Genetics*, 14(11), e1007700.

Ahmad, S. F., and Martins, C. (2019). The modern view of B chromosomes under the impact of high scale omics analyses. *Cells, 8*(2), 156.

Arai, H., Anbutsu, H., Nishikawa, Y., et al. (2023a). Combined actions of bacteriophage-encoded genes in *Wolbachia*-induced male lethality. *iScience, 26*(6), 106842.

Arai, H., Takamatsu, T., Lin, S. R., et al. (2023b). Diverse molecular mechanisms underlying microbe-inducing male killing in the moth *Homona magnanima*. *Applied and Environmental Microbiology, 89*(5), 1–15.

Bachtrog, D. (2013). Y-chromosome evolution: Emerging insights into processes of Y-chromosome degeneration. *Nature Reviews Genetics, 14*(2), 113–124.

Beukeboom, L. W. (2012). Microbial manipulation of host sex determination. *BioEssays, 34*(6), 484–488.

Borodin, P., Chen, A., Forstmeier, W., et al. (2022). Mendelian nightmares: The germline-restricted chromosome of songbirds. *Chromosome Research, 30*(2–3), 255–272.

Boveri, T. (1895). Über die Befruchtungs- und Entwicklungsfähigkeit Kernloser Seeigeleir und über die Möglichkeit ihrer Bastardierung. *Archiv für Entwicklungsmechanik, 2*, 394–443, plates xxiv–xxv.

Brockhurst, M. A., Chapman T., King K. C., Mank J. E., Paterson S., and Hurst G. D. D. (2014). Running with the Red Queen: The role of biotic conflicts in evolution. *Proceedings of the Royal Society B, 281*(1797), 20141382.

Bull, J. J. (1979). Advantage for the evolution of male haploidy and system with similar genetic transmission. *Heredity, 43*(3), 361–381.

Burga, A., Ben-David, E., and Kruglyak, L. (2020). Toxin-antidote elements across the tree of life. *Annual Review of Genetics, 54*, 387–415.

Chang, C. H., Natividad, I. M., and Malik, H. S. (2023). Expansion and loss of sperm nuclear basic protein genes in *Drosophila* correspond with genetic conflicts between sex chromosomes. *eLife 12*, 1–30.

Charlat, S., Hornett, A. A., Dyson, P. P. Y., et al. (2005). Prevalence and penetrance variation of male-killing *Wolbachia* across Indo-Pacific populations of the butterfly *Hypolimnas bolina*. *Molecular Ecology, 14*(11), 3525–3530.

Cocquet, J., Ellis, P. J. I., Mahadevaiah, S. K., Affara, N. A., Vaiman, D., and Burgoyne P. S. (2012). A genetic basis for a postmeiotic X versus Y chromosome intragenomic conflict in the mouse. *PLoS Genetics, 8*(9), e1002900.

Coleman, A. W. (1982). Sex is dangerous in a world of potential symbionts or the basis of selection for uniparental inheritance. *Journal of Theoretical Biology, 97*(3), 367–369.

Cosmides, L. M., and Tooby, J. (1981). Cytoplasmic inheritance and intragenomic conflict. *Journal of Theoretical Biology, 89*, 83–129.

Courret, C., Chang, C. H., Wei, K. H. C., Montchamp-Moreau, C., and Larracuente, A. M. (2019). Meiotic drive mechanisms: Lessons from *Drosophila*. *Proceedings of the Royal Society B, 286*(1913).

David, P., Degletagne C., Saclier, N., et al. (2022). Extreme mitochondrial DNA divergence underlies genetic conflict over sex determination. *Current Biology, 32*(10), 2325–2333.e6.

Dedukh, D., and Krasikova, A. (2022). Delete and survive: Strategies of programmed genetic material elimination in eukaryotes. *Biological Reviews, 97*(1), 195–216.

Doorn, G. S. van. 2014. Evolutionary transitions between sex-determining mechanisms: A review of theory. *Sexual Development, 8*(1–3): 7–19.

Duron, O., Bouchon, D., Boutin, S., et al. (2008). The diversity of reproductive parasites among arthropods: *Wolbachia* do not walk alone. *BMC Biology, 6*, 1–12.

Engelstädter, J., and Hurst, G. D. D. (2007). The impact of male-killing bacteria on host evolutionary processes. *Genetics, 175*(1), 245–254.

Finseth, F. (2023). Female meiotic drive in plants: Mechanisms and dynamics. *Current Opinion in Genetics and Development, 82*, 102101.

Fisher, R. A. (1930). *The Genetical Theory of Natural Selection*. Oxford University Press.

Fraïsse, C., Picard, M. A. L., and Vicoso, B. (2017). The deep conservation of the *Lepidoptera* Z chromosome suggests a non-canonical origin of the W. *Nature Communications, 8*(1).

Fujita, R., Inoue, M. N., Takamatsu, T., et al. (2021). Late male-killing viruses in *Homona magnanima* identified as Osugoroshi viruses, novel members of *Partitiviridae*. *Frontiers in Microbiology, 11*, 1–10.

Fukui, T., Kawamoto, M., Shoji, K., et al. (2015). The endosymbiotic bacterium *Wolbachia* selectively kills male hosts by targeting the masculinizing gene. *PLoS Pathogens, 11*(7), 1–14.

Gardner, A., and Ross, L. (2014). Mating ecology explains patterns of genome elimination. *Ecology Letters, 17*(12), 1602–1612.

Gardner, A., and Úbeda, F. (2017). The meaning of intragenomic conflict. *Nature Ecology and Evolution, 30*, 1807–1815.

Gemmell, N. J., Metcalf, V. J., and Allendorf, F. W. (2004). Mother's curse: The effect of mtDNA on individual fitness and population viability. *Trends in Ecology and Evolution, 19*(5), 238–244.

Geuverink, E., and Beukeboom, L. W. (2014). Phylogenetic distribution and evolutionary dynamics of the sex determination genes *doublesex* and *transformer* in insects. *Sexual Development, 8*(1–3), 38–49.

Hackstein, J. H. P., Hochstenbach, R., Hauschteck-Jungen, E., and Beukeboom, L. W. (1996). Is the Y chromosome of *Drosophila* an evolved supernumerary chromosome? *BioEssays, 18*(4), 317–323.

Haig, D. (2016). Transposable elements: Self-seekers of the germline, team-players of the soma. *BioEssays, 38*(11), 1158–1166.

Haig, D. (2021). Concerted evolution of ribosomal DNA: Somatic peace amid germinal strife. *BioEssays, 43*(12), 2100179.

Hamilton, W. D. (1967). Extraordinary sex ratios. *Science, 156*, 477–488.

Harumoto, T., Anbutsu, H., Lemaitre, B., and Fukatsu, T. (2016). Male-killing symbiont damages host's dosage-compensated sex chromosome to induce embryonic apoptosis. *Nature Communications, 7*, 12781.

Harumoto, T., and Lemaitre, B. (2018). Male-killing toxin in a bacterial symbiont of *Drosophila*. *Nature, 557*(7704), 252–255.

Havird, J. C., Forsythe, E. S., Williams, A. M., Werren, J. H., Dowling, D. K., and Sloan, D. B. (2019). Selfish mitonuclear conflict. *Current Biology, 29*(11), R496–R511.

Herbette, M., and Ross, L. (2023). Paternal genome elimination: Patterns and mechanisms of drive and silencing. *Current Opinion in Genetics and Development, 81,* 102065.

Herran, B., Geniez, S., Delaunay, C., et al. (2020). The shutting down of the insulin pathway: A developmental window for *Wolbachia* load and feminization. *Scientific Reports, 10*(1), 1–9.

Herran, B., Sugimoto, T. N., Watanabe, K., et al. (2023). Cell-based analysis reveals that sex-determining gene signals in *Ostrinia* are pivotally changed by male-killing *Wolbachia*. *PNAS Nexus, 2*(1), 1–10.

Hitchcock, T. J., and Gardner, A. (2020). A gene's-eye view of sexual antagonism. *Proceedings of the Royal Society B, 287*(1932), 20201633.

Hodson, C. N., and Ross, L. (2021). Evolutionary perspectives on germline-restricted chromosomes in flies (*Diptera*). *Genome Biology and Evolution, 13*(6), 1–19.

Hornett, E. A., Kageyama, D., and Hurst, G. D. D. (2022). Sex determination systems as the interface between male-killing bacteria and their hosts. *Proceedings of the Royal Society B, 289*(1972).

Hurst, G. D. D., Jiggins, F. M., and Pomiankowski, A. 2002. Which way to manipulate host reproduction? *Wolbachia* that cause cytoplasmic incompatibility are easily invaded by sex ratio-distorting mutants. *The American Naturalist, 160*(3), 360–373.

Imarazene, B., Du, K., Beille, S., et al. (2021). A supernumerary "B-sex" chromosome drives male sex determination in the Pachón cavefish, *Astyanax mexicanus*. *Current Biology, 31*(21), 4800–4809.e9.

Jaenike, J. (2001). Sex chromosome meiotic drive. *Annual Review of Ecology and Systematics, 32,* 25–49.

Johnson, L. J. (2008). Selfish genetic elements favor the evolution of a distinction between soma and germline. *Evolution, 62*(8), 2122–2124.

Johnson Pokorná, M., and Reifová, R. (2021). Evolution of B chromosomes: From dispensable parasitic chromosomes to essential genomic players. *Frontiers in Genetics, 12,* 1–11.

Kageyama, D., Harumoto, T., Nagamine, K., et al. (2023). A male-killing gene encoded by a symbiotic virus of *Drosophila*. *Nature Communications, 14*(1).

Kageyama, D., Ohno, M., Sasaki, T., et al. (2017). Feminizing *Wolbachia* endosymbiont disrupts maternal sex chromosome inheritance in a butterfly species. *Evolution Letters, 1*(5), 232–244.

Katsuma, S., Hirota, K., Matsuda-Imai, N., et al. (2022). A *Wolbachia* factor for male killing in Lepidopteran insects. *Nature Communications, 13*(1), 1–12.

Kern, P., Cook, J. M., Kageyama, D., and Riegler, M. (2015). Double trouble: Combined action of meiotic drive and *Wolbachia* feminization in *Eurema* butterflies. *Biology Letters, 11*(5).

Klein, S. J., and O'Neill, R. J. (2018). Transposable elements: Genome innovation, chromosome diversity, and centromere conflict. *Chromosome Research, 26*(1–2), 5–23.

Kozielska, M., Weissing, F. J., Beukeboom, L. W., and Pen, I. (2010). Segregation distortion and the evolution of sex-determining mechanisms. *Heredity, 104*(1), 100–112.

Lewis, J. J., Cicconardi, F., Martin, S. H., Reed, R. D., Danko, C. G., and Montgomery, S. H. (2021). The *Dryas iulia* genome supports multiple gains of a W chromosome from a B chromosome in butterflies. *Genome Biology and Evolution, 13*(7), 1–13.

Li, X., Visser, S., Son, J. H., et al. (2024). Divergent evolution of male-determining loci on proto-Y chromosomes of the house fly. *Nature Communications, 15*, 5984.

Ma, W. J., and Schwander, T. (2017). Patterns and mechanisms in instances of endosymbiont-induced parthenogenesis. *Journal of Evolutionary Biology, 30*(5), 868–888.

Malik, H. S. (2022). Driving lessons: A brief (personal) history of centromere drive. *Genetics, 222*(4).

McKinney, G. J., Nichols, K. M., and Ford, M. J. (2021). A mobile sex-determining region, male-specific haplotypes and rearing environment influence age at maturity in chinook salmon. *Molecular Ecology, 30*(1), 131–147.

Meiklejohn, C. D., and Tao, Y. (2009). Genetic conflict and sex chromosome evolution. *Trends in Ecology and Evolution, 25*(4), 215–223.

Metz, C. W. (1926). Genetic evidence of selective segregation of chromosomes in Sciara (Diptera). *Proceedings of the National Academy of Sciences of the United States of America, 12*(12), 690–692.

Michod, R. E. (2006). The group covariance effect and fitness trade-offs during evolutionary transitions in individuality. *Proceedings of the National Academy of Sciences of the United States of America, 103*(24), 9113–9117.

Michod, R. E. (2007). Evolution of individuality during the transition from unicellular to multicellular life. *Proceedings of the National Academy of Sciences of the United States of America, 104*(S1), 8613–8618.

Muralidhar, P., and Veller, C. (2018). Sexual antagonism and the instability of environmental sex determination. *Nature Ecology and Evolution, 2*, 343–351.

Negri, I., Franchini, A., Gonella, E., et al. (2009). Unravelling the *Wolbachia* evolutionary role: The reprogramming of the host genomic imprinting. *Proceedings of the Royal Society B, 276*(1666), 2485–2491.

Negri, I., Pellecchia, M., Mazzoglio, P. J., Patetta, A., and Alma, A. (2006). Feminizing *Wolbachia* in *Zyginidia pullula* (Insecta, Hemiptera), a leafhopper with an XX/XO sex-determination system. *Proceedings of the Royal Society B,* 273(1599), 2409–2416.

Normark, B. B. (2003). The evolution of alternative genetic systems in insects. *Annual Review of Entomology, 48,* 397–423.

Normark, B. B., and Ross, L. (2014). Genetic conflict, kin and the origins of novel genetic systems. *Philosophical Transactions of the Royal Society B,* 369(1642).

Nur, U., Werren, J. H., Eickbush, D. G., Burke, W. D., and Eickbush, T. H. (1988). A "selfish" B chromosome that enhances its transmission by eliminating the paternal genome. *Science, 240*(4851), 512–514.

Östergren, B. (1945). Parasitic nature of extra fragment chromosomes. *Botaniska Notiser, 2,* 157–163.

Parratt, S. R., Frost, C. L., Schenkel, M. A., Rice, A., Hurst, G. D. D., and King K. C. (2016). Superparasitism drives heritable symbiont epidemiology and host sex ratio in a wasp. *PLoS Pathogens, 12*(6), e1005629.

Partridge, L., and Hurst, L. D. (1998). Sex and conflict. *Science, 281*(5385), 2003–2008.

Patel, M. R., Miriyala, G. K., Littleton, A. J., et al. (2016). A mitochondrial DNA hypomorph of cytochrome oxidase specifically impairs male fertility in *Drosophila melanogaster. eLife, 5,* 1–27.

Patten, M. M., Schenkel, M. A., and Ågren, J. A. (2023). Adaptation in the face of internal conflict: The paradox of the organism revisited. *Biological Reviews, 98*(5), 1796–1811.

Perlman, S. J., Hodson, C. N., Hamilton, P. T., Opit, G. P., and Gowen B. E. (2015). Maternal transmission, sex ratio distortion, and mitochondria. *Proceedings of the National Academy of Sciences of the United States of America, 112*(33), 10162–10168.

Perlmutter, J. I., Bordenstein, S. R., Unckless, R. L., et al. (2019). The phage gene *wmk* is a candidate for male killing by a bacterial endosymbiont. *PLoS Pathogens, 15*(9), 1–29.

Perlmutter, J. I., Meyers, J. E., and Bordenstein, S. R. (2021). A single synonymous nucleotide change impacts the male-killing phenotype of prophage WO gene *wmk. eLife, 10,* 1–20.

Perrin, N. (2009). Sex reversal: A fountain of youth for sex chromosomes? *Evolution, 63*(12), 3043–3049.

Pomiankowski, A., Nöthiger, R., and Wilkins, A. (2004). The evolution of the *Drosophila* sex-determination pathway. *Genetics, 166*(4), 1761–1773.

Rice, W. R. (1987). The accumulation of sexually antagonistic genes as a selective agent promoting the evolution of reduced recombination between primitive sex chromosomes. *Evolution, 41*(4), 911–914.

Roco, Á. S., Olmstead, A. W., Degitz, S. J., Amano, T., Zimmerman, L. B., and Bullejos, M. (2015). Coexistence of Y, W, and Z sex chromosomes in *Xenopus tropicalis. Proceedings of the National Academy of Sciences of the United States of America, 112*(34), E4752–E4761.

Saunders, P. A., Perez, J., Ronce, O., and Veyrunes, F. (2022). Multiple sex chromosome drivers in a mammal with three sex chromosomes. *Current Biology, 32*(9), 2001–2010.e3.

Schenkel, M. A., Billeter, J.-C., Beukeboom, L. W., and Pen, I. (2023). Divergent evolution of genetic sex determination mechanisms along environmental gradients. *Evolution Letters, 7*(April), 132–144.

Schnable, P. S., and Wise, R. P. (1998). The molecular basis of cytoplasmic male sterility and fertility restoration. *Trends in Plant Science, 3*(5), 175–180.

Seah, B. K. B., and Swart, E. C. (2023). When cleaning facilitates cluttering—Genome editing in ciliates. *Trends in Genetics, 39*(5), 344–346.

Sellis, D., Guérin, F., Arnaiz, O., et al. (2021). Massive colonization of protein-coding exons by selfish genetic elements in *Paramecium* germline genomes. *PLoS Biology, 19*(7), 1–37.

Sharma, A., Heinze, S. D., Wu, Y., et al. (2017). Male sex in houseflies is determined by *mdmd,* a paralog of the generic splice factor gene *CWC22. Science, 356*(6338), 642–645.

Smith, J. J., Timoshevskaya, N., Ye, C., et al. (2018). The sea lamprey germline genome provides insights into programmed genome rearrangement and vertebrate evolution. *Nature Genetics, 50*(2), 270–277.

Smith, J. J., Timoshevskiy, V. A., and Saraceno, C. (2021). Programmed DNA elimination in vertebrates. *Annual Review of Animal Biosciences, 9,* 173–201.

Smith, S. H., Hsiung, K., and Böhne, A. (2023). Evaluating the role of sexual antagonism in the evolution of sex chromosomes: New data from fish. *Current Opinion in Genetics and Development, 81,* 102078.

Sobanski, J., Giavalisco, P., Fischer, A., et al. (2019). Chloroplast competition is controlled by lipid biosynthesis in evening primroses. *Proceedings of the National Academy of Sciences of the United States of America, 116*(12), 5665–5674.

Stouthamer, R., Van Tilborg, M., De Jong, J. H., Nunney, L., and Luck, R. F. (2001). Selfish element maintains sex in natural populations of a parasitoid wasp. *Proceedings of the Royal Society B, 268*(1467), 617–622.

Sugimoto, T. N., and Ishikawa, Y. (2012). A male-killing *Wolbachia* carries a feminizing factor and is associated with degradation of the sex-determining system of its host. *Biology Letters, 8*(3), 412–415.

Tennessen, J. A., Wei, N., Straub, S. C. K., Govindarajulu, R., Liston, A., and Sashman, T. L. (2018). Repeated translocation of a gene cassette drives sex-chromosome turnover in strawberries. *PLoS Biology, 16*(8), e2006062.

Togashi, T., Parker, G. A., and Horinouchi, Y. (2023). Mitochondrial uniparental inheritance achieved after fertilization challenges the nuclear-cytoplasmic conflict hypothesis for anisogamy evolution. *Biology Letters, 19*(9), 20230352.

Úbeda, F., Patten, M. M., and Wild, G. (2014). On the origin of sex chromosomes from meiotic drive. *Proceedings of the Royal Society B, 282*(1798), 20141932.

Uller, T., Pen, I., Wapstra, E., Beukeboom, L. W., and Komdeur, J. (2007). The evolution of sex ratios and sex-determining systems. *Trends in Ecology and Evolution, 22*(6), 292–297.

Vedanayagam, J., Herbette, M., Mudgett, H., et al. (2023a). Essential and recurrent roles for hairpin RNAs in silencing de novo sex chromosome conflict in *Drosophila simulans. PLoS Biology, 21*(6), 1–23.

Vedanayagam, J., Lin, C. J., Papareddy, R., et al. (2023b). Regulatory logic of endogenous RNAi in silencing de novo genomic conflicts. *PLoS Genetics, 19*(6), 1–26.

Veneti, Z., Bentley, J. K., Koana, T., Braig, H. R., and Hurst, G. D. D. (2005). A functional dosage compensation complex required for male killing in *Drosophila. Science, 307*(5714), 1461–1463.

Verhulst, E. C., Pannebakker, B. A., and Geuverink, E. (2023). Variation in sex determination mechanisms may constrain parthenogenesis-induction by endosymbionts in haplodiploid systems. *Current Opinion in Insect Science, 56,* 101023.

Verhulst, E. C., van de Zande, L., and Beukeboom, L. W. (2010). Insect sex determination: It all evolves around *transformer. Current Opinion in Genetics and Development, 20*(4), 376–383.

Verspoor, R. L., Smith, J. M. L., Mannion, N. L. M., Hurst, G. D. D., and Price, T. A. R. (2018). Strong hybrid male incompatibilities impede the spread of a selfish chromosome between populations of a fly. *Evolution Letters, 2*(3), 169–179.

Vontzou, N., Pei, Y., Mueller, J. C., et al. (2023). Songbird germline-restricted chromosome as a potential arena of genetic conflicts. *Current Opinion in Genetics and Development, 83,* 102113.

Wang, J., Veronezi, G. M. B., Kang, Y., Zagoskin, M., O'Toole, E. T., and Davis, R. E. (2020). Comprehensive chromosome end remodeling during programmed DNA elimination. *Current Biology, 30*(17), 3397–3413.

Wedell, N. (2020). Selfish genes and sexual selection: The impact of genomic parasites on host reproduction. *Journal of Zoology, 311*(1), 1–12.

Werren, J. H. (1991). The paternal-sex-ratio chromosome of *Nasonia. The American Naturalist, 137*(3), 392–402.

Werren, J. H. (2011). Selfish genetic elements, genetic conflict, and evolutionary innovation. *Proceedings of the National Academy of Sciences of the United States of America, 108*(S2), 10863–10870.

Werren, J. H., Nur, U., and Wu, C. I. (1988). Selfish genetic elements. *Trends in Ecology and Evolution, 3*(11), 297–302.

Werren, J. H., Skinner, S. W., and Charnov, E. L. (1981). Paternal inheritance of a *daughterless* sex ratio factor. *Nature, 293*(5832), 467–468.

Werren, J. H., and Stouthamer, R. (2003). PSR (paternal sex ratio) chromosomes: The ultimate selfish genetic elements. *Genetica, 117*(1), 85–101.

Wilkins, A. S. (1995). Moving up the hierarchy: A hypothesis on the evolution of a genetic sex determination pathway. *BioEssays, 17*(1), 71–77.

Yoshida, K., and Kitano, J. (2012). The contribution of female meiotic drive to the evolution of neo-sex chromosomes. *Evolution, 66*(10), 3198–3208.

Zug, R., and Hammerstein, P. (2012). Still a host of hosts for *Wolbachia:* Analysis of recent data suggests that 40% of terrestrial arthropod species are infected. *PLoS One, 7*(6), 7–10.

PART III

APPLICATIONS OF CONFLICT

Clinical Applications of the Paradox of the Organism

AMY M. BODDY AND J. ARVID ÅGREN

ABSTRACT The success of medical interventions depends on how well we understand the biological phenomenon being treated. Adopting a paradox-of-the-organism perspective that views the human body as an adaptive compromise could transform how we study, treat, and cure diseases. This is especially true for maladies that involve (the potential for) conflict between interacting entities, such as cancer, and complications stemming from pregnancy, which can include microchimerism (i.e., the presence of cells from one individual in another genetically distinct individual). In this chapter, we review the traditional ways these diseases have been conceptualized and treated, and argue that an account that takes within-organism conflicts seriously suggests new and better routes to address all conditions.

Introduction

From the outside, organisms appear to have much in common with machines. From the inside, the picture is very different. The human body, for example, is a coalition of trillions of cells coordinating their activities in their common interest, rather than acting on their own (Reynolds 2007; Sender et al. 2016). Yet, ever since its origins a billion years ago, multi-cellular life has been crucially dependent on the suppression of cell-level agency in favor of organismal-level agency (Huxley 1912/2022). This is why phenomena like cancer and other forms of cellular cheating represent such a threat to organismal fitness (Aktipis 2020).

Functioning bodies require both internal cooperation and conflict suppression. The link between internal unity and the machine-like features of organisms is an old one (Riskin 2016). Thinkers like René Descartes (1596–1650) described animals as living machines, and theologians like the Dutchman Barnard Nieuwentyt (1654–1718) and the Englishman William Paley (1743–1805) made much use of the so-called watchmaker analogy (LeMahieu 1976). In this thought exepriment, we are asked to consider what makes a watch different from a stone. The answer, the argument goes, lays in that the watch is made up of many interlocking parts, each of which can only be understood in light of the overall purpose of the watch: to tell time. Plants and animals, Nieuwentyt and Paley suggested, are more like watches than stones. Take the human eye, with its retina, lens, and cornea. Clearly, they argued, these components fitting together only made sense given the purpose of concentrating light into something we can see. From this line of reasoning, two defining features of biological design can be identified (Gardner 2009; Lewens 2019). The first is contrivance. Like watches, organisms have traits that serve specific functions; they are contrived for a goal. The second is relation. This means that all parts serve the same goal: Organisms and machines are thus characterized by an internal unity of purpose so that all parts work together for the good of the organism (cf. Okasha 2018).

We now know that the appearance of design in nature needs no divine designer (Scott and West, this volume). Even Paley's favorite example of the human eye is a result of evolutionary constraints and compromises (Lents 2018). As currently structured, the organization of multiple layers of our retina results in a blind spot in the human eye (our brain works around this blind spot). In contrast, if one were to build an eye from scratch, they would build the photoreceptor layer on top of the retina and have no blind spot (as is the case in the cephalopod lineage). The historical legacy of the human eye's development has resulted in evolutionary compromises that may increase susceptibility to diseases, such as a detached retina.

Despite these known evolutionary constraints and design flaws, we can still find plenty of traces of machine rhetoric when contemporary researchers describe biological processes. Writing in the midst of the Industrial Revolution, Paley drew heavily on the modern technology of his time and his arguments are full of references to steam engines, watches, and telescopes (Gillespie 1990). Today, we still talk of the natural world in similar ways (Bongard and Levin 2021). For example, genomes have long been described as the blueprints for building bodies (e.g., Plomin 2019). An unstated assumption in these machine metaphors is the thought that all genes are coordinating their efforts in the interest of the body, just as the different components in a blueprint or the parts of the watch all serve the purpose of constructing a building or giving the time.

The paradox of the organism demonstrates that these metaphors can sometimes be misleading. Multicellular organisms are products of major transitions in individuality, where biological entities transitioned from surviving and reproducing on their own into doing so as part of new collective wholes (Maynard Smith and Szathmáry 1995; Bourke 2011; West et al. 2015). These transitions share two fundamental characteristics. The first is the emergence of cooperation among independent entities, resulting in the creation of a new higher-level individual entity. The second is the evolution of mechanisms to suppress conflict among lower-level entities, which is critical to the functioning of this new level of individuality. However, conflict suppression is never complete (Ågren et al. 2019). As a consequence, a prominent feature of multicellular organisms is that different parts often do not always work together for the common good of the organism (Patten et al. 2023). Instead, they are often in direct conflict with each other. The unity of purpose of the individual organism— what Paley called the relations of parts—is constantly threatened from the inside by entities like cancer cells and selfish genetic elements—that is, genes that can enhance their own odds of evolutionary success at the expense of other genes (Burt and Trivers 2006; Ågren and Clark 2018). Complex multicellular organisms are a consequence of billions of years of adaptive compromises, not well-designed machines.

Taking the paradox of the organism seriously has consequences for both evolutionary theory and more practical issues. One example of the latter is how we study and treat disease. Human diseases have traditionally been classified based on their underlying causes, usually into four types: infectious, genetic, lifestyle, and environmental diseases (World Health Organization 2019). While evolutionary approaches may help with all kinds of medical issues (Natterson-Horowitz et al. 2023), the paradox perspective highlights that many different forms of diseases have in common that they involve evolutionary agents that (have the potential to) come into conflict. Here, we take a paradox-of-the-organism perspective on three such examples: cancers, pregnancy, and microchimerism. We first show how in all cases clinical complications arise when conflict overpowers cooperation, but also how recognizing this central feature opens up novel approaches to treatment. We end by considering new questions that arise from approaching these human conditions in this way.

Cancer

Cancer is a group of diseases involving the rapid growth of abnormal cells that spread to areas beyond their usual location in the body (Weinberg 2023). Every year, cancers are responsible for some ten million deaths in humans worldwide (Ferlay et al. 2021). As a malaise, cancers are genotypically and phenotypically diverse, but oncologists talk of "hallmarks," a set of fundamental characteristics typically found in cancerous cells, that unify them (Hanahan and Weinberg 2000, 2011). These include the accumulation of mutations that can stimulate tumor growth, such as interference in the process of cell growth and death through continuous stimulation of cell division, the inactivation of molecules that would normally halt such divisions, and resistance to induced cell death. Cancer cells also tend to acquire a kind of replicative immortality through the maintenance of telomeres, which prevent cell senescence. Finally, many cancer cells promote the growth of new blood vessels to support their nutrient supply and acquire the ability to invade surrounding tissues and spread to distant organs through metastasis.

It has long been acknowledged that evolutionary theory has a lot to offer the study of cancer (Cairns 1975; Nowell 1976), but recent years have seen a renewed urgency to this insight (Greaves and Maley 2012; Aktipis and Nesse 2013; Boddy 2022; Strobl et al. 2023). One reason is that we now recognize that cancer is a prevalent disease across all multicellular life (Aktipis et al. 2015), and much work has been dedicated to understanding why some species are affected more than others (Compton et al. 2025). Cancer is a disease of cellular mutations that typically arise through cellular division. A common-sense evolutionary argument may lead us to expect that larger species with longer lifespans should go through more cell divisions than smaller species with shorter lifespans, and therefore have a higher rate of cancers. However, elephants and blue whales still exist, which would not be possible if they had the same rate of cancer mutations as humans (Caulin and Maley 2011). The general lack of correlation between body size and cancer rate has been referred to as Peto's paradox (Peto et al. 1975). There are several proposed solutions to this puzzle (Nunney et al. 2015). One example, in line with a paradox-of-the-organism mindset, is the argument that species with low rates of cancers have evolved better mechanisms of internal conflict mediation (Ågren et al. 2019). Supporting this is the observation that larger species appear to have more copies of cancer suppression genes, such as *TP53*, than smaller species (Abegglen et al. 2015; Sulak et al. 2016).

The field of evolutionary oncology is still in its infancy, but evolutionary theory may already inform our strategies for treating cancer. Traditionally, the goal of cancer therapy has been to eradicate as much of the tumor as possible without harming the patient with the treatment—whether drugs, radiation, or chemotherapy. The clinical practice known as the maximum tolerated dose paradigm is based on the idea that this approach will either eradicate the tumor, or at least keep the patient alive as long as possible (DeVita and Chu 2008). While this approach has resulted in great improvements in extending the quality life of patients, the problem with this strategy has been that—just like the evolution of pesticide resistance among crop pests—cancer treatments almost inevitably stop working (Gatenby et al. 2009) (Figure 8.1a). Today, drug resistance is a leading cause of death in cancer patients (Vasan et al. 2019). Some cancer biologists have therefore argued that the key to successful cancer therapy is not to find a silver bullet of novel drugs, but to apply evolutionary theory to make decisions about how and when treatments should be administrated (West et al. 2023).

This is where the emphasis on a balance between conflict and cooperation at the heart of the paradox of organism may provide a way forward. Multicellularity has evolved independently numerous times and in all three domains of life (Grosberg and Strathmann 2007; Herron et al. 2022). It comes with several benefits, including division of labor, specialization, and protection from external threats. As in other transitions in individuality, such as the origin of genomes or eukaryotic cells, its success depends on two things: aligning the fitness interests of the constituent particles and suppressing selfish behavior at lower levels (Maynard Smith and Szathmáry 1997; Bourke 2011; West et al. 2015). As a paradox-of-the-organism perspective highlights, however, while this suppression is largely effective, it is never complete (Ågren et al. 2019; Howe et al. 2022). The result of the proliferation of cheats is cancer, and its widespread occurrence is exactly what we should expect (Aktipis 2020). The connection between the defining features of multicellularity and the hallmarks of cancer has been pointed out, and understanding the tension between cellular cooperation and conflict is therefore crucial (Aktipis et al. 2015; Nedelcu 2020).

This is where the relevance to treatment comes in (Ågren and Scott 2023). While not typically presented in the vocabulary of the major transitions, very similar concerns provide the premise of so-called adaptive therapy (Gatenby et al. 2009; West et al. 2023). Whereas the maximum tolerated dose paradigm is based on the assumption that cancer tumors are generally genetically homogenous, what has become known as adaptive therapy makes use of the recent discovery that most tumors are genetically heterogenous, with some cells being sensitive to and some being resistant to treatment. As such, treatment-sensitive and treatment-resistant cells often end up competing for resources, a competition that

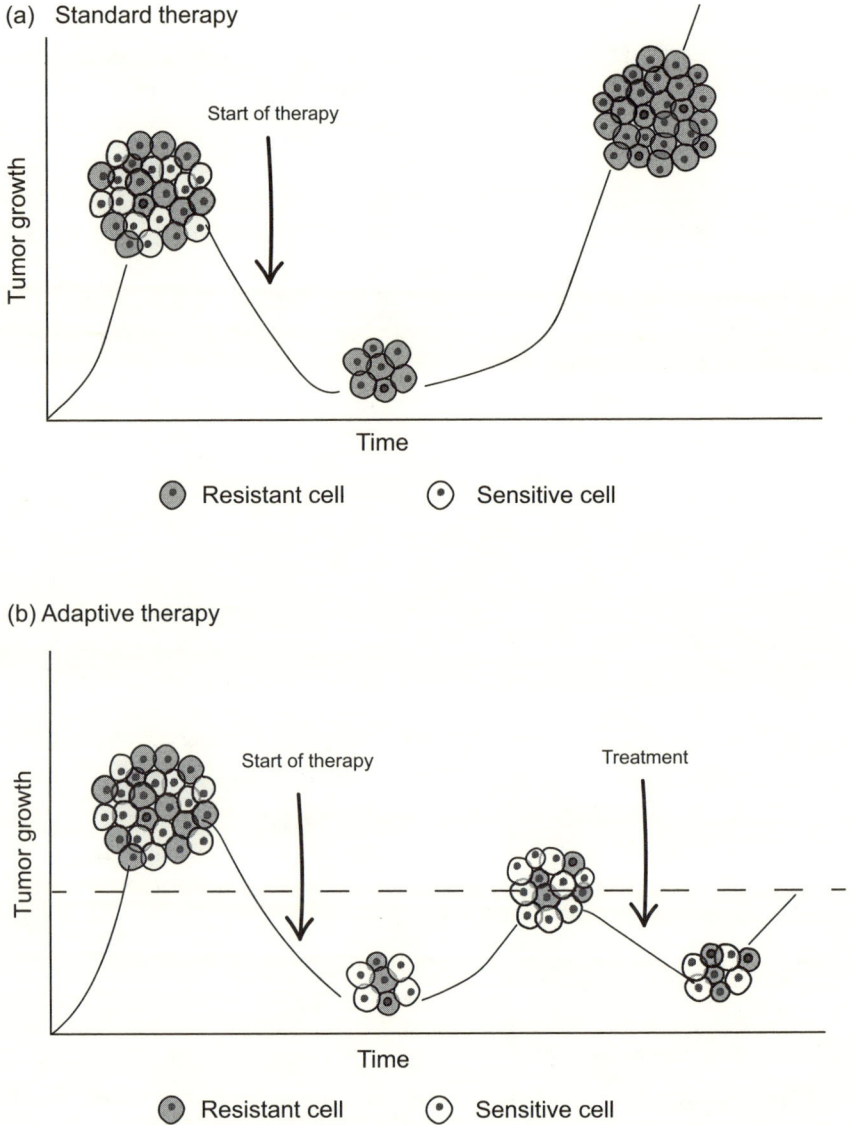

Figure 8.1. Evolutionary approaches to cancer therapy could lead to better patient outcomes. (a) The goal of standard chemotherapy treatment is to remove all cancer cells from the population. However, cells resistant to the therapy inevitably arise (dark cells), leading to competitive release and the growth of a resistant tumor population. (b) The goal of adaptative therapy, an evolutionary informed therapy, is to reduce tumor burden while keeping both drug-resistant (dark cells) and drug-sensitive (white cells) in the tumor population. Treatment is only given if tumor grows above the threshold (dotted line). These cell lineages compete, and ideally the tumor population stays sensitive to the drug for longer periods.

can be harnessed to slow down overall tumor progression. This is why traditional treatment approaches that use a predetermined treatment schedule—applying a drug or chemo as often as toxicity permits to reduce the tumor size as much as possible—almost always result in the evolution of treatment resistance (Enriquez-Navas et al. 2016) (Figure 8.1a). The problem with this strategy is that if there are treatment-resistant cells present in the tumor (or that arise de novo during treatment), they will be release from competition with treatment-sensitive cells. This a within-body example of the principle of competitive release familiar from traditional ecological theory.

Adaptive therapy was developed in response to this problem. Unlike traditional approaches, the goal is here is not to eradicate the tumor; instead, adaptive therapy tries to control it. The strategy involves using a lower dose that reduces the number of drug-sensitive cells, but stopping treatment before the drug-resistance cells have completely taken over (Figure 8.1b). This allows within-tumor competition to prevent resistant cells from completely taking over, which slows down overall tumor growth (Hansen and Read 2020). In effect, adaptive therapy aims to turn cancer into a chronic disease that you can live with, as long as treatment is regularly applied. To date, adaptive therapy has been largely a theoretical idea, but clinical trials have shown promising results in both prostate (Zhang et al. 2017) and ovarian cancers (Hockings et al. 2023), where it has increased survival times while greatly reducing drug use.

The language surrounding cancer has been shaped by metaphor more than almost any other disease (Sontag 1978). Cancer cells are said to "invade" or to "colonize" new parts of the body; treatments "kill," and tumors are "expelled"; patients "fail" to recover. All in the name of the "war on cancer." The paradox of the organism and adaptive therapy offer a new way to conceptualize cancers and highlight that they are not something foreign; conflicts are an inherited part of us. If new forms of cancer treatment and therapy allow us to live with our conflicts, new terminology surrounding cancer is needed.

Pregnancy

In *The Extended Phenotype*, Richard Dawkins describes the organism as a "physically discrete machine, usually walled off from other such machines" (Dawkins 1982, 250). However, pregnancy is a transient state during which such definition does not hold. Throughout pregnancy, the boundaries of two individual organisms are blurred, and this seems to be a special case of the paradox of an organism that is generally more intuitive to grasp. In line with this, pregnancy has often been recognized as a compromise between two genetic entities (e.g., Haig 1993, 2019). The biology of pregnancy, as well as subsequent pregnancy-associated diseases and their treatments, are generally viewed from both a conflict and cooperation framework (Boddy et al. 2015). The wider acceptance that both conflict and cooperation play a role in pregnancy is likely due to a combination of two features: (1) the transient state of gestation (ranging from nineteen days for a mouse to almost two years for an elephant); and (2) the nature of two overlapping but competing interests between mother and offspring. In other words, conflicts within the organism are intuitive for a short timeframe, even if they eventually result in two genetically distinct organisms (Díaz-Muñoz et al. 2016).

Several aspects of pregnancy are enlightened by the paradox of the organism's lens of adaptive compromises. During pregnancy, there is clearly cooperation for survival and reproductive success. At the same time, conflict arises because mothers and offspring do not share all genes (Trivers 1974; Haig 1993; Crespi and Semeniuk 2004). For example, the mother and fetus will have different optimal growth rates and gestation lengths, which leads to conflict over resources. This scenario is especially true for species with deeply invasive placental morphologies (such as humans), where there is likely more conflict due to the direct connection between shared resources. After birth, conflicts among mother and fetus may also depend on parental investment, including biparental care (Úbeda 2008).

Conflicts between mothers and offspring can lead to pathologies in pregnancy. For example, preeclampsia, which is the leading cause of maternal and perinatal mortality, is a condition during pregnancy in which the mother's blood pressure becomes dangerously elevated (Poon et al. 2019). Typically arising in the second trimester (greater than twenty weeks of gestation), preeclampsia is thought to be caused by a lack of nutrition to the fetus. In many cases, the lack of resources is due to fetal trophoblast cells having poor placental invasion and the mother developing narrow arteries that cannot properly maintain the nutritional demands of the growing fetus. This phenotype can lead to the fetus increasing maternal blood pressure to meet the demands of its rapidly growing body. Once the pregnancy is well established, the maternal body has no way to limit this resource allocation or lower blood pressure. In this evolutionary compromise scenario, the fetus has "won the upper hand" in regulating maternal resources in a way that is clearly not the optimal design from the maternal point of view. Indeed, there are species that may have evolved less invasive placentas over time, such as lemurs and loris in the primate lineage, hoofed mammals such as cows and moose, and aquatic mammals such as dolphins and whales (Wildman et al. 2006; Elliot and Crespi 2009). From a paradox-of-the-organism perspective, we may hypothesize that such adaptations are due to a resolution of conflict over resource control and distribution between the mother and baby.

Conditions during pregnancy, like preeclampsia, offer valuable case studies for learning how to manage other diseases from a paradox-of-the-organism perspective. Clinically, the *duality* of the organism, where the mother and baby are simultaneously one unit but two separate genetic individuals, forces medical professionals to make decisions on what is best for each patient (mother and offspring). In resolving these compromises, optimal solutions will often be different for the two patients. For example, while it's best for the baby to stay in utero as long as possible, preeclampsia typically leads to pre-term birth of the infant. In cases of severe preeclampsia, there are complex treatment algorithms to optimize the health of the mother and the baby. The goal is to manage maternal blood pressure, all while prolonging the pregnancy as long as possible to maximize the changes of infant survival (Dimitriadis et al. 2023). For the mother, delivery of the placenta is necessary for the maternal blood pressure to return to a state of homeostasis, yet early delivery of the baby (earlier than thirty-four weeks) puts infants at an increased risk for health complications. Understanding the conflict between mother and baby leads to a different clinical approach. In this scenario, maternal health directly conflicts with fetal health. The ultimate goal is not to "cure" the mother, as doing so could result in the baby being born too prematurely. Instead, the focus is on monitoring and managing the health of both mother and baby until a critical threshold is reached, and then in many cases of preeclampsia, doctors induce labor early for the mother's safety. What can we learn about treatment for pregnancy related conditions, such as preeclampsia, for other conditions where internal conflicts generate diseases? Disease management and monitoring can be effective strategies when a cure is not possible.

Microchimerism

Typically, the duality of pregnancy is considered to end with childbirth. However, we now know that the impact of pregnancy extends beyond gestation through microchimerism, which raises the question of how the lessons learned from pregnancy can be applied to pathologies arising from microchimerism.

Chimerism—in which an organism is composed of biological material, such as cells, from other genetically distinct entities—has appeared multiple times in multicellular organisms (e.g., as reviewed in Kapsetaki et al. 2023). Chimerism can range from large scales—such as the invasive crazy yellow ant, which is a hybrid of two separate genetic lineages (Darras et al. 2023)—to smaller cellular exchanges that occur during mammalian pregnancy, called microchimerism. This kind of biology is quite paradoxical because most of us learn that organisms with adaptive immune systems can (and should!) distinguish between self and non-self. Indeed, this is by some accounts the very definition of what an individual organism is (see Pradeu 2011 for a critical examination of this idea). Consistent with this immunological definition of an organism, many organisms cannot accept tissues or cells from another genetically distinct individual without immunological rejection. Indeed, responding to and removing cells that are not from the host germline is likely an important mechanism to maintain organismality and to protect from "cellular cheaters" that originate from genetically distinct lineages that could take over the host. However, an important corollary to this dates back to the 1950s and the work of Peter Medawar and colleagues, who were inspired by the fact that cattle twins could receive grafts from each other without rejection (Owen 1945). This was not true for humans and other species, and demonstrated that fetal exposure to non-self-antigens (cellular material) led to immunological tolerance of this genetic material in later adulthood (Billingham et al. 1953). These experiments showed how the immune system can acquire tolerance of genetically different individuals within one host if exposed early enough to the immature fetal immune system. The observation that the immune system can tolerate the "non-self" suggests that a strict immune definition of organismality is incomplete. The definition is further undermined by the discovery that an individual's immune system can be *autoreactive*—responding to self in terms of tissue healing and maintaining tissue homeostasis (Rankin and Artis 2018), all while maintaining a tolerance for what we would consider foreign genetic entities, such as the microbiome (Chu and Mazmanian 2013). The emerging, updated immunological view of the individual is therefore that an individual is made up of heterogenetic elements and the immune system must constantly redefine boundaries that distinguish the organism from its environment (Pradeu

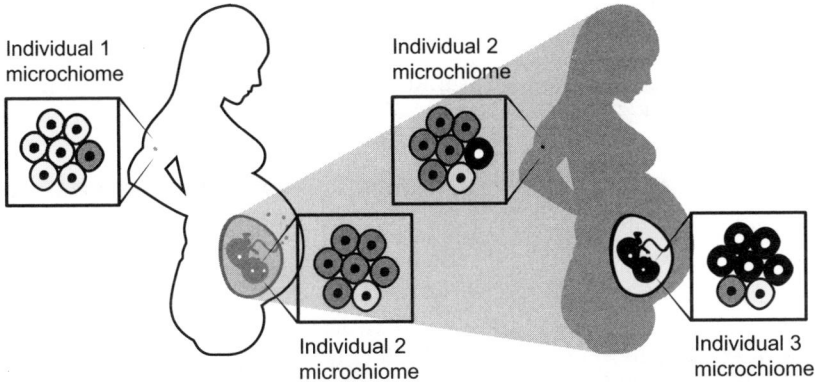

Figure 8.2. Microchimerism across the life course. During pregnancy, there is a transfer of cells between the pregnant mother and baby. The combination of cells from genetically distinct individuals is referred to as the microchiome. As depicted, an individual's microchiome changes across their life course, shown by Individual 2. Starting in fetal development, Individual 2 has cells from herself and a small amount of cells from her mother (Individual 1). When Individual 2 becomes pregnant, she receives cells from her baby (Individual 3), transiting to a more diverse microchiome.

and Cooper 2012; Pradeu 2019). Acquired tolerance, as in the case of Medawar's cattle, is likely one mechanism for how chimerism arises and explains why chimerism is widespread across multicellular life. Even so, it remains unclear whether the presence of foreign cells causes harmful conflicts with the host—and whether this tolerance is an adaptive trait or if chimeric organisms are simply "making the best of a bad situation" and whether chimeric organisms are more vulnerable to diseases, including cancer (Kapsetaki et al. 2023).

The idea of the organism as a homogenous multicellular cooperative entity is further challenged by the discovery of microchimerism, the transfer of cells between mother and baby during pregnancy, or cellular exchange during twinning. In the context of mammalian pregnancy, cells from one genetic individual are able to migrate, differentiate, and express their own genetic material in another half-genetically identical individual. This migration happens in both directions, from mother to fetus and from fetus to mother, and microchimeric cells can survive for decades, even the entire lifetime of the individual (Bianchi et al. 1996). The cellular makeup of both host cells and relative cells, including mother, grandmother, offspring, and potentially older siblings, can be referred to as the *microchiome* (Figure 8.2). The diversity of an individual's microchiome is influenced by birth order and parity across an individual's life course (Kinder et al. 2015, 2017).

Microchimeric cells can be found in almost any tissue or organ, including the heart, liver, lung, and even the brain, and can differentiate into multiple cell types (see Boddy et al. 2015; Bianchi et al. 2021). The question of whether these cellular voyagers are by-products of mammalian gestation or adaptive helpers in mammalian pregnancy remains unanswered. Many researchers propose this cellular exchange is necessary to maintain immunological tolerance to the genetically distinct (semi-allogenic) fetus during pregnancy. In other words, we, as mammals, may need this cellular exchange to generate the immunological tolerance necessary to reproduce. If true, and the swapping of cells is necessary for our fitness, how does this change our perspective of organismality? And can pathology arise from these cellular agents that may operate for their own genetic interest?

Still, while the swapping of fetal and maternal cells may help maintain a pregnancy, microchimerism has been implicated in many disease phenotypes, including autoimmune diseases (Lambert and Nelson 2003; Gadi and Nelson 2007; Gammill and Nelson 2010; Fugazzola et al. 2012). Fetal cells have been found in numerous disease tissues of the maternal body, including tissue sites impacted by systemic sclerosis (Nelson 1998), rheumatoid arthritis (Kanaan et al. 2019), and cancer (Gadi and Nelson 2007). Furthermore, females are at an increased risk for autoimmune disease compared to males (Ngo et al. 2014), leading some to speculate that this seeding of different cellular lineages during pregnancy could put some at risk for developing autoimmune disease later in life (Boddy et al. 2015; Natri et al. 2019). In contrast, other evidence from epidemiological studies support the notion that these cellular exchanges may be beneficial for women. For example, a large study of 272 Danish women reported that women with detectable levels of male microchimerism in their blood lived longer than women with no detectable microchimerism (Kamper-Jørgensen et al. 2014). Clearly, the impact microchimerism has on the host biology is complex and yet to be fully understood.

Similar to fetal cells' association with maternal physiology, there is evidence that maternal cells transferred to the developing fetus can impact health and disease. Maternal cellular exchange has been shown to be useful for the developing fetal immune system, including the transfer of maternal immune cells and protective maternal antibodies (Harrington et al. 2017; Kinder et al. 2017; Stelzer et al. 2021; Balle et al. 2022). Additionally, there is support that maternal microchimeric cells are beneficial for expanded tolerance and reproductive success of their offspring in mouse models, suggesting maternal microchimerism can have effects on offspring fitness (Kinder et al. 2015). Open questions remain regarding how microchimeric cells impact human pregnancy and whether they can influence subsequent pregnancies. Indeed, if mirochimeric cells influence pregnancy success, these reproductive benefits may outweigh the longer-term health consequences, such as autoimmune diseases.

What is the balance of conflict and cooperation when it comes to microchimeric cells? The positive and negative effects implicated in microchimerism research have led researchers to question whether these cellular voyagers are helpful or harmful (Khosrotehrani and Bianchi 2003). Conceptualizing the organism as a paradox can change how we view this biological phenomenon of transfer of cells. From the perspective of self versus non-self, these foreign cells should be considered harmful, and from a clinical point of view, they should be targeted by the host system for removal before they trigger unwanted immune responses. However, from the paradox of an organism perspective, microchimeric cells might benefit the host (e.g., enhancing reproduction) and thus are maintained in the host body. Crucially, if we accept that organisms are a messy collection of adaptive compromises, we can move away from the simplistic dichotomies of "self versus non-self" and "helpful versus harmful" to an understanding wherein we expect cooperation when fitness is aligned and conflict when fitness interests are misaligned (Boddy et al. 2015).

Moving beyond the helpful-or-harmful dichotomy also highlights that not all microchimeric cells may lead to disease, such as an autoimmune reaction, but disease may emerge only in certain contexts. Current research on fetal microchimerism and maternal autoimmune disease suggests that the interaction between the human leukocyte antigen type of both the mother and baby matters in disease risk (Kanaan et al. 2019). If we can begin to understand the context in which microchimerism positively impacts the host without a negative immune response, these mechanisms could provide useful insight into many diseases, including infertility, graft versus host disease during transplants, and the treatment of autoimmune diseases. Medawar's Noble Prize-winning study showed the transfer of non-self-material during fetal development could generate immunological tolerance; however, exposing offspring to non-self-genetic material just a few days after birth led to an immunological response. These results suggest there is a critical timepoint for the immune system to learn "what is self." If we think about the outcomes from Medawar's study in terms of microchimerism, we suggest there are different starting points for fetal and maternal microchimerism. Fetal microchimerism is introduced to the maternal body with a mature immune system, while maternal microchimerism is introduced to the fetus during development. This leads us to predict more conflict from fetal microchimerism than maternal microchimerism.

How Understanding Conflict Can Improve Medical Care

A paradox-of-the-organism perspective on medicine makes new features stand out. For example, infectious disease biology is centered on how bodies manage the threat of foreign entities. However, what really counts as a foreign cell? A traditional immunological definition of an organism is similar to what Medawar outlined in that it will readily accept grafts from other parts of its own body, but not from other bodies. Where does this leave individuals with autoimmune disease, where bodies fail to distinguish between their own cells and foreign cells and the immune system attacks the former? Are such individuals less "organismal"? Are monozygotic twins less organismal than fraternal twins? These implications seem like a stretch. Instead, the previous examples make it clear that one cannot look at a collection of cells and their genetics and know what they do or how they belong.

Instead, context matters for whether a collection of cells is in conflict or will cooperate (Díaz-Muñoz et al. 2016). Pregnancy is an obvious cooperative outcome, but one that still regularly leads to conflict. In the case of microchimerism, it is still not clear how cells from a relative can impact host physiology, but observations to date and a paradox-of-the-organism perspective leads us to suspect that fitness effects are likely to be both positive and negative. Finally, while it is clear that the uncontrollable growth of tumors generally is harmful for the body, not all tumors are malign. Therefore, much cancer work in the past decades has therefore focused on the genetic differences between benign and malign tumors. A surprising insight that has emerged from this work is that many so-called driver mutations typically found in malign tissue are also detected in normal tissue (Martincorena et al. 2018). What is more is that the benign manifestations may actually compete for space with the malign ones and reduce cancer risk.

The central conceptual lesson of the paradox of the organism is that bodies need not be fully harmonious to be functional. Conflicts are an inherent part of who we are. Recognizing this will allow us to better understand and treat pathologies—both in pregnancy, such as preeclampsia (controlling blood pressure or delivering the baby) and in cancer (adaptive therapy to control the tumor without eradicating it). In these treatment approaches we may ask several questions: Why deliver a baby prematurely? And why withhold treatment from a tumor? There are many reasons why we may be better off managing these diseases than attempting to cure them. There is no magic cure when the cells—and their conflicts—are a part of you.

References

Abegglen, L. M., Caulin, A. F., Chan, A. et al. (2015). Potential mechanisms for cancer resistance in elephants and comparative cellular response to DNA damage in humans. *Journal of the American Medical Association, 314*(17), 1850–1860.

Ågren, J. A., and Clark, A. G. (2018). Selfish genetic elements. *PLoS Genetics, 14*(11), e1007700.

Ågren, J. A., Davies, N. G., and Foster, K. R. (2019). Enforcement is central to the evolution of cooperation. *Nature Ecology and Evolution, 3*(7), 1018–1029.

Ågren, J. A., and Scott, J. G. (2023). Viruses, cancers, and evolutionary biology in the clinic: A commentary on Leeks et al. 2023. *Journal of Evolutionary Biology, 36*(11), 1587–1589.

Aktipis, A. (2020). *The Cheating Cell: How Evolution Help Us Understand and Treat Cancer.* Princeton University Press.

Aktipis, C. A., Boddy, A. M., Jansen, G., et al. (2015). Cancer across the tree of life: Cooperation and cheating in multicellularity. *Philosophical Transactions of the Royal Society B, 370*(1673), 20140219.

Aktipis, C. A., and Nesse, R. M. (2013). Evolutionary foundations for cancer biology. *Evolutionary Applications, 6*(1), 144–159.

Balle, C., Armistead, B., Kiravu, A., et al. (2022). Factors influencing maternal microchimerism throughout infancy and its impact on infant T cell immunity. *Journal of Clinical Investigation, 132*(13), e148826.

Bianchi, D. W., Zickwolf, G. K., Weil, G. J., Sylvester, S., and DeMaria, M. A. (1996). Male fetal progenitor cells persist in maternal blood for as long as 27 years postpartum. *Proceedings of the National Academy of Sciences of the United States of America, 93*(2), 705–708.

Bianchi, D. W., Khosrotehrani, K., Way, S. S., MacKenzie, T. C., Bajema, I., and O'Donoghue, K. (2021). Forever connected: The lifelong biological consequences of fetomaternal and maternofetal microchimerism. *Clinical Chemistry, 67*(2), 351–362.

Billingham, R. E., Brent, L., and Medawar, P. B. (1953). Actively acquired tolerance of foreign cells. *Nature, 172*(4379), 603–606.

Boddy, A. M. (2022). The need for evolutionary theory in cancer research. *European Journal of Epidemiology, 38,* 1259–1264.

Boddy, A. M., Fortunato, A., Sayres, M. W., and Aktipis, A. (2015). Fetal microchimerism and maternal health: A review and evolutionary analysis of cooperation and conflict beyond the womb. *BioEssays, 37*(10), 1106–1118.

Bongard, J., and Levin, M. (2021). Living things are not (20th century) machines: Updating mechanism metaphors in light of the modern science of machine behavior. *Frontiers in Ecology and Evolution, 9,* 650726.

Bourke, A. F. G. (2011). *Principles of Social Evolution.* Oxford University Press.

Burt, A., and Trivers, R. (2006). *Genes in Conflict: The Biology of Selfish Genetic Elements.* Harvard University Press.

Cairns, J. (1975). Mutation selection and the natural history of cancer. *Nature, 255*(5505), 197–200.

Caulin, A. F., and Maley, C. C. (2011). Peto's paradox: Evolution's prescription for cancer prevention. *Trends in Ecology and Evolution, 26*(4), 175–182.

Chu, H., and Mazmanian, S. K. (2013). Innate immune recognition of the microbiota promotes host-microbial symbiosis. *Nature Immunology, 14*(7), 668–675.

Compton, Z., Harris, V., Mellon, W., et al. (2025). Cancer prevalence across vertebrates. *Cancer Discovery, 15*(1), 227–244.

Crespi, B., and Semeniuk, C. (2004). Parent-offspring conflict in the evolution of vertebrate reproductive mode. *The American Naturalist, 163*(5), 635–653.

Darras, H., Berney, C., Hasin, S., Drescher, J., Feldhaar, H., and Keller, L. (2023). Obligate chimerism in male yellow crazy ants. *Science, 380*(6640), 55–58.

Dawkins, R. (1982). *The Extended Phenotype: The Gene as the Unit of Selection.* Oxford University Press.

DeVita, V. T., Jr., and Chu, E. (2008). A history of cancer chemotherapy. *Cancer Research, 68*(21), 8643–8653.

Díaz-Muñoz, S. L., Boddy, A. M., Dantas, G., Waters, C. M., and Bronstein, J. L. (2016). Contextual organismality: Beyond pattern to process in the emergence of organisms. *Evolution, 70*(12), 2669–2677.

Dimitriadis, E., Rolnik, D. L., Zhou, W., et al. (2023). Pre-eclampsia. *Nature Reviews Disease Primers, 9*(1), 8.

Elliot, M. G., and Crespi, B. J. (2009). Phylogenetic evidence for early hemochorial placentation in eutheria. *Placenta, 30*(11), 949–967.

Enriquez-Navas, P. M., Kam, Y., Das, T., et al. (2016). Exploiting evolutionary principles to prolong tumor control in preclinical models of breast cancer. *Science Translational Medicine, 8*(327), 327ra24.

Ferlay, J., Colombet, M., Soerjomataram, I., et al. (2021). Cancer statistics for the year 2020: An overview. *International Journal of Cancer, 149*(4), 778–789.

Fugazzola, L., Cirello, V., and Beck-Peccoz, P. (2012). Microchimerism and endocrine disorders. *Journal of Clinical Endocrinology and Metabolism, 97*(5), 1452–1461.

Gadi, V. K., and Nelson, J. L. (2007). Fetal microchimerism in women with breast cancer. *Cancer Research, 67*(19), 9035–9038.

Gammill, H. S., and Nelson, J. L. (2010). Naturally acquired microchimerism. *International Journal of Developmental Biology, 54*(2–3), 531–543.

Gardner, A. (2009). Adaptation as organism design. *Biology Letters, 5*(6), 861–864.

Gatenby, R. A., Silva, A. S., Gillies, R. J., and Frieden, B. R. (2009). Adaptive therapy. *Cancer Research, 69*(11), 4894–4903.

Gillespie, N. C. (1990). Divine design and the industrial revolution: William Paley's abortive reform of natural theology. *Isis, 81*(2), 214–229.

Greaves, M., and Maley, C. C. (2012). Clonal evolution in cancer. *Nature, 481*(7381), 306–313.

Grosberg, R. K., and Strathmann, R. R. (2007). The evolution of multicellularity: A minor major transition? *Annual Review of Ecology, Evolution, and Systematics, 38*(1), 621–654.

Haig, D. (1993). Genetic conflicts in human pregnancy. *Quarterly Review of Biology, 68*(4), 495–532.

Haig, D. (2019). Cooperation and conflict in human pregnancy. *Current Biology, 29*(11), R455–R458.

Hanahan, D., and Weinberg, R. A. (2000). The hallmarks of cancer. *Cell, 100*(1), 57–70.

Hanahan, D., and Weinberg, R. A. (2011). Hallmarks of cancer: The next generation. *Cell, 144*(5), 646–674.

Hansen, E., and Read, A. F. (2020). Modifying adaptive therapy to enhance competitive suppression. *Cancers, 12*(12).

Harrington, W. E., Kanaan, S. B., Muehlenbachs A., et al. (2017). Maternal microchimerism predicts increased infection but decreased disease due to plasmodium falciparum during early childhood. *Journal of Infectious Diseases, 215*(9), 1445–1451.

Herron, M. D., Conlin, P. L., and Ratcliff, W. C. (2022). *The Evolution of Multicellularity.* CRC Press.

Hockings, H., Lakatos, E., Huang, W., et al. (2023). Adaptive therapy achieves long-term control of chemotherapy resistance in high grade ovarian cancer. *bioRxiv.* https://doi.org/10.1101/2023.07.21.549688.

Howe, J., Rink, J. C., Wang, B., and Griffin, A. S. (2022). Multicellularity in animals: The potential for within-organism conflict. *Proceedings of the National Academy of Sciences of the United States of America, 119*(32), e2120457119.

Huxley, J. (1912/2022). *The Individual in the Animal Kingdom*. MIT Press.

Kamper-Jørgensen, M., Hjalgrim, H., Nybo Andersen, A.-M., Gadi, V. K., and Tjønneland, A. (2014). Male microchimerism and survival among women. *International Journal of Epidemiology, 43*(1), 168–173.

Kanaan, S. B., Sensoy, O., Yan, Z., Gadi, V. K., Richardson, M. L., and Nelson, J. L. (2019). Immunogenicity of a rheumatoid arthritis protective sequence when acquired through microchimerism. *Proceedings of the National Academy of Sciences of the United States of America, 116*(39), 19600–19608.

Kapsetaki, S. E., Fortunato, A., Compton, Z., et al. (2023). Is chimerism associated with cancer across the tree of life? *PloS One, 18*(6), e0287901.

Khosrotehrani, K., and Bianchi, D. W. (2003). Fetal cell microchimerism: Helpful or harmful to the parous woman? *Current Opinion in Obstetrics and Gynecology, 15*(2), 195–199.

Kinder, J. M., Jiang, T. T., Ertelt, J. M., et al. (2015). Cross-generational reproductive fitness enforced by microchimeric maternal cells. *Cell, 162*(3), 505–515.

Kinder, J. M., Stelzer, I. A., Arck, P. C., and Way S. S. (2017). Immunological implications of pregnancy-induced microchimerism. *Nature Reviews Immunology, 17*(8), 483–494.

Lambert, N., and Nelson, J. L. (2003). Microchimerism in autoimmune disease: More questions than answers? *Autoimmunity Reviews, 2*(3), 133–139.

LeMahieu, D. L. (1976). *The Mind of William Paley: A Philosopher and His Age.* University of Nebraska Press.

Lents, N. H. (2018). *Human Errors: A Panorama of Our Glitches, from Pointless Bones to Broken Genes.* Houghton Mifflin Harcourt.

Lewens, T. (2019). Neo-paleyan biology. *Studies in History and Philosophy of Science Part C: Studies in History and Philosophy of Biological and Biomedical Sciences, 76,* 101185.

Martincorena, I., Fowler, J. C., Wabik, A., et al. (2018). Somatic mutant clones colonize the human esophagus with age. *Science, 362*(6417), 911–917.

Maynard Smith, J., and Szathmáry, E. (1995). *The Major Transitions in Evolution.* Oxford University Press.

Natri, H., Garcia, A. R., Buetow, K. H., Trumble, B. C., and Wilson, M. A. (2019). The pregnancy pickle: Evolved immune compensation due to pregnancy underlies sex differences in human diseases. *Trends in Genetics, 35*(7), 478–488.

Natterson-Horowitz, B., Aktipis, A., Fox, M., et al. (2023). The future of evolutionary medicine: Sparking innovation in biomedicine and public health. *Frontier Science Series, 1,* 997136.

Nedelcu, A. M. (2020). The evolution of multicellularity and cancer: Views and paradigms. *Biochemical Society Transactions, 48*(4), 1505–1518.

Nelson, J. L. (1998). Microchimerism and the pathogenesis of systemic sclerosis. *Current Opinion in Rheumatology, 10*(6), 564–571.

Ngo, S. T., Steyn, F. J., and McCombe, P. A. (2014). Gender differences in autoimmune disease. *Frontiers in Neuroendocrinology, 35*(3), 347–369.

Nowell, P. C. (1976). The clonal evolution of tumor cell populations. *Science, 194*(4260), 23–28.

Nunney, L., Maley, C. C., Breen, M., Hochberg, M. E., and Schiffman, J. D. (2015). Peto's paradox and the promise of comparative oncology. *Philosophical Transactions of the Royal Society B, 370*(1673).

Okasha, S. (2018). *Agents and Goals in Evolution*. Oxford University Press.

Owen, R. D. (1945). Immunogenetic consequences of vascular anastomoses between bovine twins. *Science, 102*(2651), 400–401.

Patten, M. M., Schenkel, M. A., and Ågren, J. A.. (2023). Adaptation in the face of internal conflict: The paradox of the organism revisited. *Biological Reviews, 98*(5), 1796–1811.

Peto, R., Roe, F. J., Lee, P. N., Levy, L., and Clack J. (1975). Cancer and ageing in mice and men. *British Journal of Cancer, 32*(4), 411–426.

Plomin, R. (2019). *Blueprint: How DNA Makes Us Who We Are*. MIT Press.

Poon, L. C., Shennan, A., Hyett, J. A., et al. (2019). The International Federation of Gynecology and Obstetrics (FIGO) initiative on pre-eclampsia: A pragmatic guide for first-trimester screening and prevention. *International Journal of Gynaecology and Obstetrics, 145*(S1), 1–33.

Pradeu, T. (2011). *The Limits of the Self: Immunology and Biological Identity*. Oxford University Press.

Pradeu, T. (2019). Immunology and individuality. *eLife, 8*, e47384.

Pradeu, T., and Cooper, E. L. (2012). The danger theory: 20 years later. *Frontiers in Immunology, 3*, 287.

Rankin, L. C., and Artis, D. (2018). Beyond host defense: Emerging functions of the immune system in regulating complex tissue physiology. *Cell, 173*(3), 554–567.

Reynolds, A. (2007). The theory of the cell state and the question of cell autonomy in nineteenth and early twentieth-century biology. *Science in Context, 20*(1), 71–95.

Riskin, J. (2016). *The Restless Clock: A History of the Centuries-Long Argument over What Makes Living Things Tick*. University of Chicago Press.

Sender, R., Fuchs, S., and Milo, R. (2016). Revised estimates for the number of human and bacteria cells in the body. *PLoS Biology, 14*(8), e1002533.

Sontag, S. (1978). *Illness as Metaphor*. Farrar, Straus and Giroux.

Stelzer, I. A., Urbschat, C., Schepanski, S., et al. (2021). Vertically transferred maternal immune cells promote neonatal immunity against early life infections. *Nature Communications, 12*(1), 4706.

Strobl, M. A. R., Gallaher, J., Robertson-Tessi, M., West, J., and Anderson, A. R. A. (2023). Treatment of evolving cancers will require dynamic decision support. *Annals of Oncology, 34*(10), 867–884.

Sulak, M., Fong, L., Mika, K., et al. (2016). TP53 copy number expansion is associated with the evolution of increased body size and an enhanced DNA damage response in elephants. *eLife, 5*, e11994.

Trivers, R. L. (1974). Parent-offspring conflict. *Integrative and Comparative Biology, 14*(1), 249–264.

Úbeda, F. (2008). Evolution of genomic imprinting with biparental care: Implications for Prader-Willi and Angelman syndromes. *PLoS Biology, 6*(8), e208.

Vasan, N., Baselga, J., and Hyman, D. M. (2019). A view on drug resistance in cancer. *Nature, 575*(7782), 299–309.

Weinberg, R. A. (2023). *The Biology of Cancer.* W. W. Norton.

West, J., Robertson-Tessi, M., and Anderson, A. R. A. (2023). Agent-based methods facilitate integrative science in cancer. *Trends in Cell Biology, 33*(4), 300–311.

West, S. A., Fisher, R. M., Gardner, A., and Kiers, E. T. (2015). Major evolutionary transitions in individuality. *Proceedings of the National Academy of Sciences of the United States of America, 112*(33), 10112–10119.

Wildman, D. E., Chen, C., Erez, O., Grossman, L. I., Goodman, M., and Romero, R. (2006). Evolution of the mammalian placenta revealed by phylogenetic analysis. *Proceedings of the National Academy of Sciences of the United States of America, 103*(9), 3203–3208.

World Health Organization. (2019). *International Statistical Classification of Diseases and Related Health Problems.* 11th ed. https://icd.who.int/.

Zhang, J., Cunningham, J. J., Brown, J. S., and Gatenby, R. A. (2017). Integrating evolutionary dynamics into treatment of metastatic castrate-resistant prostate cancer. *Nature Communications, 8*(1), 1816.

CHAPTER NINE

Internal Conflict in Genomes and in Persons

SAMIR OKASHA

ABSTRACT The phenomenon of intra-genomic conflict is conceptually as well as empirically interesting, as it challenges our pre-theoretic sense that an organism is a unified entity with a single agenda or goal. A somewhat similar challenge is posed by the phenomenon of intrapersonal conflict, which challenges our pre-theoretic sense that a human being is psychologically unified, and in extreme cases has led theorists to posit the existence of multiple selves within a single human body. The analogy between the two types of internal conflict is instructive, as our philosophical experience in grappling with intrapersonal conflict in psychology provides a useful model for how to understand intra-genomic conflict in biology. Moreover, the analogy highlights a potential limitation on the widespread practice in evolutionary biology of casting adaptive explanations in a psychological idiom.

Introduction

In his 1990 article on the paradox of the organism, Dawkins offers what he describes as "almost a philosophical argument" aimed at "clarifying the meaning" of the concept of an organism (Dawkins 1990). The paradox that Dawkins identifies is that organisms appear to be reasonably cohesive and integrated entities, despite their genomes containing multiple genetic units whose interests do not always coincide. Dawkins points out that an "outlaw" gene in an organism that is able to boost its own chance of being transmitted to the next generation, even if this entails a reduction in the organism's overall reproductive output, can be favored by natural selection. In the years since Dawkins wrote this, we have learned that outlaws, or selfish genetic elements as they are usually known, are numerous in kind and found in very many taxa; examples include segregation distorters, transposons, and selfish cytoplasmic elements (Burt and Trivers 2006). This finding accentuates the paradox that Dawkins drew attention to. How is it that organisms manage to persist as stable units, with complex phenotypic adaptations, given that this requires the cooperation of many different genes? How can organismic cohesion be reconciled with internal conflict of interest?

I agree with Dawkins that this question is of philosophical as well as scientific interest. However, the philosophical significance that I see in the question is somewhat different from that which Dawkins sees. I am struck by the fact that the paradox of the organism raises an issue that, at a certain level of abstraction, is structurally akin to an issue much discussed in the philosophy of mind and psychology. To see this, consider why selfish genetic elements, and the intra-genomic conflicts to which they give rise, are so counterintuitive. It is because they show that a certain entity—the organism—that we pre-theoretically treat as a biological unity (an "individual"), and that can validly be treated as such for many purposes, is in fact somewhat disunified because it contains a multiplicity of elements with different agendas. The parallel with the psychological case is this. We typically treat a human being as a psychological unity, in the sense of corresponding to, or perhaps being identical with, a single person or self; and for most practical purposes this works well. However, internal conflict is also a reality, for people often have conflicting goals, beliefs, and intentions. Where such conflict becomes too extreme, unity gives way to multiplicity. Phenomena such as dissociative identity order and split brains suggest that, in principle, a single human body can house multiple selves who "disagree" with each other. Thus psychological unity, just like biological unity, is threatened by the existence of internal conflicts.

My aim in this chapter is to explore this parallel between intra-genomic and intrapersonal conflict. How closely does the parallel hold? What if anything can we learn from it? What are the theoretical and/or the practical implications? I do not pretend to have definitive answers to these questions, but hope to identify some points of interest.

Thematic Analogy or Causal Connection?

I am not the first author to note the parallel between intra-genomic and intrapersonal conflict. Previous authors who have been impressed by this analogy include David Haig (2006) and Robert Trivers (2009). However, Haig and Trivers argue for a different and indeed stronger connection than the one I explore here. They suggest that there might actually be a causal connection between intra-genomic conflict and internal psychological conflict in humans. It is worth briefly exploring their line of argument in order to juxtapose it with my own.

Haig and Trivers make their argument in connection with genomic imprinting—the phenomenon in which a gene's expression in its host organism depends on whether the gene was paternally or maternally inherited. In typical cases of imprinting, either the maternally or the paternally inherited copy of a gene is silenced by DNA methylation while the other copy is expressed. Haig (2002) showed how this parent-specific gene expression can give rise to intra-genomic conflict. The key point is that a paternally inherited gene in a developing organism has no evolutionary interest in future reproduction by the organism's mother, it so will seek to—that is, will be selected to—extract maximal resources from the mother. By contrast, a maternally inherited gene does have an interest in the mother's future reproduction, so it will "disagree" with the paternally inherited gene about the optimal amount of maternal provisioning. Thus the phenomenon of parent-specific gene expression, combined with basic kin selectionist logic, means that we should expect internal conflict: Genes at the imprinted locus will favor different values of a given phenotypic trait—the amount of maternal provisioning—depending on their parental origin. Haig's early work makes a convincing case that such conflicts are not just a theoretical expectation but also an empirical reality.

Building on this foundation, Haig and Trivers speculate that in humans, parent-specific gene expression may have psychological consequences, in the form of internal conflict between the maternal and paternal parts of our psyche. Thus Trivers writes, "We literally have a paternal self and a maternal self and they are often in conflict" (Trivers 2009, 163). Similarly, Haig suggests that intra-genomic conflict may lead to a "divided self," in which the human mind is inhabited by multiple personae with different agendas (Haig 2006, 2008). Dennett (2013) indicates his sympathy for this line of argument.

The Trivers–Haig hypothesis is an intriguing one. Though there is no direct evidence in its favor, so far as I know, the hypothesis seems initially plausible, both on theoretical grounds and because certain imprinted genes are known to causally influence human cognitive and behavioral traits (Crespi 2008). (Such causal influence is illustrated by the well-known cases of Prader-Willi and Angelman syndromes, which result from loss of paternal and maternal copies, respectively, of a gene on chromosome 15). It is not difficult to imagine mechanisms that could underpin the Trivers–Haig hypothesis. For example, maternally and paternally expressed genes might affect different regions of the brain that activate different impulses or motivations, leading the brain to send conflicting signals that result in the person experiencing psychological conflict. However, in a careful study of the issue, the philosopher David Spurrett (2016) has queried Trivers's and Haig's reasoning, and I find his arguments convincing.

Spurrett's most important point is that there is no particular reason why intra-genomic conflict should give rise to the sort of psychological conflict that Trivers and Haig suggest. In general, the fact that the paternally and maternally inherited copies of a gene will be selected to bring about different values of a phenotype, such as maternal provisioning, does not imply that at evolutionary equilibrium we will observe conflict over provisioning unfold within a single organism in ontogenetic time. A plausible alternative is that what evolves will be a fixed level of provisioning that represents a phenotypic "compromise" between the paternal and maternal optima. The same holds true for psychological traits. Though the paternally inherited part of our genome may "want" something different from the maternally inherited part, and thus be in conflict with it, this conflict plays out over evolutionary time. The resulting evolutionary process will not necessarily give rise to organisms that experience real-time psychological conflict, nor that have a divided self. An alternative is that an undivided self will evolve with a single consistent set of preferences that reflect a compromise between the interests of its genomic constituents.

This objection to the Trivers–Haig suggestion seems right to me, and it provides an instance of an important general point—namely, that there is no simple way of deducing what the phenotypic outcome of a given intra-genomic conflict will be. There are lots of possibilities, as evolutionary theorists recognize (e.g., Rautiala and Gardner 2023). (It depends, among other things, on which gene or genes "control" the phenotype, and on whether modifiers at other loci evolve to suppress the conflict.) So let us put to one side the idea of a direct causal connection between intra-genomic conflict and intrapersonal psychological conflict. Nonetheless, it remains true that there is an interesting parallel, or thematic analogy, between the two phenomena.

I believe that this analogy is not superficial, but rather quite instructive, for two reasons. First, the phenomenon of intra-genomic conflict is conceptually unsettling; it leaves us uncertain how to apply concepts such as individual, organism, and phenotypic adaptation, as Dawkins's own discussion illustrates. I suggest that our philosophical experience of grappling with the analogous issues surrounding unity of self could provide a useful model here. Second, there is a long-standing practice in evolutionary biology of "psychologizing" adaptive explanations—that is, casting them in an agential or intentional guise (see examples below). This practice is controversial, but if used carefully can yield heuristic insight. The analogy between intra-genomic and intrapersonal conflict helps to show the limits of this practice.

Organismic Unity and Intra-genomic Conflict

Let us start by setting human psychology aside, and concentrating instead on the connection between intra-genomic conflict and organismic unity. Dawkins posed the question of how organisms are able to exist as cohesive entities given the potential for conflict between their genomic constituents. In broad terms, the answer to this question is clear: Organisms have evolved effective ways of suppressing such conflict. Putative examples of such suppression mechanisms include Mendelian segregation, a single-celled bottleneck in the life cycle, two-step meiosis, recombination, unicellular inheritance of mitochondria, and transfer of genes from cytoplasm to nucleus (Haig and Grafen 1991; Burt and Trivers 2006; Queller and Strassman 2009). These mechanisms limit the ability of individual genes to boost their own transmission at the expense of other genes in the organism, thus serving to align genes' interests. Suppression mechanisms do not work perfectly, as evidenced by the discovery of selfish elements in a wide range of taxa. But the majority of the genes within any organism are not selfish genetic elements, and plausibly it is partly in virtue of this that organisms are able to exist as stable, cohesive entities at all (Scott and West 2019; see also Chapter 1 in this volume).

This suggests that in existing organisms, selfish genetic elements will always be in a minority, outnumbered by "well-behaved" genes that do not undermine organismic fitness—as posited by Leigh (1971) in his "parliament of the genes" metaphor. If this is true, as empirically seems to be the case, it means that internal genetic conflict can usefully be conceptualized as a disagreement between a rogue selfish genetic element, or "genomic parasite," and the rest of the genome. That is, the majority of the genes in the genome "agree" on what they want—namely, maximization of the organism's reproductive output or fitness—but the rogue element disagrees. If one dislikes the metaphor of disagreement, the point can equally be expressed in terms of "preferences" (Okasha 2018, chap. 1).

Most of the genes within the genome have identical preferences over the relevant set of phenotypic outcomes, but the rogue selfish genetic element prefers a different outcome.

When we talk about selfish genetic elements being in a "minority," the literal meaning of this is simply that such elements are outnumbered by other genes in the genome. This is an objective notion (presuming it is clear how to count the genes in a genome and how to decide whether a given gene is or is not a selfish element). But one might wonder whether this literal meaning captures the most important issue. An alternative is to consider not the number of selfish genetic elements but rather the amount of "phenotypic control" they exert relative to the rest of the genome.[1] Conceivably, a genome might contain a high number of selfish elements that have little influence on the phenotype, or vice-versa. This raises the tricky issue of how phenotypic control can be measured. I cannot do justice to this issue here, so I will stick with the literal meaning of "minority." It is an open question whether my conclusions would still hold if instead we meant degree of phenotypic control, modulo a suitable measure of this.

Now let us ask the following question (inspired by Dawkins's aim of clarifying the meaning of the organism concept). Suppose it is indeed true that organisms would be unable to exist unless the selfish genetic elements within their genomes were in a minority. Is this an empirical or a conceptual truth? That is, is the point that empirically, given what we know about how organisms and genes work, no organism would be viable if it contained too many selfish genetic elements? Or is the point that conceptually, given what we mean by the term "organism," most genes in an organism's genome must have a common interest—and thus selfish genetic elements must be in a minority—on pain of there being no organism in the first place? Dawkins hints at this latter option when he proposes to define an organism as "an entity all of whose genes share the same stochastic expectation of the distant future" (1990, S82).

I suggest that the answer is as follows. It is at most an empirical truth, not a conceptual truth, that the existence and/or cohesion of organisms requires selfish genetic elements to be in a minority. However, there is an important conceptual truth in the near vicinity of this empirical truth.

To see why it is not conceptually true that selfish genetic elements must be in a minority within an organism's genome, consider what are usually

---

[1] Thanks to Manus Patten for raising this issue.

taken to be the defining attributes of an organism. There is no precise consensus, but a standard list includes things like autonomy, functional integration, internal division of labor, having a life cycle, having an immune system, and having clear spatiotemporal boundaries (Clarke 2010). Now there may well be a close correlation, and indeed a direct causal link between possessing these attributes and successfully suppressing selfish genetic elements, such that unsuppressed selfish genetic elements will always be in a minority in the genome of any biological entity that exhibits these attributes to a high degree. But this is not a necessary connection; it is a function of the biology of our actual world. We can certainly imagine an organism lacking internal suppression mechanisms such that selfish genetic elements are not outnumbered by well-behaved genes, at least if "organism" is defined by the usual list of attributes. The point is that actual organisms are not like that.

To see the nearby conceptual truth, consider one standard way of describing genetic conflicts—namely, as the result of a selfish genetic element pursuing its own interest to the detriment of the organism's interest. Now this description is familiar, and makes good sense, but only because selfish genetic elements are outnumbered by other genes in the genome. The point is not that an organism's existence requires this outnumbering, which is at most an empirical truth, but rather that we can sensibly speak of an organism as having a single "interest" only insofar as most of the genes in its genome have a common interest. This can be best seen by adopting a strict gene's-eye viewpoint.[2] From this viewpoint, within-organism conflict arises because genes' evolutionary interests differ, with some genes enhancing their representation in the next generation at the expense of others. So, strictly speaking, the conflict of interest is between a selfish genetic element and other genes in the organism, not between the element and the organism because it is genes not organisms that get transmitted and thus have interests (in the evolutionary sense). However, an organism can be said to have an "interest" derivatively, if the majority of the genes in its genome have a common interest. Because then we can identify the organism's interest with that of its genomic majority.

---

[2] Adopting the gene's-eye viewpoint is not mandatory, but it is the simplest and most intuitive way of making sense of intra-genomic conflict. On this point, Gardner and Welch (2011, 1808) describe "the conceptual tangles of early attempts to explain the high prevalence of selfish genetic elements," before the gene's-eye view had been developed.

My claim that internal conflict is strictly speaking between a selfish genetic element and other genes, not the organism, may appear to disagree with the position taken by Patten in recent work (Patten et al. 2023; Patten 2024). Patten distinguishes two sorts of internal conflict: "part versus part" and "part versus collective." This distinction maps to a difference between two types of selfish genetic element—namely, transmission distorters and trait distorters. However, the disagreement here is more apparent than real. I agree with Patten that "pure" transmission distorters, which have no phenotypic effect on the organism, obey a different evolutionary logic to that of trait distorters, which alter the value of some phenotypic trait away from the collective's optimum, or preferred value of the trait. My point is simply that, from a strict gene's-eye perspective, the "collective's optimum" must really mean the trait value favored by (most) other genes in the genome.

If this is right, it means that the alignment of genes' interests within an organism—that is achieved by suppression mechanisms—has both an empirical and a conceptual significance. Empirically, such alignment is needed for an organism to evolve into a cohesive, functionally integrated unit with traits that contribute to the same overall goal (i.e., the organism's survival and reproduction). Conceptually, this alignment is necessary for an organism to have a single evolutionary interest—that is, to be the sort of thing that can be benefited or harmed in the evolutionary sense.

This brings us to the notion of organismic unity (which is a close relative of Queller and Strassmann's notion of "organismality") (Queller and Strassmann 2009). This unity has two distinct aspects that, although closely related, should be kept distinct. On the one hand, a typical organism is a unity in the sense of being a cohesive, functionally integrated entity. On the other hand, an organism is a unity in the sense of being an entity with a single "agenda," or evolutionary interest, as opposed to multiple agendas. Both aspects of unity are intimately related to internal genetic conflicts and their suppression, but in different ways. Too much genetic conflict means that empirically, an organism will be unlikely to display the first aspect of unity; that is, cohesion and functional integration. Too much genetic conflict means that conceptually an organism will not have a single agenda, because this requires that selfish genetic elements be in a minority, and thus that genetic conflict be the exception, not the rule.

Psychological Unity and Intrapersonal Conflict

To develop the analogy with psychology, let us start by clarifying what we mean by psychological unity. It is useful to distinguish two (related) aspects of this, which we may call "unity of self" and "unity of purpose." (Neither should be confused with what some philosophers call "unity of consciousness."[3]) Unity of self refers to the commonsense idea that each human body contains a single person, or self. "Contains" is really just a metaphor, so a more accurate formulation says that to each human body there corresponds a single person, or self. Thus where unity of self does not obtain, this means the presence of multiple selves within a single human body. The latter possibility has been defended on both abstract philosophical grounds (Rovane 1997), and on empirical grounds, as the correct interpretation of certain psychological and neuroscientific phenomena (Brook 2015).

Unity of purpose refers to the coherence of a person's mental states with each other and with their actions. It involves having broadly consistent beliefs, having intentions and goals that cohere with each other, and performing actions that conduce toward one's goals given one's beliefs (Kennett and Matthews 2003; Okasha 2018). Though she does not use the phrase "unity of purpose," the basic idea is nicely expressed by the philosopher Kathy Wilkes in a paper about multiple personality:

> [We require] enough unity amongst all a man's mental states . . . to enable us to treat the individual as a single intentional or rational system; so that, for example, most of the time his behaviour can be shown to be intelligible against the background of his past behaviour and experience . . . so that most of the time he can steer roughly consistent and autonomous path, seeking reasonable effective and practical means to an ordered set of short-term and long-term goals. (1981, 347)

Unity of purpose, in this sense, is presupposed by the standard folk-psychological practice of attributing beliefs and goals to other people and explaining their actions and behaviors by reference to what we take to their goals to be.

Note that unity of purpose comes in degrees; it is not an all-or-nothing matter. Indeed, we should probably think of complete unity of purpose

---

[3] See Bayne (2010) and Brook and Raymont (2021) for a discussion of unity of consciousness.

as an idealization that real people only approximate to a greater or lesser extent. For commonplace psychological phenomena such as cognitive dissonance (harboring inconsistent beliefs), akrasia (weakness of the will), and impulsiveness (doing things that undermine one's long-term welfare) show that people rarely achieve a complete unity of purpose. However, the departures are usually sufficiently minor that the unitary ideal holds to a decent approximation, evidenced by the fact that folk-psychological explanations often work tolerably well in daily life.

On the face of it, unity of self appears not to come in degrees. Surely a single body either houses a single person or it does not, given that "persons" come in whole number units? In a sense this is true; clearly there is no such thing as half a person (or self). However, if we grant that a single body can in principle contain multiple personae, then it seems likely that there will be a gray area, or zone of indeterminacy, where it is unclear whether a given body contains one person or two. In this zone, there may be "no fact of the matter," as philosophers say, about whether unity of self obtains or not. So although selves (or persons) do not come in degrees, this does not make unity of self into a fully determinate, yes-or-no matter.

This observation helps us to appreciate the relation between unity of self and of purpose. Minor deviations from complete unity of purpose are commonplace, and perfectly compatible with the existence of a single self. It is when the deviations are sufficiently extreme, systematic, and exhibit a pattern suggestive of more than one autonomous agency within a single human body that it becomes plausible to postulate a multiplicity of selves. In short, the presence of multiple selves within a single human represents a limit case of the breakdown of unity of purpose. Intermediate cases correspond to the gray area where it is unclear whether there is one self or more.

To illustrate these points, consider the phenomenon, much discussed in psychology and behavioral economics, of peoples' apparent inability to rationally pursue their own long-term interest in the face of short-term temptations. The starting observation here is the familiar fact that people often succumb to immediate gratification, and conversely put off doing unpleasant tasks, only to later rue their lack of self-control. This is a failure of unity of purpose—an agent's behavior does not conduce toward their own professed long-term goal. Now a standard way of modeling such behavior is to use so-called *dual-self* models, which posit the existence of a short-term and a long-term self within a single human being, whose interests do not exactly coincide (see Brocas and Carillo 2014 for an overview). In some models, the two selves engage in game-theoretic interaction with each other, each trying to further their own goal and to frustrate their opponent's. This illustrates how failure of unity of purpose can provide the motivation for positing a multiplicity of selves.

One interesting issue that dual-self models raise is whether there is any basis for identifying one of the two selves as the "real" person. Elster (2009) suggests that in many cases there is. He notes an asymmetry: The long-term self will often strategically try to frustrate (or "bind") the short-term self but not vice-versa (Elster 2009, 22). Consider for example a man who gives his wife the key to the liquor cabinet before going out for the night, to prevent his future (short-term) self from raiding the cabinet when he arrives home. However, the same man, when drunk on another occasion, is unlikely to duplicate the key and leave copies all around the house in a bid to prevent his future (long-term) self from being able to restrict access again. To Elster, this suggest that the long-term self is the "real" person.

Now admittedly, it is not clear whether the "two selves" description should be taken literally in dual-self models. Some authors suggest that it is simply a colorful gloss on a psychological/behavioral phenomenon that could be described more soberly and that, so described, would be compatible with a single, albeit perhaps irrational, self. In a similar vein, Heil (1989) argues that the idea of a divided mind, proposed by some philosophers as a way of understanding weakness of the will, should not be taken literally. However, in other contexts where a multiplicity of selves has been posited, this is intended as a literal description of the situation.

Perhaps the best-known such context is theorizing about dissociative identity disorder, formerly known as multiple personality disorder. This intriguing condition apparently involves the existence of multiple "persons" within a single body, who usually appear sequentially. Dissociative identity disorder has always been controversial, with disagreements over its correct characterization and the interpretation of the clinical findings. However, a number of careful philosophical investigations have concluded that the multiple selves interpretation is defensible and perhaps even inevitable in at least some cases (Wilkes 1981; Humphrey and Dennett 1989; Radden 2011; Brook 2015). In particular, Dennett and Humphrey write that "The end result [in dissociative identity disorder] would appear to be in many cases a person who is genuinely split. That is, the grounds for assigning several selves to such a human being can be as good as—indeed the same as—those for assigning a single self to a normal human being" (Humphrey and Dennett 1989, 890).

In cases of dissociative identity disorder, the main reason for positing a multiplicity of selves within a single human is that at different times the human exhibits quite different personae, each with their own agenda—that is, with different beliefs and goals. If we try to cleave to the "one body, one person" rule, standard folk-psychological attributions become impossible; we find ourselves unable to say what it is that "the" person believes, wants, or is trying to achieve. The most parsimonious explanation of the observed behavior (including verbal behavior) thus requires positing the existence of multiple selves, or personae. Moreover, in some cases these personae appear to be stable enough, and to appear frequently enough, that there is no clear basis on which to single out one of them as the "true" self.

A very different type case is the split-brain phenomenon (which has nothing to do with dissociative identity disorder). In such cases, a patient has their corpus callosum—a region of the cortex that connects the left and right hemispheres—surgically severed in an attempt to cure them of epilepsy. The salient empirical finding is that in carefully designed laboratory experiments, in which the visual information reaching each hemisphere is carefully controlled, a split-brain patient can sometimes be induced to behave as if they had two separate agencies within them, one in each hemisphere (Schechter 2018). The left hemisphere controls speech, while the right hemisphere controls the movement of the left hand and so is capable of non-verbal communication. In some experiments, the two agencies seem to communicate with each other, with the right hemisphere apparently trying to send a message to the left by writing with the left hand on the back of the right hand. This remarkable phenomenon has been the subject of much theoretical attention. The correct interpretation of the results is controversial, but again the option of saying that there are two people within a single body has been seriously entertained by both philosophers and neuroscientists, and is in many ways the most natural description of the experimental findings (Schechter 2018).

The points I want to emphasize here are as follows. First, it is when the unity of purpose ideal fails dramatically, and in a systematic fashion, that we are led to entertain the possibility of multiple selves within a single human. Thus there is an intimate link between unity of purpose and self. Second, in both the dissociative identity disorder and the split-brain cases, the salient points are (1) the apparent existence of a multiplicity of agents, with separate and/or conflicting agendas, within a single human; and

(2) the absence of any principled basis on which to identify one of these as the "real" self.

The Parallel Fleshed Out

Now that we have introduced biological and psychological unity and seen how each can fail, let us consider the parallel in more detail.

Complete unity of purpose in the psychological case is the analogue of the complete suppression of selfish genetic elements in the biological case—that is, all genes in the organism's genome "agree" on what the organism should do. As we have seen in the psychological case, this implies that a single self is present; in the biological case, it implies that the organism itself can be said to have a single "interest," or agenda. In the psychological case, this will manifest as behavior that makes perfect sense given the person's goals; in the biological case, it will manifest as phenotypic traits (including but not limited to behaviors) that make complete sense given the organism's goal of surviving and reproducing (modulo the usual constraints on adaptation, of course.)

In the psychological case, minor deviations from complete unity of purpose are common, as we have seen, and are compatible with the presence of a single self, or person. They will manifest as behaviors that appear irrational or hard to make sense of given the person's goals, such as succumbing to temptation. Similarly, in the biological case, when an organism's genome contains one or a few selfish genetic elements, there is less than perfect agreement among genes. This is still compatible with the organism having an overall agenda, so long as selfish genetic elements are in the minority. It will manifest as phenotypes that are hard to make sense of, or that appear suboptimal given the organism's agenda (see examples below).

In the psychological case, extreme breakdown of unity of purpose can undermine the unity of self, as the dissociative identity disorder and split-brain cases illustrate. This manifests as behavior that appears to suggest the presence (simultaneous or sequential) of more than one person within a single human, as we have seen. In the biological case, the analogue would be a case where the selfish genetic elements in an organism were not outnumbered by "well-behaved" genes in the genome. This would mean, not that the selfish genetic elements were acting to undermine the organism's agenda, but that the organism could not be said to have a single agenda at all. There are no empirical cases of this, so far as I know; however, as argued above, it is a conceptual possibility.

To further flesh out the parallel, les us consider the isomorphism, or apparent isomorphism, between adaptive explanation in biology and intentional explanation in folk psychology (Dennett 1987; Okasha 2018). In the former, we explain an organism's traits by showing how they further its evolutionary "goal" of survival and reproduction. In the latter, we explain a person's behavior by showing how it furthers whatever their goal is (which will often have nothing to do with survival and reproduction). Indicative of this isomorphism is the fact that adaptive explanations are often naturally couched in psychological language, as for example when we explain why mother rats often kill their young by saying that they know the juveniles won't survive and don't want to waste resources on them.

Opinions differ about the merits of "psychologizing" adaptive explanations in this way. Defenders regard it as a convenient shorthand; opponents say that it is anthropomorphic, risks conflating ultimate and proximate explanations, and invites naive adaptationism (Kennedy 1992). My sympathies are with the defenders; but here I want to set aside this issue (which arises quite generally and has nothing to do with intra-genomic conflict per se). Instead, the point I wish to stress is this: Just as standard intentional explanation in folk psychology presupposes that there is not too much intrapersonal conflict, so standard adaptive explanation in biology presupposes that there is not too much intra-genomic conflict.

To see this, consider what would happen if we tried to "psychologize" adaptive explanations of an organism's phenotypic traits where those traits are partly due to selfish genetic elements. For example, take the phenomenon of cytoplasmic male sterility in hermaphroditic plants in which cytoplasmic genes, which are only transmitted maternally, cause their host plant to cease making pollen. In some forms of cytoplasmic male sterility the plant still grows stamen, but they are smaller than they should be and entirely nonfunctional. Now consider a biologist observing this phenomenon for the first time and attempting to understand it in the ordinary Darwinian way. They would surely be puzzled. The plant goes to the trouble of growing stamen, but those stamen do not actually make any pollen! It is as though the plant both does and does not "want" to self-fertilize, in that some of its traits contribute toward this goal while others detract from it. Thus there is no simple answer to the question of what the plant is "trying" to achieve with its constellation of phenotypic traits because the presupposition of unity of purpose (which underpins the application of intentional terms such as "trying" to the whole organism) does not obtain.

A second example is sperm-killing in the fruitfly *Drosophila melanogaster*, in which a gene on the X chromosome kills any Y-bearing sperm. This reduces the total number of viable sperm, which in turn is thought to reduce the fly's overall reproductive success. Again, someone observing this phenomenon though a standard Darwinian lens would likely be baffled—most of the fly's traits, such as its mating dance, appear designed to maximize its reproductive output. However, other traits, such as not producing any Y-bearing sperm, appear to detract from this aim. Thus it becomes difficult to say what the fly is "trying" to achieve, with its overall constellation of phenotypic traits. The fly is akin to an irrational human, who both does and does not want to achieve a given outcome.

In both the plant and the fly examples, there is of course no mystery as to how the phenotypes in question evolved. The reason that the plant's stamen do not make pollen is that cytoplasmic genes gain by suppressing pollen production; and the reason for the fruit fly's impeded spermatogenesis is that genes on the X chromosome genes gain nothing from the production of Y-bearing sperm. The point is that these explanations do not invoke advantage to the organism, as do "ordinary" adaptive explanations, so they cannot be "psychologized" in the usual way. Intentional psychological idioms applied at the whole organism level, whether used literally to explain conscious human behavior, or in an "as if" sense to explain evolved behavior of non-human organisms, presuppose a high degree of unity of purpose. And where there is intra-genomic conflict, as in these examples, the biological analogue of complete unity of purpose—agreement between all the genes in the genome—does not obtain.

Note that in these examples, it is the biological analogue of unity of purpose, rather than unity of self, that (partially) fails.[4] This reflects the fact that the selfish genetic elements that cause impeded spermatogenesis and male sterility are outnumbered by other genes in the fly and plant genomes, respectively. Thus, in accordance with the argument made above, we can still sensibly talk about the fly's and the plant's interests, or agendas. Most of the fly's and plant's traits further their interests, but some do not. This is analogous to a person with conflicting goals who engages in irrational behavior, rather than to a case where the internal conflict is so extreme that we cannot speak of a single person at all.

---

[4] Here I correct an argument found in Okasha (2018), chap. 1, where I failed to distinguish unity of purpose from unity of self.

Conclusion

The paradox of the organism draws attention to the fact that, for the most part, organisms behave as integrated, cohesive entities despite containing multiple genomic constituents whose interests don't automatically coincide. For Dawkins, the paradox shows that we need to rethink and ultimately redefine what we mean by an organism. I have argued that this is not exactly right: The paradox shows not that "organism" needs to be redefined, but rather that an organism, in the ordinary sense, can only be said to have its own "agenda" when most of its constituent genes have their interests aligned. However, I agree with Dawkins that the paradox has philosophical significance, though I locate it in a slightly different place.

There is a striking parallel between intra-genomic conflict in biology and intrapersonal conflict in psychology. Both phenomena, as well as being unexpected, call into the question a pre-theoretic assumption of unity. In the biological case, intra-genomic conflict undermines organismic unity in two related ways. Unless most selfish genetic elements are effectively suppressed, then empirically an organism is unlikely to exhibit a high degree of internal cohesion; and conceptually, an organism will not have a single "agenda" because this requires that most of its genes have their evolutionary interests aligned. In the psychological case, intrapersonal conflict undermines the unity of purpose presumption on which folk-psychological attribution depends. In extreme cases, this may undermine unity of self—that is, there may be multiple selves within a single body. Complete unity of purpose in the psychological case, which implies unity of self, corresponds to complete suppression of selfish genetic elements in the biological case, which implies complete alignment of genes' interests and the existence of a single organismic agenda.

The parallel as developed here is thematic rather than causal, in contrast with the Trivers–Haig hypothesis that intra-genomic conflict has led to a divided self in humans. Like any thematic parallel, whether one regards it as significant or not is partly a subjective matter. In this case I do regard the parallel as significant, given the isomorphism of adaptive explanation in biology with intentional explanation in psychology. A standard adaptive explanation of an organismic trait shows how the trait furthers the organism's goal or agenda; this presupposes that the organism does have an agenda and that the trait in question contributes to it. A standard intentional explanation of a human behavior proceeds by showing that the behavior furthers the person's goal; this presupposes that there is a single person and that they exhibit sufficient unity of purpose to be truly credited with the goal in question. In both cases, the explanations only work to the extent that internal conflict is the exception rather than the rule.

Acknowledgments

Thanks to Manus Patten, Topaz Halperin, Ulrich Stegmann, Arvid Ågren, Nick Shea, David Haig, and the late Daniel Dennett for discussion of this material. Thanks to Manus Patten, Stu West, Tom Scott, and Philippe Huneman for comments on an earlier version. This chapter is part of a project that has received funding from the European Research Council (ERC) under the European Union's Horizon 2020 research and innovation program (grant no. 101018523).

References

Bayne, T. (2010). *The Unity of Consciousness*. Oxford University Press.

Brocas, I., and Carrillo, J. D. (2014). Dual-process theories of decision making: A selective survey. *Journal of Economic Psychology 41*, 45–54.

Brook, A. (2015). Disorders of unified consciousness: Brain bisection and dissociative identity disorder. In *Disturbed Consciousness: New Essays on Psychopathology and Theories of Consciousness,* edited by R. J. Gennaro, 209–225. MIT Press.

Brook, A., and Raymont, P. (2021). The unity of consciousness. In *The Stanford Encyclopedia of Philosophy,* edited by Edward N. Zalta, Summer. https://plato .stanford.edu/archives/sum2021/entries/consciousness-unity/.

Burt, A., and Trivers, R. (2006). *Genes in Conflict: The Biology of Selfish Genetic Elements*. Harvard University Press.

Crespi, B. (2008). Genomic imprinting in the development and evolution of psychotic spectrum conditions. *Biological Reviews, 83,* 441–493.

Dawkins, R. (1990). Parasites, desiderata lists and the paradox of the organism. *Parasitology, 100,* S63–S73.

Dennett, D. C. (1987). *The Intentional Stance*. MIT Press.

Dennett D. C. (2013). The normal well-tempered mind: A conversation with Daniel C. Dennet. *Edge,* January 8. http://edge.org/conversation/the-normal -well-tempered-mind/.

Elster, J. (2009). *Ulysses Unbound.* Cambridge University Press.

Gardner, A., and Welch, J. J. (2011). A formal theory of the selfish gene. *Journal of Evolutionary Biology, 24,* 1801–1813.

Haig, D. (2002). *Genomic Imprinting and Kinship.* Rutgers University Press.

Haig, D. (2006). Intrapersonal conflict. In *Conflict,* edited by M. K. Jones and A. C. Fabian, 8–22. Cambridge University Press.

Haig, D. (2008). Conflicting messages: Genomic imprinting and internal communication. In *Sociobiology of Communication: An Interdisciplinary Perspective,* edited by P. d'Ettorre and D. P. Hughes, 209–224. Oxford University Press.

Haig, D., and Grafen, A. (1991). Genetic scrambling as a defense against meiotic drive. *Journal of Theoretical Biology, 153*(4), 531–558.

Heil, J. (1989). Minds divided. *Mind, 98*(392), 571–583.

Humphrey, N., and Dennett, D. C. (1989). Speaking for ourselves: An assessment of multiple personality disorder. *Raritan, 9*(1), 68–98.

Kennedy, J. S. (1992). *The New Anthropomorphism.* Cambridge University Press.

Kennett, J., and Matthews, S. (200). The unity and disunity of agency. *Philosophy, Psychiatry, and Psychology, 10*(4), 305–312.

Leigh, E. G. (1971). *Adaptation and Diversity: Natural History and the Mathematics of Evolution.* Cooper.

Okasha, S. (2018). *Agents and Goals in Evolution.* Oxford University Press.

Queller, D. C., and Strassmann, J. E. (2009). Beyond society: The evolution of organismality. *Philosophical Transactions of the Royal Society B, 364*(1533), 3143–3155.

Radden, J. (2011). Multiple selves. In *The Oxford Handbook of the Self,* edited by S. Gallagher, 547–570. Oxford University Press.

Rautiala, P., and Gardner, A. (2023). The geometry of evolutionary conflict. *Proceedings of the Royal Society B, 290*(1992), https://doi.org/10.1098/rspb.2022.2423.

Rovane, C. (1997). *The Bounds of Agency.* Princeton University Press.

Schechter, E. (2018). *Self-Consciousness and "Split" Brains: The Minds' I.* Oxford University Press.

Scott, T. W., and West, S. A. (2019). Adaptation is maintained by the parliament of genes. *Nature Communications, 10*(5163).

Spurrett, D. (2016). Does intragenomic conflict predict intrapersonal conflict? *Biology and Philosophy, 31,* 313–333.

Trivers, R. (2009). Genetic conflict within the individual. *Sonderdruck der Berliner-Brandenburgische Akademie der Wissenschschaften 14,* 149–199. Akademic Verlag.

Wilkes, K. V. (1981). Multiple personality and personal identity. *British Journal for the Philosophy of Science 32*(4), 331–348.

CHAPTER TEN

Host-Symbiont Conflict in Reef-Building Corals and the Risk of Coral Bleaching

DAKOTA E. McCOY, BRENDAN CORNWELL, SÖNKE JOHNSEN, AND JENNIFER A. DIONNE

ABSTRACT Tropical coral reef ecosystems depend on the solar-powered symbioses between coral animals and photosynthetic dinoflagellate microalgae, often considered a mutualistic relationship. However, corals can experience conflict with their microalgae, potentially destabilizing their symbiosis compared to more mutualistic partnerships. Corals may parasitize their symbionts, or vice versa. We propose a continuum from mutualistic to parasitic interactions, from cooperation to conflict, among reef-building corals and other photosymbiotic creatures. We outline five axes of quantifiable conflict: (1) reproductive mode, (2) host sex, (3) resource distribution, (4) solar irradiation, and (5) symbiont housing. For each of these five traits we present metrics of conflict and examples from coral and other symbiotic relationships. Host-symbiont conflict in corals may cause coral bleaching, a catastrophic breakdown of symbiosis that can destroy reef ecosystems. We hypothesize that evolutionary host-symbiont conflict makes some coral-symbiont systems unstable—prone to bleaching—in the face of even small temperature fluctuations. The proposed evolutionary conflict perspective opens new avenues for ecological interventions to protect reefs. Additionally, corals and other photosymbiotic animals are a useful case study to illuminate a fundamental question: How do symbioses evolve from conflict to cooperation and vice versa? Finally, coral reefs offer a striking example of how evolutionary conflicts at small scales may have dire consequences for entire ecosystems.

Introduction

Coral Reefs Are Threatened by Paradoxical Bleaching, Perhaps Explainable by Host-Symbiont Conflict. Tropical reefs teem with life. They sustain countless marine species, shore up coastlines against erosion and rising seas, and feed millions of humans. A coral colony is animal, vegetable, and mineral (Figure 10.1). A thin layer of coral animal tissue, punctuated by polyps, spreads over the surface of a calcareous skeleton. Single-celled plants, microalgae often called zooxanthellae (Box 10.1), live inside coral gastrodermal cells where they photosynthesize to provide energy. Diverse

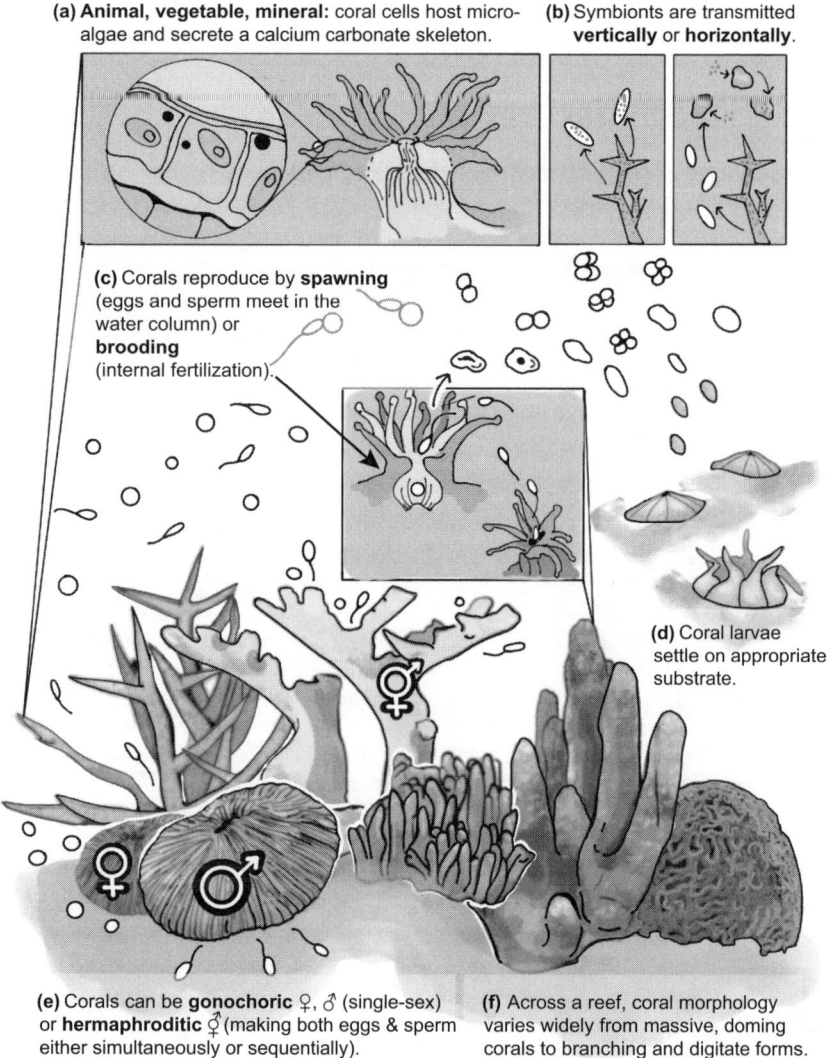

(a) Animal, vegetable, mineral: coral cells host micro-algae and secrete a calcium carbonate skeleton.

(b) Symbionts are transmitted **vertically** or **horizontally**.

(c) Corals reproduce by **spawning** (eggs and sperm meet in the water column) or **brooding** (internal fertilization).

(d) Coral larvae settle on appropriate substrate.

(e) Corals can be **gonochoric** ♀, ♂ (single-sex) or **hermaphroditic** ⚥ (making both eggs & sperm either simultaneously or sequentially).

(f) Across a reef, coral morphology varies widely from massive, doming corals to branching and digitate forms.

Figure 10.1. Overview of coral life history and reproduction.

bacteria, fungi, and viruses dwell in nooks and crannies to round out the coral holobiont. Many other animals house photosynthetic symbionts, from worms and sponges to clams and sea slugs.

Most tropical corals build their own skeleton through calcification, which forms a habitat for diverse assemblages of fish and marine invertebrates. A single coral polyp grows and divides to become a large colony made up of hundreds or thousands of genetically identical polyps. Coral

BOX 10.1

The Dinoflagellate Microalgae Symbiont

Corals host unicellular microalgae inside their cells. In general, scientists know less about these symbionts than about coral reef ecology and coral biology (Nitschke et al. 2022; Matz 2024). These eukaryotic microalgae in corals, often termed zooxanthellae, are dinoflagellates (order Dinophyceae; family Symbiodiniaceae; formerly, all genus *Symbiodinium*) (LaJeunesse 2020; Nitschke et al. 2022). Symbiodiniaceae originated in the Jurassic Period about 160 million years ago and diversified alongside a major radiation of reef-building corals (LaJeunesse et al. 2018). Across Symbiodiniaceae species, and even within a single species, these microalgae vary tremendously in their growth rate, tolerance to heat stress, photosynthetic efficiency, physiology, and more (Parkinson et al. 2025).

Some Symbiodiniaceae species live exclusively in a free-living form, while the famous symbiotic forms can live either inside the host or freely in the environment. Inside a host, members of Symbiodiniaceae measure 5–13 μm in diameter, are spherical, lack flagella, and have cellulose cell walls (Sheppard et al. 2018; LaJeunesse 2020). In the little-understood free-living state, Symbiodiniaceae seem to alternate between an immobile spherical shape and a mobile "dinomastigote" shape with two round halves and two flagella that allow them to swim (Sheppard et al. 2018). These free-living symbionts live in the water column, on sediments, and on macroalgae surfaces (Fujise et al. 2021). Symbiodiniaceae species can reproduce asexually through mitosis from their spherical form, or sexually, but we know little about their sexual reproduction (Figueroa et al. 2021).

These dinoflagellates photosynthesize to produce and transfer large amounts of glucose, amino acids, and other organic compounds to their host. In turn, the host feeds on particulate matter and provides nutrients to the symbiont. Symbiodiniaceae radiated to specialize within various animal hosts—including corals, sponges, clams, anemones, jellyfish, and flatworms (Thornhill et al. 2014)—and unicellular hosts including ciliates and foraminifera (Parkinson et al. 2025).

Most corals acquire symbionts from the environment, although some vertically transmit them through eggs (see Box 10.2, Figure 10.1, and Figure 10.2). To survive, corals must acquire symbionts during the planula or newly settled polyp stage (Sheppard et al. 2018).

Scientists do not fully understand how corals recruit, select, and acquire symbionts, but it starts when the coral cell envelops the symbiont through phagocytosis (Davy et al. 2012). Corals attract symbionts by emitting nitrogen-containing compounds such as ammonium (Fitt 1985), and through green fluorescence that attracts symbiotic microalgae (Aihara et al. 2019). The microalgae actively swim toward the coral, even against a current (Pasternak et al. 2006). Once acquired, symbionts live in the corals' gastrodermal cells, which border the gastrodermal cavity of the host (Davy et al. 2012). Corals usually host one dominant symbiont species that arises from a single cell dividing mitotically (LaJeunesse 2020; Parkinson et al. 2025).

After a bleaching event, no one knows for sure how the ejected Symbiodinaceae fare. In some cases, they probably die due to broken membranes or destroyed photosystems (Franklin et al. 2004). Further work is needed. Certain Symbiodiniaceae species resist environmental stress, like the heat- and cool-tolerant *Durusdinium trenchii,* which can remain in coral after bleaching events (Nitschke et al. 2022). Conversely, the genus *Cladocopium* dominates coral ecosystems but seems more susceptible to heat. As Parkinson and colleagues (2025) point out, however, these microalgae are so diverse that it is difficult to generalize at the genus level or even the species level.

Does the coral or the dinoflagellate control the symbiosis? It is not clear (LaJeunesse 2020). The coral exerts some control by enclosing the symbiont within not only cytoplasm and host membrane but also a specialized "symbiosome membrane" that cuts it off from outside nutrients. As for who controls bleaching events— do corals expel algae? Do algae flee?—both sides may exert influence, and the term "bleaching" encompasses a wide variety of events (Parkinson et al. 2025) that likely fall under different regulatory controls. However, the coral does seem to control symbiont populations through cellular processes such as autophagy, cell-cycle arrest, and apoptosis (Gorman et al. 2025).

colonies and the reef ecosystem depend on their solar-powered symbioses with the photosynthetic microalgae. The coral–microalgae relationship is classically considered a mutualistic symbiosis that benefits both partners. The symbionts receive nutrients (carbon dioxide, ammonium, and key micronutrients) and shelter; the corals receive energy (in the form of sugars) that is hard to come by in the characteristically nutrient-poor waters where they live.

As waters warm, the symbiosis typically fails. Corals bleach. They lose their symbionts, turn white, and usually die. It is not clear whether the coral or the microalgae controls bleaching or, indeed, what fate befalls the expelled microalgae (LaJeunesse 2020; Parkinson et al. 2025) (see Box 10.1). However, we know that the coral requires symbionts to survive but microalgae can live on their own. Sometimes, a colony can recover from a bleaching event by recruiting new symbionts or allowing remaining symbionts to reproduce and spread within the coral (Nakamura et al. 2003; Rodrigues and Grottoli 2007; Seneca and Palumbi 2015; Thomas and Palumbi 2017; Cornwell et al. 2021; Galindo-Martínez et al. 2022; Walker et al. 2023). Some colonies and reefs are resilient in the face of climate change (Bay and Palumbi 2014; Palumbi et al. 2014; Humanes et al. 2022). But with rapid climate change, corals' natural mechanisms of resilience are inadequate (e.g., Ainsworth et al. 2016; but see Lachs et al. 2023). We have lost over half of the world's coral reefs since 1950 (Eddy et al. 2021). Coral bleaching is an urgent climate catastrophe (Carpenter et al. 2008; Knowlton et al. 2021).

But coral bleaching is also a biological paradox. Why do corals eject the microalgae partners they need to survive? Why does a mere 1°C of warming kill many reef-building coral species but spare other corals and analogous species of photosymbiotic sponges, bivalves, and worms? We hypothesize that the coral symbiosis itself must be subject to countervailing evolutionary pressures that prevent it from becoming more stable. One such evolutionary pressure may be conflicts of interest between the coral host and the symbiotic microalgae.

We propose that host-symbiont conflict drives coral bleaching, which we view as a breakdown of symbiosis. Bleaching-prone reef corals may cooperate less with their symbionts than do hardier corals and other photosymbiotic animals. Less cooperative symbioses are unstable because hosts and parasites rapidly coevolve in an arms race (Woolhouse et al. 2002; Buckingham and Ashby 2022). An arms race is like a tug-of-war; if both sides are evenly matched, the rope is stable. But a small slip of the hands—small mutations or environmental changes—sends the rope flying (Moore and Haig 1991; Haig 1993, 2019; Brandvain and Haig 2018; McCoy and Haig 2020). Alternatively, many scientists consider bleaching to be a maladaptive side-effect of climate change on a fully mutualistic symbiosis. The distinction matters not only to help us understand evolution but also to open new avenues for conserving and rebuilding reefs, for example by engineering "cooperative" microalgae or corals.

To test whether bleaching-prone corals have more parasitic symbioses, we propose strategies to quantify the degree of conflict in photosynthetic symbioses.

Corals and Other Symbiotic Partnerships Vary from Parasitic to Mutualistic. Many seeming mutualisms conceal conflicts. Hosts and symbionts clash over who gets which nutrients and how they are divided; who gets to reproduce and how often; who controls basic biological processes in the symbiotic organism; and more. When plants host nitrogen-fixing bacteria and fungi (Heath and Tiffin 2007), they sanction selfish partners (Kiersi and Denison 2008). When *Wolbachia* bacteria inhabit invertebrates (Werren et al. 2008), they bias their own transmission by feminizing or killing males because they transmit only through the egg (Stouthamer et al. 1999; Werren et al. 2008; Toft and Andersson 2010; Cordaux et al. 2011). Mutualisms benefit the direct fitness of both partners, while parasitisms benefit one partner at a direct fitness cost to the other (West et al. 2007b). As plants and *Wolbachia* have taught us, most symbioses sit somewhere in between. But so far, the coral–microalgae symbiosis—widely considered a mutualism—has largely escaped such scrutiny.

Corals may parasitize their symbiotic microalgae more than other animals. Indeed, there is little evidence that coral–microalgae symbioses are mutualistic at all (Wooldridge 2010; Blackstone and Golladay 2018; Matz 2024). Corals certainly benefit from symbionts, who photosynthesize inside corals and provide energy-rich sugars to their hosts. But do symbionts benefit from corals? Symbionts receive nutrition, and perhaps protection. However, corals restrict symbiont reproduction, trap them in regions of high irradiation, expel symbionts who do not produce and share enough photosynthate, and isolate symbionts singly inside coral cells. Corals even feed on their symbionts (Titlyanov et al. 1996; Wiedenmann et al. 2023) and withhold nitrogen to restrict symbiont growth (Matz 2024). Perhaps symbionts get less out of the supposed mutualism than corals do. In other animals that host photosymbionts, the symbionts are often housed communally, allowed to reproduce, granted a greater portion of photosynthate, and protected from irradiation (Venn et al. 2008).

Corals and other photosymbiotic animals seem to occupy a variety of positions along the parasitism-mutualism spectrum. But to take this idea one step further we must solidify the sometimes nebulous idea of conflict. What does it mean for an animal or endosymbiont to receive a benefit from a symbiotic relationship (Douglas and Smith 1989; Law and Dieckmann 1998)? We can test whether an organism performs better in symbiosis or in isolation, but challenges arise. What does it mean to perform better? For symbionts that can also live freely, we have little information about how successful their life is outside of a host (Box 10.1); for symbionts that cannot live freely, we cannot distinguish between benefit and dependence (Douglas and Smith 1989). Perhaps the free-living ancestors of some symbionts had greater fitness than their dependent, domesticated descendants.

We propose five trait categories to quantify conflict in corals and other photosynthetic animals (Table 10.1). These joint phenotypes are loci of conflict where trait changes help one partner but harm the other (Queller 2014; Queller and Strassmann 2018): (1) reproductive mode, (2) host sex, (3) resource sharing, (4) solar irradiation, and (5) symbiont housing. When the coral or microalgae adopt selfish strategies at the expense of the other, this internal conflict threatens the form and fitness of the whole organism (Schenkel et al. 2024). To borrow Patten and colleagues' (2023) classification of organismal conflicts over transmission versus traits, we offer two categories of transmission distorters—(1) reproductive mode and (2) host sex—and three categories of trait distorters—(3) resource sharing, (4) solar irradiation, and (5) symbiont housing. Each trait varies among coral species and even within individual coral colonies over time.

We predict that coral-microalgae conflict in these domains will correlate with bleaching risk. That is, a changing environment may particularly harm higher-conflict symbioses. Further, by quantifying these traits in animal-microbe photosymbioses, we can ask evolutionary questions such as whether mutualism tends to evolve from parasitism or vice versa.

Reproductive Mode: Symbionts Are Passed Down or Acquired from the Environment

Hypothesis: Animal hosts that reproduce together with their symbionts through vertical transmission have more mutualistic symbioses because host and symbiont fates are tied.

Synopsis: Many reef-building corals reproduce independently of their symbionts, a predictor of evolutionary conflict. When larvae

Table 10.1. Five Domains of Quantifiable Conflict in Coral-Microalgae Symbiosis

Trait	Theoretical predictions	
	High-conflict state	Low-conflict state
Reproductive mode *Corals pass down microalgae through eggs or get them from the environment (vertical or horizontal transmission).*	Coral reproduces without symbionts (horizontal transmission), so symbionts meet an evolutionary dead end and have little incentive to improve host fitness.	Coral and symbiont reproduce together (vertical transmission), so their evolutionary interests are aligned. Symbionts are incentivized to improve host fitness.
Host sex *Male coral polyps are an evolutionary dead end for microalgae, who only transmit through the egg in female vertical transmitters.*	Symbionts would rather be in a female. Symbionts can bias their own transmission at the cost of host fitness by feminizing hosts, killing males, and more (selfish strategies are not yet known in coral but widespread in other creatures).	Symbionts reproduce with the host and do not need to alter sex determination in hosts (e.g., in hermaphroditic corals, at least some of the symbionts can reproduce through the female line).
Resource sharing *Coral hosts extract more or less photosynthate from microalgae.*	Photosynthate is inequitably shared (e.g., some corals extract 95% of the photosynthate from microalgae).	Photosynthate is more equitably shared; microalgae retain a substantial portion for their own cellular and reproductive needs.
Solar irradiation *Corals irradiate symbionts to increase productivity or screen light and protect microalgae from stress.*	Corals expose symbionts to irradiation to maximize photosynthetic production but therefore hold symbionts under high stress levels.	Symbionts are protected from harmful UV radiation and intense light; they are maintained at lower, safer levels of productivity.
Symbiont housing *Symbionts may be housed in large groups or isolated within individual membranes inside coral cells.*	Corals isolate approximately one symbiont per host cell to restrict nutrients, restrict reproduction, and detect cheaters.	Symbionts are housed communally and/or flexibly, with freedom to move, reproduce, and interact with environmental nutrients.

BOX 10.2

Coral Reproduction and Life Cycles

Corals are diverse. Like many marine invertebrates, corals have biphasic life histories that can vary in (a)sexual reproduction, parental investment in offspring, pelagic larval durations, and timing of sexual reproduction, among many other factors (Sheppard et al. 2018). Here we present a brief overview of these traits and outline how they affect their partnership with members of the Symbiodiniaceae.

Most corals can reproduce asexually and sexually (Richmond 1997). Most corals are also colonial, meaning that they are composed of many genetically identical polyps. Colonies grow when new polyps bud off from older polyps, and polyps across the colony can communicate with one another and share resources. However, polyps do not need to be physically connected to survive, which means that if the colony is broken into pieces, each piece can continue growing. Therefore, among corals growing on a reef, many individuals could be genetically identical. Corals can also undergo sexual reproduction where individuals produce either eggs, sperm, or both (Figure 10.2).

Coral reproduce sexually through brooding or, as is more common (Baird et al. 2009; Hartmann et al. 2017), through broadcast spawning (Figures 10.1 and 10.2). Spawning colonies release gametes into the water column, often coordinating with exquisite precision so that all colonies of a species across a region spawn at the same hour on a single night each year. After fertilization, the embryo will develop into a planula larva that will reside in the water column for hours or weeks before it tries to settle on suitable substrate (Figure 10.1). Planula larvae do not feed, but vertical transmitters' larvae have symbionts who provision them with photosynthate.

Alternatively, corals can brood their larvae: After fertilization the planula develop inside the parent colony, typically spend a limited time in the plankton (if any), and are therefore less likely to travel long distances before settling on a hard substrate.

Finally, symbionts can either be maternally provisioned or environmentally acquired (i.e., vertically or horizontally transmitted; see Figure 10.2) (Baird et al. 2009; Hartmann et al. 2017).

These two categories describe how symbionts are passed down from generation to generation of coral hosts (Figure 10.1). Maternal (vertical) transmission typically involves the parent preloading eggs with symbionts that will then give rise to the next generation of corals with those same symbionts. Environmental (horizontal) acquisition tends to occur near the end of the larval stage around settlement. In this case, the planula develops in the plankton without symbiont partners and then acquires symbionts from its environment (likely from symbiont populations in the benthos, although much less is known about this stage of the symbiont life cycle). Hosts and symbionts are significantly more likely to coevolve in vertical transmitters (Swain et al. 2018). In these cases, the symbiont's fitness partially relies on the reproductive success of their coral host, which propagates the symbiont to offspring. Consequently, vertical transmitters are expected to exhibit less conflict, as the fitness of both partners aligns with the host's reproductive success (Swain et al. 2018; Drew et al. 2021).

pick up symbionts from their environment ("horizontal transmission"), the symbiont has little evolutionary incentive to improve the reproductive fitness of the host. In contrast, "vertical transmitters," who transmit their symbionts from parent to egg, have more tightly aligned evolutionary interests. What is good for the goose is good for the gosling—and for all the gosling's vertically inherited symbionts. Generally, vertical transmission selects for cooperation while horizontal transmission selects for conflict (Drew et al. 2021). We hypothesize that vertically transmitting corals have a more mutualistic, stable symbiosis with, e.g., equitable resource sharing.

In Corals: Most Reef-Building Corals Horizontally Acquire Symbionts, a Seeming Paradox

Most scleractinian coral species are horizontal transmitters who reproduce independently of their symbionts. Estimates vary from approximately 71% (Fadlallah 1983), to as many as 85% (Harrison and Wallace 1990). Transmission mode is tightly tied to reproductive strategy (Figure 10.2). Corals either spawn (males and females broadcast eggs and

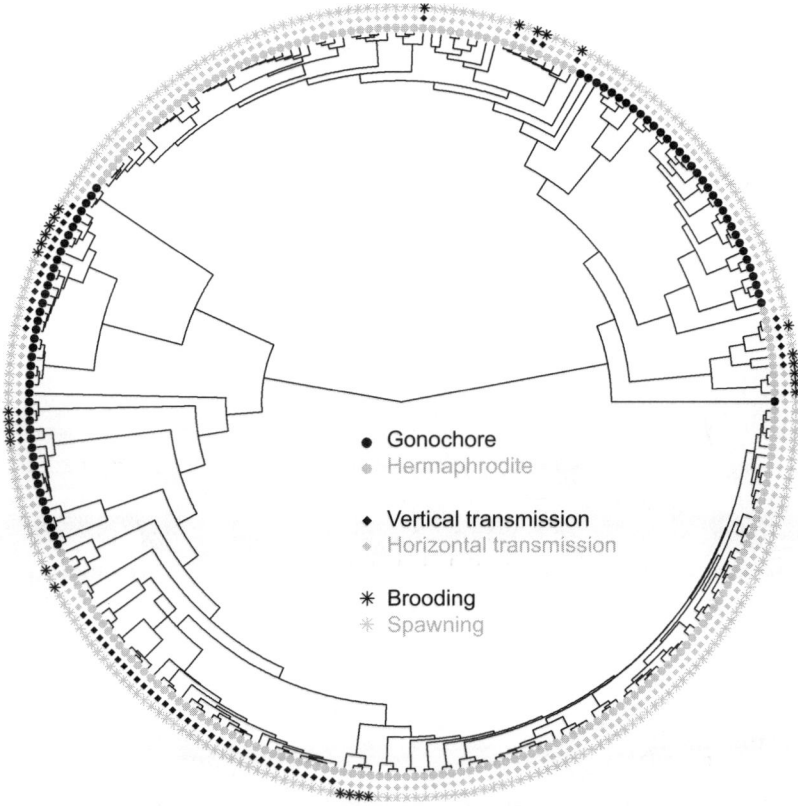

Figure 10.2. Scleractinian corals vary in sexual system, symbiont transmission mode, and reproductive strategy. Data sources: Gault, J. et al. (2021a). Lineage-specific variation in the evolutionary stability of coral photosymbiosis. [Dataset]. Dryad; Gault, J. et al. (2021b). Lineage-specific variation in the evolutionary stability of coral photosymbiosis. *Science Advances*, 7(39); Hartmann, A. C., et al. (2017). The paradox of environmental symbiont acquisition in obligate mutualisms. *Current Biology*, 27(23), 3711–3716; Madin, J. S., et al. (2016). The Coral Trait Database, a curated database of trait information for coral species from the global oceans. *Scientific Data*, 3(1), 160017.

sperm) or brood (eggs are internally fertilized by sperm, develop in the polyp, then disperse as motile larva) (Baird et al. 2009). The majority of spawners horizontally acquire symbiotic microalgae (75%), while the majority of brooders vertically transmit their symbionts (~90%) (Baird et al. 2009) (Figure 10.2). Corals that must acquire microalgae after fertilization pick up symbionts in the planula stage or as a recently settled polyp (Box 10.2) (Sheppard et al. 2018). Interestingly, corals have convergently evolved vertical transmission at least four times, and horizontal and vertical transmitters may even coexist within a genus (Baird et al.

2009; Hartmann et al. 2017). Likewise, brooding, which is most often found in vertical transmitters, has arisen multiple independent times among both gonochoric and hermaphroditic lineages; brooding is much more common in the Atlantic (48%) than in the Indo-Pacific (10%) (Kerr et al. 2010). Some corals categorized as "vertical transmitters" also recruit symbionts from the environment through horizontal transmission. For example, scientists have shown that larvae of the brooding and horizontally transmitting coral *Seriatopora hystrix* only inherited 33% of the symbiont genetic variability from parents, indicating substantial horizontal transmission as well (Quigley et al. 2018).

Vertically Transmitting Corals May Be More Robust to Environmental Perturbation and Have Stabler Symbiotic Associations. Scientists predict that vertical transmitters will have stabler symbioses than horizontal transmitters (Swain et al. 2018; Drew et al. 2021). Some vertically transmitting coral are more resistant to environmental change (van Woesik et al. 2011; Putnam et al. 2012) and have more consistent symbiont populations over time compared to horizontal transmitters. For example, vertically transmitting *Porites* corals may be more stable than horizontally transmitting acroporids.

During bleaching events, some vertical transmitters seem more resilient and stable (Thornhill et al. 2006). For example, three brooding Atlantic corals (*Agaricia agaricites, Porites astreoides,* and *Siderastrea radians*) did not change their symbiont communities during environmental fluctuations and a moderate bleaching event (Thornhill et al. 2006). In other studies, researchers directly compared brooders, typically vertical transmitters, with spawners, which are typically horizontal transmitters. Brooders outperformed spawners in Caribbean reefs after overfishing, heat waves, and outbreaks of disease in the 1980s and 1990s (Hughes 1994; Knowlton 2001). After a mass bleaching in Belize, brooders outnumbered spawners among new coral recruits (McField 1999). Brooders survived at higher levels (73%) than spawners (29%) through the Oligocene-Miocene extinction, a time of great global cooling during which about half of all Caribbean corals went extinct (Edinger and Risk 1995). However, after the 1998 mass bleaching event on the Great Barrier Reef, brooders recovered more slowly than spawners (Emslie et al. 2008).

Kenkel and Bay (2018) tested how thermal stress affected cooperativity in three vertically transmitting coral species and their closest horizontally transmitting relatives. They quantified cooperation as the percent of photosynthetically fixed carbon produced by the symbiont and translocated to the host; parasitism was defined as the difference in that value between heat-exposed and control fragments. They found no clear link between parasitism and vertical transmission, but parasitism correlated with symbiont diversity during bleaching—suggesting, as some theory predicts, that genetic homogeneity of symbionts (rather than vertical transmission per se) drives mutualism (Kenkel and Bay 2018; see also Thornhill et al. 2006 and "Theory Predicts that Mutualism Is Tied to Vertical Transmission"). Perhaps genetically homogenous coral-symbiont-holobionts operate more effectively, like a superorganism, with less intrinsic conflict.

Corals May Horizontally Transmit Symbionts to Protect Sea-Surface-Going Larvae or Cope with Environmental Change. Why do so many corals horizontally transmit their symbionts, if theory predicts that selection should favor reproducing together? A recent phylogenetic study proposes an answer: Spawning helps dispersal, but the hot, bright sea-surface conditions necessarily encountered by dispersing coral larvae and gametes are too harmful for symbionts (Hartmann et al. 2017). High levels of light and ultraviolet (UV) radiation harm the microalgae's photosystem and cause them to release toxic factors into the coral larvae (Hartmann et al. 2017). In the rare spawning corals that also vertically transmit symbionts, eggs are smaller and have less wax ester (possibly adaptations to allow the eggs to sink away from the surface more quickly), as well as higher amounts of antioxidants and photoprotective compounds (independent of the parental light environment) compared to eggs of symbiont-free relatives. In a separate study, *Acropora intermedia* coral larvae with symbionts indeed face greater stress and mortality in simulated ocean surface conditions (Yakovleva et al. 2009).

Other factors favor horizontal transmission in corals. In the short term, corals that can acquire symbionts from the environment may be more flexible in the face of environmental change (Lesser et al. 2013; Bennett and Moran 2015), and capable of swapping out weaker for tougher symbionts (Herre et al. 1999). After mass bleaching, symbiont communities often change dramatically toward heat-tolerant clades (Quigley et al. 2022; Palacio-Castro et al. 2023). In horizontally transmitting species, host and symbiont independently adapt to local conditions based on gene flow rates, effective population size, and selection strength. Coral

may have limited local adaptation because larvae are planktonic, so gene
flow is high between distant locations with different environments (Cornwell 2020; Fuller et al. 2020). Conversely, symbiont populations tend to
be highly genetically structured (Santos et al. 2003; Andras et al. 2013;
Cornwell and Hernández 2021), suggesting that microalgae can adapt to
local conditions. Therefore, coral hosts who horizontally acquire symbionts can enjoy the benefits of a locally adapted symbiont. Finally, coral
transmission mode correlates with life history (e.g., competitive, weedy,
stress-tolerant, or generalist) (Darling et al. 2012).

In sum, many factors favor horizontal symbiont transmission in most
reef-building corals. But as a consequence, the evolutionary fates of many
coral species are less tied to those of their symbionts. Therefore, we predict higher levels of host-symbiont conflict in horizontally transmitting
corals.

Beyond Corals: Theory, Experiment, and Observation Show That Vertical Transmission Is Tied to Cooperativity

Theory Predicts That Mutualism Is Tied to Vertical Transmission. When
hosts vertically transmit their symbionts from parent to offspring, symbionts who help their host are also helping themselves. Horizontally
transmitted symbionts benefit less from improving their host's fitness
(Ewald 1987). Theory predicts that in mutualistic symbioses, hosts
should vertically transmit their symbionts (Anderson and May 1982;
Frank 1996; Sachs et al. 2004; Lesser et al. 2013; Hartmann et al.
2017). Indeed, along the evolutionary spectrum from parasitism to mutualism, many scholars identify the origin of vertical transmission as a
key step—perhaps even an essential one—in the origin of mutualism
(Ewald 1987; Toft and Andersson 2010). Four arguments support this
view (Wilkinson and Sherratt 2001): (1) shared reproduction reduces
the incentive to exploit a host (Sachs et al. 2004); (2) offspring need not
spend energy finding and recruiting symbionts; (3) genetically uniform
symbionts are cheaper (Douglas 1994), less prone to cheating (Maynard
Smith 1998), and less competitive with one another (Frank 1996); and
(4) hosts can inherit communities of coadapted symbionts, experimentally engineering a better internal ecosystem (Wilkinson and Sherratt
2001). However, horizontally transmitting symbioses are common and
can be stable and mutualistic, particularly if symbionts are genetically

homogenous, cheaters can be punished, or there are few alternatives to symbiosis (Herre et al. 1999).

Experiments Show That Horizontal Transmission Increases Parasitism in Non-Coral Symbioses. In the laboratory, symbionts forced to transmit horizontally tend to become more parasitic (Bull et al. 1991; Sachs and Wilcox 2005; Stewart et al. 2005; Dusi et al. 2015). In one study of bacteria and symbiotic filamentous bacteriophages, selection favored benevolent symbionts when transmission was limited to vertical but not when horizontal transmission was allowed (Bull et al. 1991). The newly vertically transmitting bacteriophages lost the ability to transmit horizontally; that is, they became dependent on the symbiosis.

Scientists have experimentally evolved jellyfish *Cassiopea xamachana* and their symbiont *Symbiodinium microadriaticum* to enforce either vertical or horizontal transmission. Horizontally transmitted symbionts became more parasitic; they reproduced and dispersed more per host body mass and they decreased host fitness and growth (Sachs and Wilcox 2005). However, the smaller hosts also restricted symbiont fitness (Sachs and Wilcox 2005). Parasitism may lead to pyrrhic victories!

In experimentally evolved barley stripe mosaic virus and its barley host *Hordeum vulgare*, vertical transmission is associated with mutualism. Over four generations of horizontal transmission, virulence tripled, as measured by lifelong viable seed output; after three subsequent generations of vertical transmission, virulence decreased by 40% (Stewart et al. 2005).

Researchers experimentally evolved the protozoan *Paramecium caudatum* and its parasite *Holospora undulata* to show that a parasite with both vertical and horizontal transmission can evolve into a less-harmful symbiont with vertical transmission (Dusi et al. 2015).

While comparable studies of transmission mode and parasitism are rarer in wild populations (Kenkel and Bay 2018), Herre (1993) examined eleven species of fig wasp (*Pegocapsus* or *Tetrapus* spp.) and their corresponding nematode parasites (*Parasitodiplogaster* spp.). Sometimes one wasp infects a fig, allowing only vertical transmission; other times multiple wasps swap nematodes inside the same fig. In these wasps, vertical transmission is negatively correlated with virulence.

Overall, vertical transmission seems to increase mutualism. But nature has many exceptions. For example, in humans, some vertically transmitted human pathogens may be less virulent to the mother but devastating to the child, like rubella, syphilis, Zika, and more. Broadly, infectious disease research can help us understand the parasitism-mutualism continuum of symbiosis.

Predictions: Vertically Transmitting Corals Will Be More Mutualistic and Stable

We predict that vertically transmitting corals have a more mutualistic, stable symbiosis. That is, we predict that they will bleach less easily, share more nutrients with their symbionts, subject their symbionts to less extreme radiation, and more. We expect more beneficial clades of Symbiodiniaceae to be associated with vertically transmitting corals (Lesser et al. 2013). Vertical transmitters that do not brood and horizontal transmitters that do not spawn deserve further study to determine the significance of the transmission mode itself. Corals could be compared to the well-studied system of bioluminescent *Vibrio* bacteria in bobtail squid and other fauna (McFall-Ngai 1999, 2008; Nyholm and McFall-Ngai 2004, 2021; McFall-Ngai and Ruby 2021).

Are mutualisms more likely to evolve from parasitisms, and vertical transmission from horizontal transmission, or vice versa? In bacteria, most mutualists evolved from parasitic rather than free-living ancestors, and mutualists have low diversification rates—suggesting that mutualism most often arises from parasitism (Sachs et al. 2014). Vertical transmission can evolve from horizontal even in an exploitative symbiosis (Moran 2007), which some authors describe as making the best of a bad situation (Law and Dieckmann 1998; Wilkinson and Sherratt 2001). Drew et al. (2021) reported many cases where symbionts evolved to be more mutualistic but nearly as many cases show the reverse. It is not easy to determine which transition is more common. One cannot always detect when symbioses dissolve (Sachs and Simms 2006), or when species go extinct. Mutualism and parasitism are themselves fluid and changing categories.

Host Sex: A Symbiont Would Rather Find Itself in a Female

Hypothesis: Symbiotic microalgae inside vertically transmitting corals will evolve strategies to feminize their host or feminize the sex ratio of subsequent generations.

Synopsis: Symbionts who find themselves inside a male have met an evolutionary dead end; they will not be transmitted to future generations. For symbiotic microalgae offered the opportunity to reproduce together with their coral host, they are transmitted through the egg. The sperm, which is produced by the father, typically does not contribute any cytoplasmic ingredients to animal offspring (for some fascinating exceptions, including freshwater mussels, see Bright and Bulgheresi 2010). Therefore, strong selective pressure favors a symbiont that can feminize its host or feminize the sex ratio of its host's offspring. In turn, the host is selected to evolve new, resistant means of sex determination and new techniques to separate symbionts from offspring sex determination. We predict that coral and symbiotic microalgae will show genetic or phenotypic signatures of symbiont attempts to feminize the host through parthenogenesis, male killing, phenotypic feminization of genetic males, or cytoplasmic incompatibility. Further, we predict that conflict over host sex will lead corals to rapidly change systems of sex determination and reproductive strategies.

In Corals: Very Little Is Known About Sex Determination or Any Feminizing Influence of Symbionts

To our knowledge, little is known about coral sex determination. Likewise, researchers have not documented any feminizing influence of Symbiodiniaceae. Below, we describe some interesting features of coral life history that hint at future research into host-symbiont conflict, sex determination, and reproductive strategies.

Corals Vary Between Gonochorism and Hermaphroditism, Within Species and Across the Phylogeny. Hermaphroditism can arise from genetic conflicts within animals (e.g., in scale insects; Gardner and Ross 2011). A selfish, feminizing endosymbiont could be predicted to favor the evolution of hermaphroditism in the host.

Some corals are hermaphrodites and some are gonochores; even closely related species can have opposite reproductive strategies (Figure 10.2; see Baird et al. 2009). Across the coral phylogeny, corals tend to transition from gonochorism to hermaphroditism; conversely, gaining a gonochoric life history is difficult and limited to brooders (Kerr et al. 2010). Hermaphroditism has evolved independently at least three times in primarily reef-building clades (Kerr et al. 2010). In addition, gonochorism is highly favored to be lost, and yet it is the predicted ancestral state of the common ancestor to scleractinian corals.

Further, certain coral species considered to be gonochoric can also be hermaphroditic. After eutrophication in Barbados, 2.7% of colonies (*n* = 10) of the mounding coral *Porites porites* showed hermaphroditism (Tomascik and Sander 1987). Nine of these colonies were mostly male with one to three ovaries while one was mostly female with one testis and three spermaries. At that reef, the male-to-female sex ratio among the otherwise gonochoric colonies was 2:1. In another surprising example, the supposedly gonochoric Indo-Pacific *Diploastrea heliopora* had both male and female polyps within a single colony, and even some hermaphroditic polyps with co-occurring mature ovaries and immature spermaries (Guest et al. 2012). Its polyps seemed to change sexes over time.

Might these examples of hermaphroditism represent evolutionary responses to a feminizing symbiont? If a host coral evolves to be hermaphroditic, symbionts found within that genetic individual are in luck: they can be passed to the next generation through female gametes. Therefore, symbionts would face lower selective pressure to exert a feminizing influence. Hermaphroditism may therefore be a strategy to limit the evolutionary incentives on symbionts to manipulate host sex. For now, this is just speculation. Many other selective forces favor hermaphroditism (Jarvis et al. 2022).

Parthenogenesis Can Occur in Multiple Species of Coral. Parthenogenesis, by which females produce offspring with no contribution from a male, is one host trait often induced by selfish endosymbionts. Parthenogenesis may be more widespread in corals than we currently suspect.

Parthenogenesis (and/or asexual planulae production) has been frequently described in *Pocillopora damicornis* (Ayre and Miller 2004). In one study, 94% of larvae were produced parthenogenetically, but ten of thirteen colonies also produced some sexual larvae (Combosch and Vollmer 2013). Likewise, twenty-four experimental colonies of the Florida coral *Porites astreoides* produced larvae exclusively through parthenogenesis (Vollmer 2018). The gonochoristic coral *Porites divaricata* reproduced asexually across a mangrove population, almost certainly due to parthenogenesis (Lord et al. 2023).

We raise two further comments of interest. First, the strange reproductive strategies of sea anemones, a close relative to corals, deserve further comparative study (Schlesinger et al. 2010; Armoza-Zvuloni et al. 2014). And second, in marine invertebrates, including cnidarians, parthenogenesis significantly coincides with brooding (Lively and Johnson 1997).

Corals Who Can Switch Symbiont Lineages May Be Able to Resist Any Feminizing Influence. Many coral species can switch symbiont lineages later in life, perhaps when they encounter environmental change or a better symbiont (Byler et al. 2013). Indeed, all reef-building corals may have this capacity. In one survey of fifty-four coral species and 106 symbiont types, no instances were found of a single host species interacting exclusively with a single symbiont lineage (Fabina et al. 2012). A dynamic, heterogenous coral holobiont might spare the host from the negative effects of a strictly vertically transmitted symbiont.

Beyond Corals: Endosymbionts and Parasites Manipulate Host Sex to Bias Their Own Transmission, Complicate the Evolution of Sex Determination in Hosts, and Reduce Host Fitness

Across the tree of life, many symbionts and parasites adopt nefarious strategies to maximize their own transmission at the cost of host fitness. Scientists have learned much from the model system of *Wolbachia,* a bacterial associate of insects, arthropods, worms, and more. *Wolbachia* ranges from parasitic to mutualistic (Werren et al. 2008). It is transmitted only through female eggs, and it has evolved a bewildering array of selfish strategies to help itself reproduce (Stouthamer et al. 1999; Werren et al. 2008; Toft and Andersson 2010). Four main strategies help selfish symbionts like *Wolbachia* bias their own transmission (Toft and Andersson 2010; Cordaux et al. 2011): parthenogenesis, killing males, feminizing males, and causing cytoplasmic incompatibility. As a result, hosts rapidly evolve new systems of sex determination and may face fitness costs or even extinction risk. Below, we give a brief overview of *Wolbachia* and other parasites/symbionts. We hope these biological stories will help scientists investigate similar, as-yet-unknown phenomena in corals and their microalgae.

Symbionts and Parasites Adopt Selfish Strategies to Increase Their Transmission: Parthenogenesis, Male Killing, Feminization, and Cytoplasmic Incompatibilities. Scientists describe four main strategies by which symbionts or parasites increase their transmission.

First, symbionts induce parthenogenesis, where mothers produce offspring with no male involvement. Symbionts will then occur in all descendants. *Wolbachia* disrupts early cell divisions in some mites, hymenopterans, and thrips—clades where males typically develop from unfertilized eggs. For example, in the wasp *Leptopilina clavipes,* *Wolbachia* disrupts anaphase of asexual embryonic development to produce one diploid female rather than two haploid males (Pannebakker et al. 2004).

Second, symbionts can simply kill any male offspring. Males are a dead end, while females carry symbionts that they can then transmit through the egg. Scientists do not fully understand how *Wolbachia* kills males. In the WZ/WW corn borer moths *Ostrinia* spp., *Wolbachia* apparently causes dosage compensation to fail in male embryos due to a downregulated zinc finger protein gene *Masc* (Fukui et al. 2015).

Third, symbionts can turn genetic males into functional females who produce symbiont-containing eggs. *Wolbachia* feminizes some isopod and insect males (Werren et al. 2008; Wedell 2020), which is surprising because sex determination occurs in a totally different manner in isopods (circulating hormones) and insects (typically heterogamy) (Cordaux et al. 2011). *Wolbachia* feminizes common pillbugs, *Armadillidium vulgare* (Leclercq et al. 2016; Cormier et al. 2021); the common grass yellow butterfly, *Eurema hecabe* (Hiroki et al. 2002); and—by disrupting male imprinting—leafhoppers, *Zyginidia pullula* (Negri et al. 2006, 2009). Far across the tree of life, eukaryotic microsporidia feminize isopod hosts such as *Gammarus roeselii* (Cormier et al. 2021) and *G. duebeni* (Dunn et al. 2020).

Fourth, through cytoplasmic incompatibility, symbionts can turn infected sperm into a weapon that poisons eggs devoid of symbionts, or even eggs devoid of the *right* symbionts (Verspoor et al. 2020). *Wolbachia* causes cytoplasmic incompatibility in several arachnids, isopods, and insects (Werren et al. 2008): if sperm from *Wolbachia*-infected males encounters uninfected eggs, development fails. For example, *Wolbachia* infects the tiny parasitoid wasp *Nasonia vitripennis,* desynchronizing development of paternal relative to maternal pronuclei to abort uninfected embryos (Tram and Sullivan 2002).

Selfish Parasites/Symbionts May Complicate Host Sex Determination, Disrupt Sexual Reproduction, and Harm Host Fitness. Hosts rapidly evolve countermeasures to selfish symbionts. This arms race leaves signatures in a clade's phylogeny and life history.

First, selfish symbionts can cause the host to have varied and young sex-determining chromosomes (Rigaud et al. 1997; Cordaux et al. 2011;

Becking et al. 2017). Isopods are stolid bearers of reproductive parasites across many species and over many generations (Rigaud et al. 1997; Cordaux et al. 2011; Becking et al. 2017). Their sex chromosomes are unusual for three reasons: (1) sex chromosomes look morphologically similar to other chromosomes; (2) closely related species vary in male or female heterogamety, having transitioned between XX/XY and ZZ/ZW three to thirteen times since the late Cretaceous (Becking et al. 2017); and (3) WW females and YY males can be healthy and fertile (Cordaux et al. 2011).

Second, hosts can evolve parthenogenesis, lose sexual reproduction, or both (Stouthamer et al. 2010; but see Birky and Gilbert 1971). This has occurred in more than twenty-five species of arthropods (Kageyama et al. 2012). It is possible that haplodiploidy, where males are haploid (have one set of chromosomes) and females are diploid (have two sets of chromosomes), evolved to "rescue" haploids as viable males from endosymbionts that killed males by destroying paternal chromosomes (Normark, 2004; Cordaux et al., 2011). Similarly, a host with male-killing symbionts could "rescue" its males through hermaphroditism.

Third, selfish symbionts harm host fitness in the short and long term. With cytoplasmic incompatibility (Verspoor et al. 2020), infected males often produce fewer gametes (Price and Wedell 2008; Zanders and Unckless 2019) and have less competitive sperm (Champion de Crespigny and Wedell 2006). Over evolutionary time, as the host genome evolves to counteract the selfish endosymbiont, populations can become reproductively incompatible (Verspoor et al. 2020). Crossbreeding worsens these effects as selfish elements invade a susceptible genome. Further, scientists predict that spermatogenesis will become more complex to oppose the effects of the endosymbiont (Verspoor et al. 2020).

Finally, sex-ratio distortions with too few males can lead to extinction (Wedell 2020). Such extinctions are difficult to detect in the fossil record. However, males are an evolutionary waste if they do not successfully mate, as is the case for many males of many species. From this perspective, a species with a 60:40 female-to-male sex ratio may be less extinction-prone than one with a 50:50 ratio.

Predictions: Coral's Microalgae May Exert a Feminizing Influence on Corals, Causing Complex and Labile Sex Determination and Contributing to Extinction Risk

How can we detect feminizing effects of symbionts in corals, if indeed such effects are present? We predict that some symbiotic microalgae will have evolved techniques to alter host sex through parthenogenesis, male killing, feminization, or cytoplasmic incompatibility—all effects that have been described widely in other endosymbionts, primarily in arthropods.

Further, we predict that any efforts by symbionts to alter host sex will have cascading effects on the phylogeny and evolution of corals, as has been shown in so many arthropods and other clades (Kageyama et al. 2012, 2017; Leclercq et al. 2016; Cordaux and Gilbert 2017; Wedell 2020). Specifically, these effects may (1) drive rapid evolution of sex determination systems, (2) eliminate or alter sexual reproduction, and (3) harm male fitness or burden males with costly counteradaptations, and more. We predict that if microalgae feminize corals, then coral sex-determination systems will be labile, polymorphic, and complex (Cordaux et al. 2011). When host sex determination evolves quickly, we predict further downstream consequences that could harm the ecosystem. Gynandromorphs may become more common or species may split into reproductively incompatible species (Wedell 2020).

Currently, almost nothing is known about such effects in corals or other photosymbiotic animals. We hope this section may inspire some to study the selfish effects of symbiotic microalgae further.

Resources: Coral Hosts Take More or Less Photosynthate from Their Symbionts

Hypothesis: Animal hosts which compel their symbionts to give more photosynthate experience higher levels of conflict and are more susceptible to bleaching.

Synopsis: The same species of microalgae symbionts that live inside coral also live freely in the ocean (Box 10.1). Indeed, many photosymbiotic animals must acquire their symbionts from the environment as they develop. But the lives of these symbiotic microalgae look very different outside of an animal host. Free-living microalgae use the products of photosynthesis for their own purposes, such as building cellular components or reproducing. They only leak about 5% of the energy-rich sugars (i.e., photosynthate). However, inside corals and other animals, symbiotic microalgae transfer 20%–95% of the photosynthate to their host, apparently because the host causes them to do so, though we do not yet fully understand how. Indeed, symbionts leak more energy-rich sugars when their animal host is starving or otherwise stressed, at the expense of their own energy budget. We argue that the amount of photosynthate given to a host is one metric of host-symbiont conflict. Early experiments show that hosts vary significantly in the amount of photosynthate they receive from their symbiont, and we predict that this quantity varies alongside other metrics of   . conflict. Further, we predict that some symbionts may have co-evolved resistance to "greedy" hosts that cause high amounts of photosynthate to leak from the symbionts.

In Corals and Symbiodiniaceae: Animal Hosts Cause Symbionts to Leak 20%–95% of Their Photosynthate and More Under Stressful Conditions

From the symbiont's evolutionary perspective, selection favors adaptations to use photosynthate for its own reproduction and health (Blackstone and Golladay 2018). Corals, and other animal hosts, cause their symbionts to leak substantial amounts of photosynthate for use by the animal host. Further, hosts regulate symbionts by controlling how much nitrogen they share with them (Xiang et al. 2020), and how much carbon they concentrate and deliver to them (Wooldridge 2010). We consider corals that cause symbionts to leak more photosynthate to be more selfish and less mutualistic. Conversely, we consider symbiont clades that fix more carbon, share more photosynthate, and promote more growth in the coral host to be more mutualistic and less selfish. Resource sharing is one useful metric of the degree of conflict in a symbiosis.

Much work has been done to determine the degree of mutualism in various clades of microalgae symbionts (Table 10.2) (Lesser et al. 2013). In the alphabet soup of Symbiodiniaceae clades, some seem generous but weak, while others are selfish but tough. Photosynthetic rate can inversely vary with heat tolerance (Cunning et al. 2015), as can the amount of photosynthate a symbiont shares and coral growth rates (see Matsuda et al. 2023). For example, clade-C1 symbionts share more photosynthate and promote more calcification in corals, while rarer thermally tolerant clade-D symbionts seem to be less generous partners who are more resilient to stress. The more parasitic clade-A symbionts associate with sicker coral, fix less carbon, and share less photosynthate with their host (Stat et al. 2008). However, these microalgae are so diverse within and across

Table 10.2. Broadly, Symbiodiniaceae Vary from Selfish and Hardy to Mutualistic and Less Hardy (Clade D vs. Clade C1)

Symbiont type	Animal host	Study findings
Mutualistic clade C1 versus selfish clade D	Juvenile branching corals *Acropora millepora*	Mutualistic clade C1 symbionts translocated photosynthate to the host at 121% the rate of the more selfish clade D symbionts and had 87% greater rates of electron transport through photosystem II (Cantin et al. 2009).
Selfish clade D	Caribbean corals	An opportunistic, thermally tolerant, and genetically homogenous clade D endosymbiont lineage *Durusdinium trenchii* from the Indo-Pacific invaded the Caribbean after mass bleaching in 2005 (LaJeunesse et al. 2014).
Selfish clade D	*Orbicella* spp. corals	*Durusdinium trenchii* photosynthesizes at normal rates compared to other symbionts but promotes up to 50% less coral calcification (Pettay et al. 2015).
Mutualistic clade C1 versus selfish clade D	Coral *Acropora tenuis* experimentally inoculated	Thermally tolerant clade D acquired less nitrogen than clade C1 at normal temperatures but fixed 35% more carbon than clade C1 at higher temperatures. Clade C1 outcompetes clade D at nitrogen acquisition under normal temperatures, which may explain why the less thermally tolerant clade C1 dominates globally (Baker et al. 2013).

Note: Symbiodiniaceae vary tremendously both within and across clades, complicating attempts to generalize.

species that it is difficult to generalize (Parkinson et al. 2025). In Table 10.2 we report a selection of studies on symbiont mutualism (Cantin et al. 2009; Baker et al. 2013; LaJeunesse et al. 2014; Pettay et al. 2015).

Corals presumably also vary along a parasitic-to-mutualistic continuum, but comparatively little is known. One metric of parasitism in corals is how much photosynthate they force their microalgae to leak. Up to 95% of carbon fixed by symbionts is leaked to the coral hosts (Muscatine et al. 1983, 1997). Scientists can trace the fate of carbon to determine translocation rate. Less directly, scientists can expose different strains of microalgae to tissue homogenate from different animal hosts, then measure how much photosynthate the microalgae leak in the presence of that tissue. That leakage rate can be interpreted as a metric of host greediness; the more leakage, the greedier the host. Additionally, symbionts may have coevolved resistance to the particular greedy strategies of their typical host; if this is true, we might expect giant clam symbionts to be ill-prepared to handle coral tissue homogenate and leak photosynthate profusely. Studies of photosynthate leakage offer tantalizing hints at host-symbiont coevolution and variable greediness across hosts. For example, reef-building corals seem greedier than other animals (Table 10.3) (Muscatine 1967; Sutton and Hoegh-Guldberg 1990; Masuda et al. 1994). It is important to note that symbionts inside a coral

Table 10.3. Microalgae Symbionts Leak Different Amounts of Photosynthate in the Presence of Tissue from Different Animal Hosts, a Metric of Animal Greediness

Animal host(s)	Results
Scleractinian coral *Plesiastrea versipora,* zoanthid *Zoanthus robustus,* nudibranch *Pteraeolidia ianthina,* and soft coral *Capnella gaboensis*	Tissue from the scleractinian coral caused microalgae to leak up to 42% of the photosynthate, but tissue from the zoanthid and nudibranch caused very little photosynthate leakage. Tissue from the soft coral lysed microalgae, a somewhat hard-to-interpret result (Sutton and Hoegh-Guldberg 1990).
Giant clam *Tridacna crocea* and scleractinian coral *Pocillopora damicornis*	Symbionts isolated from a giant clam and a scleractinian coral were exposed to tissue homogenate from their own, and each other's, hosts. With their own hosts, each leaked ~37% and 38%, respectively. With each other's hosts in two exposures, the clam's symbionts leaked 22% and 23% and the coral's symbionts leaked 29% and 5% (Muscatine 1967).
Giant clams Tridacnidae (*T. derasa, T. crocea, T. maxima, T. squamosa, Hippopus hippopus*) cockles Cardiidae (*Fragum fragum, F. mundum, F. unedo*), non-photosymbiotic bivalves (*Mytilus edulis, Meretrix lusoria, Ruditapes philippinarum*), and gastropods (*Umbonium giganteum, Turbo argyrostoma*)	When symbionts from giant clams *T. derasa* were tested, giant clam (*Tridacninae*) tissue homogenate promoted more translocation of photosynthate than did tissue from *Cardiidae* cockles or non-photosymbiotic bivalves and gastropods (Masuda et al. 1994). The researchers also showed that *T. derasa* tissue homogenate caused photosynthate leakage in certain free-living algae within Dinophyceae but not in other clades.

do seem to produce more photosynthate overall compared to in free-living forms; for further discussion, see "Solar Irradiation: Corals Can Screen and Protect Symbionts or Expose Symbionts to Irradiation to Grow Quickly."

Further, animal hosts seem to take more photosynthate from their symbionts when they are stressed or starved. Microalgae from healthy anemone host *Exaiptasia pallida* released 14% of photosynthate, while those from starved anemones released 25%, despite lower photosynthesis rates in the starved condition (Davy and Cook 2001). The authors saw similar results for the same microalgae incubated with tissue homogenate from the coral *Montastrea annularis*. For the coral *Stylophora pistillata*, translocation was significantly higher for corals living in lower-pH waters (80% translocation at pH 7.2; 60% at pH 8.1), allowing the coral to receive the same amount of translocated carbon (5.5–6.1 µg C cm^{-2}h^{-1}) in both scenarios even though it represented a greater portion of the symbiont's available carbon (Tremblay et al. 2013).

Beyond Corals: In Plant-Rhizobium Symbioses, Resource Sharing Varies with the Degree of Conflict

In many other symbioses, scientists use resource sharing between host and symbiont as a metric of cooperation or conflict. Phenotypes can be considered more- or less-cooperative strategies to maximize payoffs in resource markets (Noë and Hammerstein 1994, 1995; Kiersi and Denison 2008; Werner et al. 2014). In the famous plant-rhizobium system, plants create root nodules to house bacterial partners in the right conditions to convert atmospheric nitrogen into ammonia (van Rhijn and Vanderleyden 1995). The plants invest energy by giving photosynthate and phosphorus to their bacterial partners (Ma and Chen 2021). Plants can also house mycorrhizal fungi in their roots, which uptake phosphorus and mitigate drought stress (Smith and Smith 2012). These plant symbioses determine success across local (Porter and Rice 2013) and global (Afkhami et al. 2014) habitats amid environmental change.

Both plants and fungi adjust the amount of resources they give their partner depending on how generous their partner is. As Kiers et al. (2011) demonstrated, this bidirectional system of control prevents the fungal symbionts from being exploited, and therefore stabilizes the symbiosis. They found that plant *Medicago truncatula* transfers more carbon resources to a more cooperative fungal partner, *Glomus intraradices*, compared to the less-cooperative close relatives *G. custos* and *G. aggregatum*. Next, they manipulated how much phosphorus and carbon the fungi and plants could give, respectively, ultimately finding that plants gave more carbon to generous fungi that provided more phosphorus. Likewise, fungi gave more phosphorus to generous plants.

Theoretical models show that plants should sanction rhizobia that fix less nitrogen, which is predicted to stabilize symbiotic interactions (West et al. 2002). Yet these sanctions cannot fully purge the population of poor performers, which persist at low frequencies (Frederickson 2013). Heath and Tiffin (2009) found that plants appear to select for the most beneficial partners, which increase in frequency across generations, leading to increased fitness for both hosts and symbionts. In the same study, the optimal symbiont changed depending on the host's genotype. In short, there is no universal best symbiont. Furthermore, the order in which symbionts colonize a plant matters: a less beneficial symbiont may not be actively sanctioned if it arrives after a well-performing symbiont partner in the nodule. Perhaps these priority effects prevent poorly performing genotypes from being selected out of the population (Boyle et al. 2021).

The plant-mycorrhizal fungi symbiosis is a biological market in which each partner regulates its provision of nutrients in response to the other (Kiers et al. 2016). Fungal partners can even control the market value of the phosphorus they provide by compensating with alternate sources during shortages and storing surpluses, as van't Padje et al. (2021) showed in their study of the fungus *Rhizophagus irregularis* living in carrot root *Daucus carota*. More broadly, biological market theory clarifies the evolution of symbioses. For example, Werner et al. (2014) identified six strategies microbes may pursue to improve their performance in biological resource markets: "(i) avoid bad trading partners; (ii) build local business ties; (iii) diversify or specialize; (iv) become indispensable; (v) save for a rainy day; and (vi) eliminate the competition."

These experimental and theoretical studies show how we can use resource sharing to quantify cooperation and conflict.

Predictions: Corals and Symbionts Will Co-Evolve in an Arms Race over Resource Sharing, and Faster-Growing Corals Will Extract Greater Amounts of Photosynthate from Their Symbionts

We predict that as coral hosts evolve to be more demanding, symbionts will evolve resistance. Perhaps symbionts from very demanding host species would share far less if exposed to tissue from an undemanding host. Resource sharing can be seen as a tug-of-war, where both sides are pulling hard but, if equally matched, the rope does not move.

Additionally, we predict that corals with a fast-growing lifestyle will extract a greater proportion of photosynthate from their symbionts. Further, we propose that corals generally extract greater amounts of photosynthate than other animal hosts—such as sponges, bivalves, and acoels—and that the rate of photosynthate seizure will correlate with other metrics of conflict. Hosts that require, and receive, higher amounts of photosynthate will be more susceptible to changing environmental conditions that impact the symbiont photosystem.

Solar Irradiation: Corals Can Screen and Protect Symbionts or Expose Symbionts to Irradiation to Grow Quickly

Hypothesis: Animal hosts that irradiate their symbionts will experience higher levels of conflict than hosts with photoprotective adaptations; the irradiators will be more susceptible to bleaching.

Synopsis: Animal hosts face a photonic tradeoff between efficiency and protection. They can concentrate light to enhance photosynthesis, or they can screen light to protect their symbionts. In particular, they can specifically screen out UV wavelengths that tend to be the most harmful to host and symbiont, though this depends strongly on depth. We propose that one metric of host-symbiont conflict is the degree to which hosts irradiate their symbionts. Corals expertly irradiate their symbionts to "farm" microalgae with efficient light-concentrating adaptations (Hoogenboom et al. 2008; Barott et al. 2015; Kramer et al. 2022). Corals can build gigantic reefs and support entire ecosystems because they invented intensive agriculture. But intensive agriculture has downsides. In corals that irradiate their symbionts with more and higher-energy photons, symbionts must contend with constant levels of light stress. And when global warming adds heat stress to the mix, they bleach. Shallow corals, and corals at the reef crest, expose their microalgae to higher heat stress and light stress than deeper corals. We hypothesize that corals run microalgae's photosynthetic machinery at a rate that is beneficial to them as hosts but stressful for the symbionts. Thus even a small amount of heat stress can destabilize this light-stressed system and cause catastrophic bleaching. Across corals and other photosymbiotic animals, we predict that highly irradiating animal hosts will bleach more easily and severely.

In Corals: Corals Are Ultraefficient Farmers of Photosynthetic Symbionts

Reef-building corals farm their microalgae symbionts (Wooldridge 2010; Matz 2024). Corals house their symbionts singly, each within an animal cell, where the microalgae cannot move to escape irradiation. Indeed, being housed singly places many restrictions on microalgae (see "Symbiont Housing: Symbionts May Be Housed in Large Groups or Isolated"). In addition, from the nanoscale to the macroscale, corals manipulate incident light with their highly scattering tissue and skeleton to irradiate their symbionts and maximize photosynthetic output. Together, the coral approaches theoretical limits for photosynthetic efficiency (approaching eight photons to fix one CO_2 molecule; see Brodersen et al. 2014). Coral reefs are among the most productive and diverse ecosystems on earth despite occurring in nutrient-poor waters (Sheppard et al. 2018).

Symbionts living inside a coral experience a manyfold increase in irradiation compared to free-living forms. In the outer layers of coral tissue, corals concentrate light to irradiate their symbionts (Wangpraseurt et al. 2012). In the coral *Porites branneri,* symbionts living within coral absorb two to five times more light than those living outside (Enríquez et al. 2005). In general, symbionts living within corals have a higher photosynthetic efficiency (Wangpraseurt et al. 2016). Free-living Symbiodiniaceae are motile and can move away from regions where irradiation is too high (Box 10.1). Generally, heat stress and light stress act through separate but related pathways to damage the photosystem of microalgae and often kill them (Downs et al. 2013).

Corals have many adaptations to irradiate their symbiont, but these adaptations can backfire, causing or amplifying bleaching. Corals are highly efficient light collectors (Stambler and Dubinsky 2005). Their skeletons scatter light efficiently, further irradiating their symbionts (Enríquez et al. 2005; Kaniewska et al. 2011). Coral tissue itself is highly scattering and works in concert with the skeleton to propagate and homogenize the light field across symbiont-bearing areas (Wangpraseurt et al. 2016). Skeletons vary in their scattering efficiency between species, a variation that correlates with bleaching risk (Marcelino et al. 2013; Swain et al. 2016; see also Enríquez et al. 2017). Simulations show that irradiance experienced by coral photosymbionts increases exponentially as a coral bleaches (Terán et al. 2010). In general, these photonic adaptations explain why light stress drives photoinhibition and causes or amplifies coral bleaching (see Brown et al. 1994, 1999; Lesser and Farrell 2004; Anthony and Kerswell 2007; Castro-Sanguino et al. 2024).

Corals have some apparently photoprotective adaptations, including fluorescent pigments (Salih et al. 2000; Dove et al. 2001; Bollati et al. 2020) and an additional algal symbiont called *Ostreobium* (Rodríguez-Román et al. 2006; Yamazaki et al., 2008; del Campo et al. 2017; Galindo-Martínez et al. 2022). These photoprotective adaptations may benefit the host, the symbiont, or both. The green fluorescent proteins of corals are upregulated during bleaching events (Bollati et al. 2020) and may dissipate harmful energy from ultraviolet radiation to reduce light stress (Salih et al. 2000; Dove et al. 2001). In an experimental study of *Acropora palifera,* non-fluorescent color morphs were significantly more photo-inhibited by full sunlight than their fluorescent morph peers (Salih et al. 2000). Similarly, the endolithic algal symbiont of many corals (*Ostreobium* spp.) seems to protect some bleaching corals by screening sunlight. For example, in several bleaching-resistant individuals of *Orbicella faveolata,* the algal symbiont *Ostreobium* bloomed near the skeletal surface during bleaching events, absorbing a substantial proportion of incident light and therefore relieving the coral of light stress (Galindo-Martínez et al. 2022). Their bleaching-prone peers had no such algal bloom.

Coral fluorescence may itself hint at underlying host-symbiont conflict. Fascinatingly, a genetic analysis suggests that a coevolutionary arms race between corals and their symbiotic dinoflagellates caused fluorescent colors to diverge in corals. Specifically, Field et al. (2006) showed that adaptive evolution caused cyan and red fluorescence to evolve from the ancestral green color in the great star coral *Montastrea cavernosa.* Surprisingly, they showed that an intermolecular binding interface that is not related to color production was also subjected to positive natural selection. They performed reverse mutagenesis to experimentally test which mutations affected color phenotype, resurrecting ancestral versions of the proteins inferred from statistical genetic analysis. Through this work, they confirmed positive selection at sites that do not seem to impact the color phenotype. Further, these sites appeared to have a binding function and showed signatures of continuous readjustment—a well-known hall-mark of a coevolutionary chase, where "one of the interacting partners benefits from maximizing the interaction and the other from avoiding or restricting it" (Field et al. 2006). Field et al. noted that such signatures are seen in systems such as sperm-egg and host-parasite recognition. They proposed that the fluorescent proteins therefore somehow regulate the symbiotic dinoflagellates, acting as regulatory photosensors analogous to phytochromes in plants. Further work is needed to follow up on these results.

Beyond Corals: Many Photosymbiotic Creatures Protect Symbionts from Light Stress

Bivalves, acoels, foraminifera, and other hosts have evolved photonic adaptations to protect their symbionts from light stress.

Giant clams engage in "photonic cooperation" with their symbionts: Layered iridocytes enhance photosynthetic capabilities while screening damaging UV light (Rossbach et al. 2020). Specifically, the clams' mantles forward-scatter photosynthetically active radiation, back-reflect nonproductive wavelengths, and absorb and fluoresce ultraviolet light (Holt et al. 2014; Ghoshal et al. 2016; Rossbach et al. 2020; Holt 2024). Similarly, heart cockles (*Corculum cardissa* and spp.) transmit greater than 30% of visible sunlight but screen ultraviolet radiation (McCoy et al. 2024), through transparent windows in their otherwise opaque shell (Kawaguti 1950; Carter and Schneider 1997; Farmer et al. 2001). The heart cockle's windows are composed of aragonite fiber-optic cable bundles, proposed to be an adaptation to transmit photosynthetically active radiation while scattering and absorbing ultraviolet light (McCoy et al. 2024).

The photosymbiotic acoelomorph flatworm *Convolutriloba longifissura* has two types of pigment cell that seem to protect algal symbionts from light stress (Hirose and Hirose 2007).

Remarkably, the photosymbiotic foraminifera *Marginopora vertebralis* protects its photosynthetic symbionts by shuttling them away from the bright sun, deeper into the cavities of the foram's body. Apparently, stress response by the symbiont is interpreted as a signal by the host, who actively responds and contracts actin filaments to move symbionts deeper into the calcium carbonate structure (Petrou et al. 2017).

Fascinatingly, sunlight plays a role in the only example of a photosymbiosis that transitions from mutualistic to parasitic as the animal host grows up (Melo Clavijo et al. 2018): the queen conch *Strombus gigas* (Banaszak et al. 2013). In their larval stage, the conch have what seems to be a mutualistic relationship with their *Symbiodinium* symbionts, which photosynthesize because sunlight can penetrate the larval gastropod. But in adults, which have a thick shell that blocks all but 0.1%–20% of sunlight, photosynthesis is less effective and it seems that the adult snail must provide nutrition to their symbionts without receiving much, if anything, in return (Banaszak et al. 2013).

A large body of research investigates how plants protect their photosystems against light stress (Schwenkert et al. 2022; Shi et al. 2022), another fruitful source of comparison to corals. Red and brown algae produce mycosporine-like amino acids that protect tissue from UV irradiation (Frohnmeyer and Staiger 2003; Rui et al. 2019).

Predictions: Photonic Adaptations to Irradiate Symbionts Correlate with Bleaching Risk and Other Metrics of Conflict

We predict that corals will consistently irradiate their symbionts at higher levels than non-coral photosymbiotic animals, and that across species, irradiation will correlate with bleaching risk. Further, we anticipate that levels of irradiation align with metrics of nutrient sharing and transmission mode. Because the available downwelling irradiance varies dramatically with depth, scientists could take advantage of this natural experiment to compare levels of conflict in shallow versus mesophotic coral reefs. We predict that shallow-sea corals, which have the opportunity for intensive agriculture with greater downwelling irradiance, will experience higher levels of conflict than the deeper, mesophotic corals. Irradiation generally represents a farm-and-discard strategy rather than a protect-and-coexist strategy. We suspect that reef building corals so dramatically irradiate their symbionts because they are in direct competition with seaweeds and other corals for space in the narrow band of seas with appropriate temperature and sunlight. Global real estate for reefs is limited, and these creatures largely cannot move; they can only grow. Therefore, they face intense competition for growth. In contrast, the deep-sea houses scleractinian corals with no seaweed competition. Scleractinian corals likely arose in the deep sea at depths of 200–2,300 m, without sunlight and photosymbionts (Campoy et al. 2023). To better understand the evolutionary history of host-symbiont conflict over irradiation, scientists could compare the ecology of deep-sea to shallow-sea corals as well comparing symbiotic to aposymbiotic species.

Symbiont Housing: Symbionts May Be Housed in Large Groups or Isolated

Hypothesis: Symbionts living communally in animal hosts have more power and less conflict with their host compared to those isolated in individual coral cells, which can be controlled and coerced.

Synopsis: Corals typically isolate each symbiont inside a membrane inside a coral cell. Unlike corals, other animals house their symbiotic microalgae in communal spaces extracellularly, such as those found in heart cockles (*Corculum* spp.) and giant clams Tridacninae, or in higher numbers intracellularly, such as in sponges (*Cliona* spp.). Symbionts housed communally have more power because hosts cannot detect cheaters in a shared physiological space. The host detects only total photosynthate produced, allowing individual symbionts to contribute less and potentially even reproduce unnoticed. Corals seem to have solved this problem by housing symbionts within single cells, individually monitoring their photosynthetic performance and individually providing them with the carbon needed to survive and photosynthesize (or withholding carbon to punish cheaters). We predict that photosymbionts vary from singly-housed-intracellular to multiply-housed-intracellular to communally-housed-extracellular; along that spectrum, we predict symbionts to reproduce more and provide less photosynthate when they are more communally housed.

In Corals: Intracellular Symbionts May Be Restricted from Reproduction and Monitored for Defection

Corals Generally Have One Symbiont Per Cell, Enabling Close and Specific Monitoring. In most animals that host photosymbionts, including corals, symbionts are intracellular and surrounded by a host-derived membrane called the symbiosomal membrane (Venn et al. 2008). The animal host provides carbon dioxide, water, and other important materials to the symbiont and therefore controls most aspects of the metabolic partnership (Venn et al. 2008; Wooldridge 2010; Blackstone and Golladay 2018). Animal hosts seem to restrict symbiont reproduction, just as we restrict cells within our own body from dividing freely and developing into cancer (see Okasha 2021).

Corals host one or only a few symbionts per host cell (Blackstone and Golladay 2018). Across thirty-three species of anthozoan, including twenty scleractinian corals, most animals host only one symbiont per cell (average symbionts per cell = 1.54 ± 0.30; range $1.11-2.19$; see Muscatine et al. 1998). High nutrient levels seem to prompt symbionts to reproduce. When eleven species were flooded with nutrients, the number of symbionts per cell increased from an average of 1.46 ± 0.21 to 1.77 ± 0.28 (Muscatine et al. 1998). In other species, symbionts are often housed in more communal situations (see "Beyond Corals: Housing Varies Dramatically Across Photosymbioses, and Plant Hosts Monitor the Specific Cooperativity of Rhizobium Symbionts").

Because corals house symbionts singly, they can detect cheaters. Evolutionary theorists predict the evolution of defector symbionts that will divide more quickly and share less photosynthate than the host prefers (Blackstone and Golladay 2018). These defectors could hide from host oversight if they lived in communal pools, but singly housed symbionts cannot hide. For example, in a coral relative, the sea anemone *Aiptasia pulchella,* the host preferentially expels rapidly dividing symbionts (Baghdasarian and Muscatine 2000).

Perhaps one-on-one cellular symbioses could also create a stronger mutualism, in which policing enables both parties to exchange more materials without the looming threat of cheating (by either party). This idea deserves further consideration.

Environmental Stress Causes Bleaching by Disrupting Corals' Innate Mechanisms for Detecting and Expelling Cheater Symbionts. Corals' cheater-detection mechanisms may malfunction under environmental stress, causing them to eject both defecting and healthy symbionts (Blackstone and Golladay 2018). Defectors that fail to export photosynthate cause product inhibition and collapse carbon concentration mechanisms, resulting in electron backup in photosystems I and II. Then, reactive oxygen species form, potentially causing damage and programmed cell death. High heat and light levels also trigger these processes, not because the symbionts are defecting per se, but because environmental stress reduces the rate of photosynthesis (Blackstone and Golladay 2018). Under heat stress, cooperators look an awful lot like defectors, causing corals' cheater-detection systems to overreact and drive massive coral bleaching.

Host carbon supply to symbionts depends on their photosynthesis, which serves as a checkpoint against cheaters. Host enzymes concentrate carbon to provide $CO_{2(aq)}$ to microalgae. However, corals need to receive photosynthate from symbionts to activate carbon concentration. Heat or light stress disrupts symbiont photosynthesis and causes a vicious cycle: Less CO_2 is provided to the symbionts, they photosynthesize less, and so on (Wooldridge 2010).

Indeed, in six scleractinian coral species—*Acropora selago, Acropora muricata, Heliofungia actiniformis, Ctenactis echinata, Oxypora lacera,* and *Pocillopora eydouxi*—corals expelled mostly degraded symbiont cells under normal conditions but released many healthy symbiont cells during heat stress (Fujise et al. 2013). One thermally tolerant species *P. eydouxi* did not change the number of symbionts expelled between the two conditions.

Symbionts Reproduce Less Inside Corals Compared to in Culture; Little Is Known About Free-Living Symbiodiniaceae. Because corals host symbionts singly within cells, they can restrict symbiont reproduction. Generally, we expect that symbionts will be allowed to reproduce less inside a host than in their free-living state, particularly in the case of non-cooperative symbioses. While little is known, studies have shown that symbionts double in number far more quickly outside of a coral host than within one (Wooldridge 2010). In a host, Symbiodiniaceae typically double every 70–100 days (Wilkerson et al. 1988); in culture, symbionts double approximately every three days (Fitt and Trench 1983).

Symbiont reproduction rates vary across coral hosts, potentially representing degrees of host-symbiont conflict. Across nine different coral species (*Madracis mirabilis, Acropora cervicornis, Agaricia agaricites, Acropora palmata, Porites astreoides, Montastraea annularis, Montastraea cavernosa, Eusmilia fastigiata,* and *Dendrogyra cylindrus*), the symbiont mitotic index—a metric of reproduction—varied from 1.1% to 14.1%. Certain corals had far greater rates of microalgae reproduction (specifically, *Eusmilia fastigiata* and *Dendrogyra cylindrus*). Within the coral *Acropora cervicornis,* the mitotic index was higher at branch tips than branch base, suggesting that the coral allows or facilitates reproduction when it must populate growing tissue with symbionts. Finally, the mitotic index was greater in deeper habitats (Wilkerson et al. 1988).

The counterpoint to reproduction is, of course, predation. Free-living microalgae get to reproduce more, but is that offset by greater predation risk? Many authors hypothesize that corals protect their symbionts from predators, but to our knowledge no study has shown this. Generally, more research is needed on Symbiodiniaceae reproduction and predation inside and outside of a host (Box 10.1).

Beyond Corals: Housing Varies Dramatically Across Photosymbioses, and Plant Hosts Monitor the Specific Cooperativity of Rhizobium Symbionts

Symbionts Are Singly Intracellular, Multiply Intracellular, Extracellular, and Everything in Between. Symbionts occupy every imaginable position on the spectrum from extracellular to intracellular. Smith et al. (1997) illustrated this diversity, from the mostly extracellular algal symbiont in lichens with only 20% cell surface contact, to acoelomorph worm symbionts confusingly described as either intra- or extracellular, to fully intracellular *Chlorella* symbionts in green hydra. How a host organizes and houses symbionts is a key nexus of coevolution through cooperation and conflict (Fronk and Sachs 2022).

In photosymbiotic animals, most symbionts are either intracellular or localized within specific regions of the host (Venn et al. 2008; Rumpho et al. 2011; Melo Clavijo et al. 2018). We will mention a few interesting cases described by Melo Clavijo et al. (2018) that hint at variable rates of cheating and conflict within photosymbiotic clades.

Most colonial ascidians (family Didemnidae) have extracellular symbioses with cyanobacteria *Prochloron* or *Synechocystis trididemni,* but at least one species, *Lissoclinum punctatum,* has an intracellular *Procholon* symbiont (Hirose et al. 2009; Hirose 2015). In the salamander *Ambystoma maculatum,* eggs can bear *Oophila* algae intracellularly, extracellularly in the egg capsule, or not at all (Burns et al. 2017). Giant clams (Norton et al. 1992; Lucas 2014; Fatherree 2023) and heart cockles (Kawaguti 1950; Farmer et al. 2001; Li et al. 2020) are an interesting exception to the intracellular symbiosis rule (Venn et al. 2008): In these bivalves, *Symbiodinium* live extracellularly in specialized tubes that connect to the stomach and wind through the soft tissue to sun-exposed upper mantle.

Sponges can have either intracellular or extracellular photosymbionts (Trautman and Hinde 2001). For example, the bioeroding sponge *Cliona orientalis* (part of the *Cliona viridis* species complex) houses *Gerakladium* dinoflagellates (Symbiodiniaceae) at varying densities within its cells (Achlatis et al. 2019). The sponge *Haliclona cymaeformis* has an extracellular symbiosis with the red macroalga *Ceratodictyon spongiosum* (Pile et al. 2003).

Too little is known about the number of symbionts per cell in intra-cellular associations. Usually, symbiont density is calculated in ways that do not control for the number of host cells. But Muscatine et al. (1998) reported the number of symbionts per cell in a sample of marine organisms, such as the hydroid *Myrionema amboinense* (ten to twelve symbionts in the cytoplasm) (Fitt and Cook 1990), and green hydra (fifteen to twenty symbiont cells per vacuole) (Muscatine and Pool 1979; McAuley 1982, 1984). Muscatine et al. further cite research showing that supplementing hydra with extra nutrition increases the symbiont count in the vacuoles to fifty to sixty symbionts and causes the symbiosis to break down (Muscatine and Neckelmann 1981; Neckelmann and Muscatine 1983; Blank and Muscatine 1987).

Taken together, these examples suggest that symbiont housing relates tightly to the equity of nutrient sharing.

Plant-Rhizobium Symbioses: Theoretical and Experimental Evidence Shows That Nutrients Are Not Public Goods but Are Allocated Based on Specific Partner Generosity. Theoretical work suggests that persistent, stable symbioses are more likely to evolve when nutrients are not public goods; that is, when partners can discriminate and adjust resource availability (West et al. 2007a; Frank 2010; Ghoul et al. 2014; Kiers et al. 2016). In biology, hosts often harbor multiple different strains of symbiont mixed together at fine spatial scales. Theory predicts that the hosts should monitor each partner to provide more resources to the generous. Indeed, in plant-rhizobia symbioses, plants penalize partners that are artificially manipulated to not fix nitrogen; indeed, such "cheating" rhizobia are 50% less reproductively successful (Kiers et al. 2003). Using soybean *Glycine max* with *Bradyrhizobium japonicum,* Kiers et al. (2003) showed that the plants detected and punished cheaters across three spatial scales: whole plant, half root, and individual nodules.

Plants appear to selectively favor generous symbionts over cheaters even within the same nodule. Researchers studied the pea plant *Mimosa pudica* and two variants of bacterial symbiont *Cupriavidus taiwanensis,* one that can fix nitrogen, and one that cannot (Daubech et al. 2017). After the symbionts invaded root nodules, they initially reproduced at the same rates, but then the fixers began outcompeting the non-fixers even within the same nodule. Within twenty days, nodules containing non-fixers degenerated (Daubech et al. 2017).

Predictions: Singly Housed Symbionts Receive Fewer Nutrients, Share More Nutrients, and Reproduce Less

We predict that the number of symbionts housed per cell represents the freedom afforded the symbiont by the host and will therefore correlate with other metrics of conflict: reproductive rate of the symbiont, nutrients received by the symbiont, and more. Across photosymbiotic animals, corals, with their singly housed intracellular symbionts, seem to sit at the high-conflict, high-control end of the spectrum. Symbionts housed in groups will be freer to reproduce more and share less. Further, we predict that evolutionary transitions will tend to proceed from extracellular symbionts housed communally along the spectrum to intracellular symbionts housed singly.

Conclusion

Coral reef ecosystems depend on the dynamic interplay between corals and photosynthetic microalgae. While traditionally viewed as mutualistic, we suggest that animal hosts and their symbiotic microalgae experience a spectrum of interactions ranging from parasitic to mutualistic. Parasitic relationships may destabilize the symbiotic equilibrium, driving coral bleaching. Our proposed framework offers five avenues for quantifying conflict within these symbioses: reproductive mode, host sex determination, resource allocation, solar irradiation, and detecting cheaters. By quantifying these conflicts, we aim to deepen our understanding of how cooperative symbioses evolve and shed light on the mechanisms underlying coral bleaching. By doing so, perhaps we can design new ecological interventions to help vulnerable reef ecosystems. Taken together, we emphasize the role of evolutionary conflict in shaping entire ecosystems, as exemplified by the dance of conflict and cooperation within coral reefs.

Acknowledgments and Funding

We thank Steve Palumbi, Manus Patten, Ellen Clarke, and Gabby Mansilla for their useful comments. Dakota McCoy was supported by the Stanford Science Fellowship and the National Science Foundation postdoctoral research fellowships in biology PRFB Program, grant 2109465.

We gratefully acknowledge support from the Chan Zuckerberg Biohub, San Francisco.

References

Achlatis, M., Schönberg, C. H. L., van der Zande, R. M., LaJeunesse, T. C., Hoegh-Guldberg, O., and Dove, S. (2019). Photosynthesis by symbiotic sponges enhances their ability to erode calcium carbonate. *Journal of Experimental Marine Biology and Ecology, 516,* 140–149.

Afkhami, M. E., McIntyre, P. J., and Strauss, S. Y. (2014). Mutualist-mediated effects on species' range limits across large geographic scales. *Ecology Letters, 17*(10), 1265–1273.

Aihara, Y., Maruyama, S., Baird, A. H., Iguchi, A., Takahashi, S., and Minagawa, J. (2019). Green fluorescence from cnidarian hosts attracts symbiotic algae. *Proceedings of the National Academy of Sciences of the United States of America, 116*(6), 2118–2123.

Ainsworth, T. D., Heron, S. F., Ortiz, J. C., et al. (2016). Climate change disables coral bleaching protection on the Great Barrier Reef. *Science, 352*(6283), 338–342.

Anderson, R. M., and May, R. M. (1982). Coevolution of hosts and parasites. *Parasitology, 85*(2), 411–426.

Andras, J. P., Rypien, K. L., and Harvell, C. D. (2013). Range-wide population genetic structure of the Caribbean sea fan coral, *Gorgonia ventalina*. *Molecular Ecology, 22*(1), 56–73.

Anthony, K., and Kerswell, A. (2007). Coral mortality following extreme low tides and high solar radiation. *Marine Biology, 151,* 1623–1631.

Armoza-Zvuloni, R., Kramarsky-Winter, E., Loya, Y., Schlesinger, A., and Rosenfeld, H. (2014). Trioecy, a unique breeding strategy in the sea anemone *Aiptasia diaphana* and its association with sex steroids. *Biology of Reproduction, 90*(6), 122, 1–8.

Ayre, D., and Miller, K. (2004). Where do clonal coral larvae go? Adult genotypic diversity conflicts with reproductive effort in the brooding coral *Pocillopora damicornis*. *Marine Ecology Progress Series, 277,* 95–105.

Baghdasarian, G., and Muscatine, L. (2000). Preferential expulsion of dividing algal cells as a mechanism for regulating algal-cnidarian symbiosis. *The Biological Bulletin, 199*(3), 278–286.

Baird, A. H., Guest, J. R., and Willis, B. L. (2009). Systematic and biogeographical patterns in the reproductive biology of scleractinian corals. *Annual Review of Ecology, Evolution, and Systematics, 40*(1), 551–571.

Baker, D. M., Andras, J. P., Jordán-Garza, A. G., and Fogel, M. L. (2013). Nitrate competition in a coral symbiosis varies with temperature among *Symbiodinium* clades. *The ISME Journal, 7,* 1248–1251.

Banaszak, A. T., García Ramos, M., and Goulet, T. L. (2013). The symbiosis between the gastropod *Strombus gigas* and the dinoflagellate *Symbiodinium*: An ontogenic journey from mutualism to parasitism. *Journal of Experimental Marine Biology and Ecology, 449,* 358–365.

Barott, K. L., Venn, A. A., Perez, S. O., Tambutté, S., and Tresguerres, M. (2015). Coral host cells acidify symbiotic algal microenvironment to promote photosynthesis. *Proceedings of the National Academy of Sciences of the United States of America, 112*(2), 607–612.

Bay, R. A., and Palumbi, S. R. (2014). Multilocus adaptation associated with heat resistance in reef-building corals. *Current Biology, 24*(24), 2952–2956.

Becking, T., Giraud, I., Raimond, M. et al. (2017). Diversity and evolution of sex determination systems in terrestrial isopods. *Scientific Reports, 7*(1).

Bennett, G. M., and Moran, N. A. (2015). Heritable symbiosis: The advantages and perils of an evolutionary rabbit hole. *Proceedings of the National Academy of Sciences of the United States of America, 112*(33), 10169–10176.

Birky, C. W., Jr., and Gilbert, J. J. (1971). Parthenogenesis in rotifers: The control of sexual and asexual reproduction. *American Zoologist, 11*(2), 245–266.

Blackstone, N. W., and Golladay, J. M. (2018). Why do corals bleach? Conflict and conflict mediation in a host/symbiont community. *BioEssays, 40*(8), e1800021.

Blank, R., and Muscatine, L. (1987). How do combinations of nutrients cause symbiotic chlorella to overgrow hydra? *Symbiosis, 3,* 123–134.

Bollati, E., D'Angelo, C., Alderdice, R., Pratchett, M., Ziegler, M., and Wiedenmann, J. (2020). Optical feedback loop involving dinoflagellate symbiont and scleractinian host drives colorful coral bleaching. *Current Biology, 30*(13), P2433–P2445.

Boyle, J. A., Simonsen, A. K., Frederickson, M. E., and Stinchcombe, J. R. (2021). Priority effects alter interaction outcomes in a legume–rhizobium mutualism. *Proceedings of the Royal Society B, 288*(1946), 20202753.

Brandvain, Y., and Haig, D. (2018). Outbreeders pull harder in a parental tug-of-war. *Proceedings of the National Academy of Sciences of the United States of America, 115*(45), 11354–11356.

Bright, M., and Bulgheresi, S. (2010). A complex journey: Transmission of microbial symbionts. *Nature Reviews Microbiology, 8*(3), 218–230.

Brodersen, K. E., Lichtenberg, M., Ralph, P. J., Kühl, M., and Wangpraseurt, D. (2014). Radiative energy budget reveals high photosynthetic efficiency in symbiont-bearing corals. *Journal of the Royal Society Interface, 11*(93), 20130997.

Brown, B., Dunne, R., Scoffin, T., and Le Tissier, M. (1994). Solar damage in intertidal corals. *Marine Ecology Progress Series, 105,* 219–230.

Brown, B. E., Ambarsari, I., Warner, M. E., et al. (1999). Diurnal changes in photochemical efficiency and xanthophyll concentrations in shallow water reef corals: Evidence for photoinhibition and photoprotection. *Coral Reefs, 18*(2), 99–105.

Buckingham, L. J., and Ashby, B. (2022). Coevolutionary theory of hosts and parasites. *Journal of Evolutionary Biology, 35*(2), 205–224.

Bull, J. J., Molineux, I. J., and Rice, W. R. (1991). Selection of benevolence in a host-parasite system. *Evolution, 45*(4), 875–882.

Byler, K. A., Carmi-Veal, M., Fine, M., and Goulet, T. L. (2013). Multiple symbiont acquisition strategies as an adaptive mechanism in the coral *Stylophora pistillata*. *PLoS One, 8*(3), e59596.

Campoy, A. N., Rivadeneira, M. M., Hernández, C. E., Meade, A., and Venditti, C. (2023). Deep-sea origin and depth colonization associated with phenotypic innovations in scleractinian corals. *Nature Communications, 14*(1), 7458.

Cantin, N. E., van Oppen, M. J. H., Willis, B. L., Mieog, J. C., and Negri, A. P. (2009). Juvenile corals can acquire more carbon from high-performance algal symbionts. *Coral Reefs, 28*(2), 405–414.

Carpenter, K. E., Abrar, M., Aeby, G., et al. (2008). One-third of reef-building corals face elevated extinction risk from climate change and local impacts. *Science, 321*(5888), 560–563.

Carter, J. G., and Schneider, J. A. (1997). Condensing lenses and shell microstructure in *Corculum* (Mollusca: Bivalvia). *Journal of Paleontology, 71*(1), 56–61.

Castro-Sanguino, C., Stick, D., Duffy, S., Grimaldi, C., Gilmour, J., and Thomas, L. (2024). Differential impacts of light on coral phenotypic responses to acute heat stress. *Journal of Experimental Marine Biology and Ecology, 581*, 152057.

Champion de Crespigny, F. E., and Wedell, N. (2006). *Wolbachia* infection reduces sperm competitive ability in an insect. *Proceedings of the Royal Society B, 273*(1593), 1455–1458.

Combosch, D. J., and Vollmer, S. V. (2013). Mixed asexual and sexual reproduction in the Indo-Pacific reef coral *Ocillopora damicornis*. *Ecology and Evolution, 3*(10), 3379–3387.

Cordaux, R., Bouchon, D., and Grève, P. (2011). The impact of endosymbionts on the evolution of host sex-determination mechanisms. *Trends in Genetics, 27*(8), 332–341.

Cordaux, R., and Gilbert, C. (2017). Evolutionary significance of *Wolbachia*-to-animal horizontal gene transfer: Female sex determination and the *f* element in the isopod *Armadillidium vulgare*. *Genes, 8*(7), 186.

Cormier, A., Chebbi, M. A., Giraud, I., et al. (2021). Comparative genomics of strictly vertically transmitted, feminizing microsporidia endosymbionts of amphipod crustaceans. *Genome Biology and Evolution, 13*(1), evaa245.

Cornwell, B. H. (2020). Gene flow in the anemone *Anthopleura elegantissima* limits signatures of local adaptation across an extensive geographic range. *Molecular Ecology, 29*(14), 2550–2566.

Cornwell, B. H., Armstrong, K., Walker, N. S., et al. (2021). Widespread variation in heat tolerance and symbiont load are associated with growth tradeoffs in the coral *Acropora hyacinthus* in Palau. *eLife, 10,* e64790.

Cornwell, B. H., and Hernández, L. (2021). Genetic structure in the endosymbiont Breviolum "muscatinei" is correlated with geographical location, environment and host species. *Proceedings of the Royal Society B, 288*(1946), 20202896.

Cunning, R., Silverstein, R. N., and Baker, A. C. (2015). Investigating the causes and consequences of symbiont shuffling in a multi-partner reef coral symbiosis under environmental change. *Proceedings of the Royal Society B, 282*(1809), 20141725.

Darling, E. S., Alvarez-Filip, L., Oliver, T. A., McClanahan, T. R., and Côté, I. M. (2012). Evaluating life-history strategies of reef corals from species traits. *Ecology Letters, 15*(12), 1378–1386.

Daubech, B., Remigi, P., Doin de Moura, G., et al. (2017). Spatio-temporal control of mutualism in legumes helps spread symbiotic nitrogen fixation. *eLife, 6,* e28683.

Davy, S. K., Allemand, D., and Weis, V. M. (2012). Cell biology of cnidarian-dinoflagellate symbiosis. *Microbiology and Molecular Biology Reviews, 76*(2), 229–261.

Davy, S. K., and B. Cook, C. (2001). The influence of "host release factor" on carbon release by zooxanthellae isolated from fed and starved *Aiptasia pallida* (Verrill). *Comparative Biochemistry and Physiology Part A: Molecular and Integrative Physiology, 129*(2), 487–494.

del Campo, J., Pombert, J.-F., Šlapeta, J., Larkum, A., and Keeling, P. J. (2017). The "other" coral symbiont: Ostreobium diversity and distribution. *The ISME Journal, 11*(1), 296–299.

Douglas, A. E. (1994). *Symbiotic Interactions.* Oxford University Press.

Douglas, A. E., and Smith, D. C. (1989). Are endosymbioses mutualistic? *Trends in Ecology and Evolution, 4*(11), 350–352.

Dove, S. G., Hoegh-Guldberg, O., and Ranganathan, S. (2001). Major colour patterns of reef-building corals are due to a family of GFP-like proteins: *Coral Reefs, 19*(3), 197–204.

Downs, C. A., McDougall, K. E., Woodley, C. M., et al. (2013). Heat-stress and light-stress induce different cellular pathologies in the symbiotic dinoflagellate during coral bleaching. *PLoS One, 8*(12), e77173.

Drew, G. C., Stevens, E. J., and King, K. C. (2021). Microbial evolution and transitions along the parasite–mutualist continuum. *Nature Reviews Microbiology, 19*(10), 623–638.

Dunn, A. M., Rigaud, T., Ford, A. T., Cothran, R., and Thiel, M. (2020). Environmental influences on crustacean sex determination and reproduction: Environmental sex determination, parasitism and pollution. *The Natural History of the Crustacea: Reproductive Biology, 6,* 394–428.

Dusi, E., Gougat-Barbera, C., Berendonk, T. U., and Kaltz, O. (2015). Long-term selection experiment produces breakdown of horizontal transmissibility in parasite with mixed transmission mode. *Evolution, 69*(4), 1069–1076.

Eddy, T. D., Lam, V. W. Y., Reygondeau, G., et al. (2021). Global decline in capacity of coral reefs to provide ecosystem services. *One Earth, 4*(9), 1278–1285.

Edinger, E. N., and Risk, M. J. (1995). Preferential survivorship of brooding corals in a regional extinction. *Paleobiology, 21*(2), 200–219.

Emslie, M., Cheal, A., Sweatman, H., and Delean, S. (2008). Recovery from disturbance of coral and reef fish communities on the Great Barrier Reef, Australia. *Marine Ecology Progress Series, 371*, 177–190.

Enríquez, S., Méndez, E. R., Hoegh-Guldberg, O., and Iglesias-Prieto, R. (2017). Key functional role of the optical properties of coral skeletons in coral ecology and evolution. *Proceedings of the Royal Society B, 284*(1853), 20161667.

Enríquez, S., Méndez, E. R., and Prieto, R. I. (2005). Multiple scattering on coral skeletons enhances light absorption by symbiotic algae. *Limnology and Oceanography, 50*(4), 1025–1032.

Ewald, P. W. (1987). Transmission modes and evolution of the parasitism-mutualism continuum. *Annals of the New York Academy of Sciences, 503*(1), 295–306.

Fabina, N. S., Putnam, H. M., Franklin, E. C., Stat, M., and Gates, R. D. (2012). Transmission mode predicts specificity and interaction patterns in coral-symbiodinium networks. *PLoS One, 7*(9), e44970.

Fadlallah, Y. H. (1983). Sexual reproduction, development and larval biology in scleractinian corals: A review. *Coral Reefs, 2*(3), 129–150.

Farmer, M. A., Fitt, W. K., and Trench, R. K. (2001). Morphology of the symbiosis between *Corculum cardissa* (Mollusca: Bivalvia) and *Symbiodinium corculorum* (Dinophyceae). *The Biological Bulletin, 200*(3), 336–343.

Fatherree, J. W. (2023). *The Giant Clams: All About These Beautiful, Unique, and Fascinating Reef Animals.* Bowker.

Field, S. F., Bulina, M. Y., Kelmanson, I. V., Bielawski, J. P., and Matz, M. V. (2006). Adaptive evolution of multicolored fluorescent proteins in reef-building corals. *Journal of Molecular Evolution, 62*(3), 332–339.

Figueroa, R. I., Howe-Kerr, L. I., and Correa, A. M. S. (2021). Direct evidence of sex and a hypothesis about meiosis in Symbiodiniaceae. *Scientific Reports, 11*(1), 18838.

Fitt, W., and Cook, C. (1990). Some effect of host feeding on growth of zooxanthellae in the marine hydroid *Myrionema amblonense* in the laboratory and in nature. In *Endocytobiology IV: Proceedings of the Fourth International Colloquium on Endocytobiology and Symbiosis, Lyon, July 4–8, 1989,* edited by P. Nardon, V. Gianinazzi-Pearson, Grenier, L. Margulis, and D. C. Smith, 281–284. INRA.

Fitt, W. K. (1985). Chemosensory responses of the symbiotic dinoflagellate *Symbiodinium microadriaticum* (dinophyceae). *Journal of Phycology, 21*(1), 62–67.

Fitt, W. K., and Trench, R. K. (1983). The relation of diel patterns of cell division to diel patterns of motility in the symbiotic dinoflagellate *Symbiodinium micro-adria ticum* Freudenthal in culture. *New Phytologist, 94*(3), 421–432.

Frank, S. A. (1996). Models of parasite virulence. *Quarterly Review of Biology, 71*(1), 37–78.

Frank, S. A. (2010). A general model of the public goods dilemma. *Journal of Evolutionary Biology, 23*(6), 1245–1250.

Franklin, D. J., Hoegh-Guldberg, O., Jones, R. J., and Berges, J. A. (2004). Cell death and degeneration in the symbiotic dinoflagellates of the coral *Stylophora pistillata* during bleaching. *Marine Ecology Progress Series, 272,* 117–130.

Frederickson, M. E. (2013). Rethinking mutualism stability: Cheaters and the evolution of sanctions. *Quarterly Review of Biology, 88*(4), 269–295.

Frohnmeyer, H., and Staiger, D. (2003). Ultraviolet-B radiation-mediated responses in plants. Balancing damage and protection. *Plant Physiology, 133*(4), 1420–1428.

Fronk, D. C., and Sachs, J. L. (2022). Symbiotic organs: The nexus of host–microbe evolution. *Trends in Ecology and Evolution, 37*(7), 599–610.

Fujise, L., Suggett, D. J., Stat, M., et al. (2021). Unlocking the phylogenetic diversity, primary habitats, and abundances of free-living Symbiodiniaceae on a coral reef. *Molecular Ecology, 30*(1), 343–360.

Fujise, L., Yamashita, H., Suzuki, G., and Koike, K. (2013). Expulsion of zooxanthellae (*Symbiodinium*) from several species of scleractinian corals: Comparison under non-stress conditions and thermal stress conditions. *Galaxea, Journal of Coral Reef Studies, 15*(2), 29–36.

Fukui, T., Kawamoto, M., Shoji, K., et al. (2015). The endosymbiotic bacterium *Wolbachia* selectively kills male hosts by targeting the masculinizing gene. *PLoS Pathogens, 11*(7), e1005048.

Fuller, Z. L., Mocellin, V. J. L., Morris, L. A., et al. (2020). Population genetics of the coral *Acropora millepora*: Toward genomic prediction of bleaching. *Science, 369*(6501), eaba4674.

Galindo-Martínez, C. T., Weber, M., Avila-Magaña, V., et al. (2022). The role of the endolithic alga *Ostreobium* spp. during coral bleaching recovery. *Scientific Reports, 12*(1).

Gardner, A., and Ross, L. (2011). The evolution of hermaphroditism by an infectious male-derived cell lineage: An inclusive-fitness analysis. *The American Naturalist, 178*(2), 191–201.

Gault, J., Bentlage, B., Huang, D., and Kerr, A. (2021a). Lineage-specific variation in the evolutionary stability of coral photosymbiosis. [Dataset]. *Dryad,* July 29. https://doi.org/10.5061/dryad.tdzo8kqop.

Gault, J., Bentlage, B., Huang, D., and Kerr, A. M. (2021b). Lineage-specific variation in the evolutionary stability of coral photosymbiosis. *Science Advances, 7*(39), eabh4243.

Ghoshal, A., Eck, E., Gordon, M., and Morse, D. E. (2016). Wavelength-specific forward scattering of light by Bragg-reflective iridocytes in giant clams. *Journal of the Royal Society Interface, 13*(120), 20160285.

Ghoul, M., Griffin, A. S., and West, S. A. (2014). Toward an evolutionary definition of cheating. *Evolution, 68*(2), 318–331.

Gorman, L. M., Tivey, T. R., Raymond, E. H., et al. 2025. Stability of the cnidarian–dinoflagellate symbiosis is primarily determined by symbiont cell-cycle arrest. *Proceedings of the National Academy of Sciences of the United States of America, 122*(14), e2412396122.

Guest, J. R., Baird, A. H., Goh, B. P. L., and Chou, L. M. (2012). Sexual systems in scleractinian corals: An unusual pattern in the reef-building species *Diploastrea heliopora. Coral Reefs, 31*(3), 705–713.

Haig, D. (1993). Genetic conflicts in human pregnancy. *Quarterly Review of Biology, 68*(4), 495–532.

Haig, D. (2019). Cooperation and conflict in human pregnancy. *Current Biology, 29*(11), R455–R458.

Harrison, P. L., and Wallace, C. C. (1990). Reproduction, dispersal and recruitment of scleractinian corals. In *Coral Reefs,* edited by Z. Dubinsky, 133–207. Elsevier Science Publishers.

Hartmann, A. C., Baird, A. H., Knowlton, N., and Huang, D. (2017). The paradox of environmental symbiont acquisition in obligate mutualisms. *Current Biology, 27*(23), 3711–3716.

Heath, K. D., and Tiffin, P. (2007). Context dependence in the coevolution of plant and rhizobial mutualists. *Proceedings of the Royal Society B, 274*(1620), 1905–1912.

Heath, K. D., and Tiffin, P. (2009). Stabilizing mechanisms in a legume-rhizobium mutualism. *Evolution, 63*(3), 652–662.

Herre, E. A. (1993). Population structure and the evolution of virulence in nematode parasites of fig wasps. *Science, 259*(5100), 1442–1445.

Herre, E. A., Knowlton, N., Mueller, U. G., and Rehner, S. A. (1999). The evolution of mutualisms: Exploring the paths between conflict and cooperation. *Trends in Ecology and Evolution, 14*(2), 49–53.

Hiroki, M., Kato, Y., Kamito, T., and Miura, K. (2002). Feminization of genetic males by a symbiotic bacterium in a butterfly, *Eurema hecabe* (Lepidoptera: Pieridae). *Naturwissenschaften, 89*(4), 167–170.

Hirose, E. (2015). Ascidian photosymbiosis: Diversity of cyanobacterial transmission during embryogenesis. *Genesis, 53*(1), 121–131.

Hirose, E., and Hirose, M. (2007). Body colors and algal distribution in the acoel flatworm *Convolutriloba longifissura:* Histology and ultrastructure. *Zoological Science, 24*(12), 1241–1246.

Hirose, E., Neilan, B. A., Schmidt, E. W., and Murakami, A. (2009). Enigmatic life and evolution of prochloron and related cyanobacteria inhabiting colonial ascidians. In *Handbook on Cyanobacteria,* edited by P. Gault, 161–189. Nova Science Publishers.

Holt, A. L. (2024). Simple mechanism for optimal light-use efficiency of photo-synthesis inspired by giant clams. *PRX Energy, 3*(2).

Holt, A. L., Vahidinia, S., Gagnon, Y. L., Morse, D. E., and Sweeney, A. M. (2014). Photosymbiotic giant clams are transformers of solar flux. *Journal of the Royal Society Interface, 11*(101), 20140678.

Hoogenboom, M. O., Connolly, S. R., and Anthony, K. R. N. (2008). Interactions between morphological and physiological plasticity optimize energy acquisition in corals. *Ecology, 89*(4), 1144–1154.

Hughes, T. P. (1994). Catastrophes, phase shifts, and large-scale degradation of a Caribbean coral reef. *Science, 265*(5178), 1547–1551.

Humanes, A., Lachs, L., Beauchamp, E. A., et al. (2022). Within-population variability in coral heat tolerance indicates climate adaptation potential. *Proceedings of the Royal Society B, 289*(1981), 20220872.

Jarvis, G. C., White, C. R., and Marshall, D. J. (2022). Macroevolutionary patterns in marine hermaphroditism. *Evolution, 76*(12), 3014–3025.

Kageyama, D., Narita, S., and Watanabe, M. (2012). Insect sex determination manipulated by their endosymbionts: Incidences, mechanisms and implications. *Insects, 3*(1).

Kageyama, D., Ohno, M., Sasaki, T., et al. (2017). Feminizing *Wolbachia* endo-symbiont disrupts maternal sex chromosome inheritance in a butterfly species. *Evolution Letters, 1*(5), 232–244.

Kaniewska, P., Magnusson, S. H., Anthony, K. R. N., Reef, R., Kühl, M., and Hoegh-Guldberg, O. (2011). Importance of macro- versus microstructure in modulating light levels inside coral colonies. *Journal of Phycology, 47*(4), 846–860.

Kawaguti, S. (1950). Observations on the heart shell, *Corculum. cardissa* (L.), and its associated zooxanthellae. *Pacific Science, 4*(1), 43–49.

Kenkel, C. D., and Bay, L. K. (2018). Exploring mechanisms that affect coral cooperation: Symbiont transmission mode, cell density and community composition. *PeerJ, 6,* e6047.

Kerr, A. M., Baird, A. H., and Hughes, T. P. (2010). Correlated evolution of sex and reproductive mode in corals (Anthozoa: Scleractinia). *Proceedings of the Royal Society B, 278*(1702), 75–81.

Kiers, E. T., Duhamel, M., Beesetty, Y., et al. (2011). Reciprocal rewards stabilize cooperation in the mycorrhizal symbiosis. *Science, 333*(6044), 880–882.

Kiers, E. T., Rousseau, R. A., West, S. A., and Denison, R. F. (2003). Host sanctions and the legume–rhizobium mutualism. *Nature, 425*(6953), 78–81.

Kiers, E. T., West, S. A., Wyatt, G. A. K., Gardner, A., Bücking, H., and Werner, G. D. A. (2016). Misconceptions on the application of biological market theory to the mycorrhizal symbiosis. *Nature Plants, 2*(5), 1–2.

Kiersi, E. T., and Denison, R. F. (2008). Sanctions, cooperation, and the stability of plant-rhizosphere mutualisms. *Annual Review of Ecology, Evolution, and Systematics, 39,* 215–236.

Knowlton, N. (2001). The future of coral reefs. *Proceedings of the National Academy of Sciences of the United States of America, 98*(10), 5419–5425.

Knowlton, N., Corcoran, E., Felis, T., et al. (2021). *Rebuilding Coral Reefs: A Decadal Grand Challenge.* International Coral Reef Society and Future Earth Coasts. https://doi.org/10.53642/NRKY9386.

Kramer, N., Guan, J., Chen, S., Wangpraseurt, D., and Loya, Y. (2022). Morpho-functional traits of the coral *Stylophora pistillata* enhance light capture for photosynthesis at mesophotic depths. *Communications Biology, 5*(1), 1–11.

Lachs, L., Donner, S. D., Mumby, P. J., et al. (2023). Emergent increase in coral thermal tolerance reduces mass bleaching under climate change. *Nature Communications, 14*(1), 4939.

LaJeunesse, T. C. (2020). Zooxanthellae. *Current Biology, 30*(19), R1110–R1113.

LaJeunesse, T. C., Parkinson, J. E., Gabrielson, P. W., et al. (2018). Systematic revision of Symbiodiniaceae highlights the antiquity and diversity of coral endosymbionts. *Current Biology, 28*(16), 2570–2580.

LaJeunesse, T. C., Wham, D. C., Pettay, D. T., Parkinson, J. E., Keshavmurthy, S., and Chen, C. A. (2014). Ecologically differentiated stress-tolerant endosymbionts in the dinoflagellate genus *Symbiodinium* (Dinophyceae) Clade D are different species. *Phycologia, 53*(4), 305–319.

Law, R., and Dieckmann, U. (1998). Symbiosis through exploitation and the merger of lineages in evolution. *Proceedings of the Royal Society B, 265*(1402), 1245–1253.

Leclercq, S., Thézé, J., Chebbi, M. A., et al. (2016). Birth of a W sex chromosome by horizontal transfer of *Wolbachia* bacterial symbiont genome. *Proceedings of the National Academy of Sciences of the United States of America, 113*(52), 15036–15041.

Lesser, M. P., and Farrell, J. H. (2004). Exposure to solar radiation increases damage to both host tissues and algal symbionts of corals during thermal stress. *Coral Reefs, 23*(3), 367–377.

Lesser, M. P., Stat, M., and Gates, R. D. (2013). The endosymbiotic dinoflagellates (*Symbiodinium* sp.) of corals are parasites and mutualists. *Coral Reefs, 32*(3), 603–611.

Li, J., Lemer, S., Kirkendale, L., Bieler, R., Cavanaugh, C., and Giribet, G. (2020). Shedding light: A phylotranscriptomic perspective illuminates the origin of photosymbiosis in marine bivalves. *BMC Evolutionary Biology, 20*(50).

Lively, C. M., and Johnson, S. G. (1997). Brooding and the evolution of parthenogenesis: Strategy models and evidence from aquatic invertebrates. *Proceedings of the Royal Society B, 256*(1345), 89–95.

Lord, K. S., Lesneski, K. C., Buston, P. M., Davies, S. W., D'Aloia, C. C., and Finnerty, J. R. (2023). Rampant asexual reproduction and limited dispersal in a mangrove population of the coral *Porites divaricata*. *Proceedings of the Royal Society B, 290*(2002), 20231070.

Lucas, J. S. (2014). Giant clams. *Current Biology, 24*(5), R183–R184.

Ma, Y., and Chen, R. (2021). Nitrogen and phosphorus signaling and transport during legume–rhizobium symbiosis. *Frontiers in Plant Science, 12*, 683601.

Madin, J. S., Anderson, K. D., Andreasen, M. H., et al. (2016). The Coral Trait Database, a curated database of trait information for coral species from the global oceans. *Scientific Data, 3*(1), 160017.

Marcelino, L. A., Westneat, M. W., Stoyneva, V., et al. (2013). Modulation of light-enhancement to symbiotic algae by light-scattering in corals and evolutionary trends in bleaching. *PLoS One, 8*(4), e61492.

Masuda, K., Miyachi, S., and Maruyama, T. (1994). Sensitivity of zooxanthellae and non-symbiotic microalgae to stimulation of photosynthate excretion by giant clam tissue homogenate. *Marine Biology, 118*(4), 687–693.

Matsuda, S. B., Opalek, M. L., Ritson-Williams, R., Gates, R. D., and Cunning, R. (2023). Symbiont-mediated tradeoffs between growth and heat tolerance are modulated by light and temperature in the coral *Montipora capitata*. *Coral Reefs, 42*(6), 1385–1394.

Matz, M. V. (2024). Not-so-mutually beneficial coral symbiosis. *Current Biology, 34*(17), R798–R801.

Maynard Smith, J. (1998). *Evolutionary Genetics*. 2nd ed. Oxford University Press.

McAuley, P. (1984). Variation in green hydra. A description of three cloned strains of *Hydra viridissima* Pallas 1766 (Cnidaria: Hydrozoa) isolated from a single site. *Biological Journal of the Linnean Society, 23*(1), 1–13.

McAuley, P. J. (1982). Temporal relationships of host cell and algal mitosis in the green hydra symbiosis. *Journal of Cell Science, 58*, 423–431.

McCoy, D. E., Burns, D. H., Klopfer, E., et al. (2024). Heart cockle shells transmit sunlight for photosynthesis using bundled fiber optic cables and condensing lenses. *Nature Communications, 15*, 9445.

McCoy, D. E., and Haig, D. (2020). Embryo selection and mate choice: Can "honest signals" be trusted? *Trends in Ecology and Evolution, 35*(4), 308–318.

McFall-Ngai, M. (2008). Hawaiian bobtail squid. *Current Biology, 18*(22), R1043–R1044.

McFall-Ngai, M. J. (1999). Consequences of evolving with bacterial symbionts: Insights from the squid-vibrio associations. *Annual Review of Ecology, Evolution, and Systematics, 30*(30), 235–256.

McFall-Ngai, M., and Ruby, E. (2021). Getting the message out: The many modes of host–symbiont communication during early-stage establishment of the squid-vibrio partnership. *mSystems, 6*(5).

McField, M. D. (1999). Coral response during and after mass bleaching in Belize. *Bulletin of Marine Science, 64*(1), 155–172.

Melo Clavijo, J., Donath, A., Serôdio, J., and Christa, G. (2018). Polymorphic adaptations in metazoans to establish and maintain photosymbioses. *Biological Reviews, 93*(4), 2006–2020.

Moore, T., and Haig, D. (1991). Genomic imprinting in mammalian development: A parental tug-of-war. *Trends in Genetics, 7*(2), 45–49.

Moran, N. A. (2007). Symbiosis as an adaptive process and source of phenotypic complexity. *Proceedings of the National Academy of Sciences of the United States of America, 104*(S1), 8627–8633.

Muscatine, L. (1967). Glycerol excretion by symbiotic algae from corals and tridacna and its control by the host. *Science, 156*(3774), 516–519.

Muscatine, L., Falkowski, P., and Dubinsky, Z. (1983). Carbon budgets in symbiotic associations. In *Intracellular Space as Oligogenetic Ecosystem: Proceedings,* edited by H. E. A. Schenk and W. Schwemmler, 649–658. De Gruyter.

Muscatine, L., Falkowski, P. G., Porter, J. W., Dubinsky, Z., and Smith, D. C. (1997). Fate of photosynthetic fixed carbon in light- and shade-adapted colonies of the symbiotic coral *Stylophora pistillata. Proceedings of the Royal Society B, 222*(1227), 181–202.

Muscatine, L., Ferrier-Pagès, C., Blackburn, A., Gates, R. D., Baghdasarian, G., and Allemand, D. (1998). Cell-specific density of symbiotic dinoflagellates in tropical anthozoans. *Coral Reefs, 17*(4), 329–337.

Muscatine, L., and Neckelmann, N. (1981). Regulation of numbers of algae in the hydra-chlorella symbiosis. *Berichte der Deutschen Botanischen Gesellschaft, 94*(1), 571–582.

Muscatine, L., and Pool, R. R. (1979). Regulation of numbers of intracellular algae. *Proceedings of the Royal Society B, 204*(1155), 131–139.

Nakamura, T., Yamasaki, H., and Woesik, R. van. (2003). Water flow facilitates recovery from bleaching in the coral *Stylophora pistillata. Marine Ecology Progress Series, 256,* 287–291.

Neckelmann, N., and Muscatine, L. (1983). Regulatory mechanisms maintaining the hydra-chlorella symbiosis. *Proceedings of the Royal Society B, 219*(1215), 193–210.

Negri, I., Franchini, A., Gonella, E., et al. (2009). Unravelling the *Wolbachia* evolutionary role: The reprogramming of the host genomic imprinting. *Proceedings of the Royal Society B, 276*(1666), 2485–2491.

Negri, I., Pellecchia, M., Mazzoglio, P. J., Patetta, A., and Alma, A. (2006). Feminizing *Wolbachia* in *Zyginidia pullula* (Insecta, Hemiptera), a leafhopper with an XX/Xo sex-determination system. *Proceedings of the Royal Society B, 273*(1599), 2409–2416.

Nitschke, M. R., Rosset, S. L., Oakley, C. A., et al. (2022). The diversity and ecology of *Symbiodiniaceae*: A traits-based review. In *Advances in Marine Biology*, edited by C. Sheppard, 55–127. Academic Press.

Noë, R., and Hammerstein, P. (1994). Biological markets: Supply and demand determine the effect of partner choice in cooperation, mutualism and mating. *Behavioral Ecology and Sociobiology, 35*(1), 1–11.

Noë, R., and Hammerstein, P. (1995). Biological markets. *Trends in Ecology and Evolution, 10*(8), 336–339.

Normark, B. B. (2004). Haplodiploidy as an outcome of coevolution between male-killing cytoplasmic elements and their hosts. *Evolution, 58*(4), 790–798.

Norton, J. H., Shepherd, M. A., Long, H. M., and Fitt, W. K. (1992). The zoo-xanthellal tubular system in the giant clam. *The Biological Bulletin, 183*(3), 503–506.

Nyholm, S. V., and McFall-Ngai, M. (2004). The winnowing: Establishing the squid–vibrio symbiosis. *Nature Reviews Microbiology, 2*(8), 632–642.

Nyholm, S. V., and McFall-Ngai, M. J. (2021). A lasting symbiosis: How the Hawaiian bobtail squid finds and keeps its bioluminescent bacterial partner. *Nature Reviews Microbiology, 19*(10), 666–679.

Okasha, S. (2021). Cancer and the levels of selection. *British Journal for the Philosophy of Science, 75*(3), 537–560.

Palacio-Castro, A. M., Smith, T. B., Brandtneris, V., et al. (2023). Increased dominance of heat-tolerant symbionts creates resilient coral reefs in near-term ocean warming. *Proceedings of the National Academy of Sciences of the United States of America, 120*(8), e2202388120.

Palumbi, S. R., Barshis, D. J., Traylor-Knowles, N., and Bay, R. A. (2014). Mechanisms of reef coral resistance to future climate change. *Science, 344*(6186), 895–898.

Pannebakker, B. A., Pijnacker, L. P., Zwaan, B. J., and Beukeboom, L. W. (2004). Cytology of *Wolbachia*-induced parthenogenesis in *Leptopilina clavipes* (Hymenoptera: Figitidae). *Genome, 47*(2), 299–303.

Parkinson, J. E., Peixoto, R. S., and Voolstra, C. R. (2025). Symbiodiniaceae. In *Coral Reef Microbiome*, edited by Raquel S. Peixoto and Christian R. Voolstra, 9–23. Springer Nature Switzerland.

Pasternak, Z., Blasius, B., Abelson, A., and Achituv, Y. (2006). Host-finding behaviour and navigation capabilities of symbiotic zooxanthellae. *Coral Reefs, 25*(2), 201–207.

Patten, M. M., Schenkel, M. A., and Ågren, J. A. (2023). Adaptation in the face of internal conflict: The paradox of the organism revisited. *Biological Reviews, 98*(5), 1796–1811.

Petrou, K., Ralph, P. J., and Nielsen, D. A. (2017). A novel mechanism for host-mediated photoprotection in endosymbiotic foraminifera. *The ISME Journal, 11*(2).

Pettay, D. T., Wham, D. C., Smith, R. T., Iglesias-Prieto, R., and LaJeunesse, T. C. (2015). Microbial invasion of the Caribbean by an Indo-Pacific coral zooxanthella. *Proceedings of the National Academy of Sciences of the United States of America, 112*(24), 7513–7518.

Pile, A. J., Grant, A., Hinde, R., and Borowitzka, M. A. (2003). Heterotrophy on ultraplankton communities is an important source of nitrogen for a sponge–rhodophyte symbiosis. *Journal of Experimental Biology, 206*(24), 4533–4538.

Porter, S. S., and Rice, K. J. (2013). Trade-offs, spatial heterogeneity, and the maintenance of microbial diversity. *Evolution, 67*(2), 599–608.

Price, T. A. R., and Wedell, N. (2008). Selfish genetic elements and sexual selection: Their impact on male fertility. *Genetica, 134*(1), 99–111.

Queller, D. C. (2014). Joint phenotypes, evolutionary conflict and the fundamental theorem of natural selection. *Philosophical Transactions of the Royal Society B, 369*(1642), 20130423.

Queller, D. C., and Strassmann, J. E. (2018). Evolutionary conflict. *Annual Review of Ecology, Evolution, and Systematics, 49,* 73–93.

Quigley, K. M., Ramsby, B., Laffy, P., Harris, J., Mocellin, V. J. L., and Bay, L. K. (2022). Symbioses are restructured by repeated mass coral bleaching. *Science Advances, 8*(49), eabq8349.

Quigley, K. M., Warner, P. A., Bay, L. K., and Willis, B. L. (2018). Unexpected mixed-mode transmission and moderate genetic regulation of *Symbiodinium* communities in a brooding coral. *Heredity, 121*(6), 524–536.

Richmond, R. H. (1997). Reproduction and recruitment in corals: Critical links in the persistence of reefs. In *Life and Death of Coral Reefs,* edited by C. Birkeland, 175–197. Chapman & Hall.

Rigaud, T., Juchault, P., and Mocquard, J.-P. (1997). The evolution of sex determination in isopod crustaceans. *BioEssays, 19*(5), 409–416.

Rodrigues, L. J., and Grottoli, A. G. (2007). Energy reserves and metabolism as indicators of coral recovery from bleaching. *Limnology and Oceanography, 52*(5), 1874–1882.

Rodríguez-Román, A., Hernández-Pech, X., E. Thome, P., Enríquez, S., and Iglesias-Prieto, R. (2006). Photosynthesis and light utilization in the Caribbean coral *Montastraea faveolata* recovering from a bleaching event. *Limnology and Oceanography, 51*(6), 2702–2710.

Rossbach, S., Subedi, R. C., Ng, T. K., Ooi, B. S., and Duarte, C. M. (2020). Iridocytes mediate photonic cooperation between giant clams (Tridacninae) and their photosynthetic symbionts. *Frontiers in Marin Science, 7,* 465.

Rui, Y., Zhaohui, Z., Wenshan, S., Bafang, L., and Hu, H. (2019). Protective effect of MAAs extracted from *Porphyra tenera* against UV irradiation-induced photoaging in mouse skin. *Journal of Photochemistry and Photobiology B: Biology, 192,* 26–33.

Rumpho, M. E., Pelletreau, K. N., Moustafa, A., and Bhattacharya, D. (2011). The making of a photosynthetic animal. *Journal of Experimental Biology, 214*(2), 303–311.

Sachs, J. L., Mueller, U. G., Wilcox, T. P., and Bull, J. J. (2004). The evolution of cooperation. *Quarterly Review of Biology, 79*(2), 135–160.

Sachs, J. L., and Simms, E. L. (2006). Pathways to mutualism breakdown. *Trends in Ecology and Evolution, 21*(10), 585–592.

Sachs, J. L., Skophammer, R. G., Bansal, N., and Stajich, J. E. (2014). Evolutionary origins and diversification of proteobacterial mutualists. *Proceedings of the Royal Society B, 281*(1775), 20132146.

Sachs, J. L., and Wilcox, T. P. (2005). A shift to parasitism in the jellyfish symbiont *Symbiodinium microadriaticum. Proceedings of the Royal Society B, 273*(1585), 425–429.

Salih, A., Larkum, A., Cox, G., Kühl, M., and Hoegh-Guldberg, O. (2000). Fluorescent pigments in corals are photoprotective. *Nature, 408*(6814).

Santos, S. R., Gutiérrez-Rodríguez, C., Lasker, H. R., and Coffroth, M. A. (2003). *Symbiodinium* sp. associations in the gorgonian *Pseudopterogorgia elisabethae* in the Bahamas: High levels of genetic variability and population structure in symbiotic dinoflagellates. *Marine Biology, 143*(1), 111–120.

Schenkel, M. A., Patten, M. M., and Agren, J. A. (2024). Quantifying internal conflicts and their threats to organismal form and fitness. *bioRxiv.* https://doi.org/10.1101/2024.02.05.578856.

Schlesinger, A., Kramarsky-Winter, E., Rosenfeld, H., Armoza-Zvoloni, R., and Loya, Y. (2010). Sexual plasticity and self-fertilization in the sea anemone *Aiptasia diaphana. PLoS One, 5*(7), e11874.

Schwenkert, S., Fernie, A. R., and Geigenberger, P. (2022). Chloroplasts are key players to cope with light and temperature stress. *Trends in Plant Science, 27*(6), 577–587.

Seneca, F. O., and Palumbi, S. R. (2015). The role of transcriptome resilience in resistance of corals to bleaching. *Molecular Ecology, 24*(7), 1467–1484.

Sheppard, C., Davy, S., Pilling, G., and Graham, N. (2018). *The Biology of Coral Reefs.* 2nd ed. Oxford University Press.

Shi, Y., Ke, X., Yang, X., Liu, Y., and Hou, X. (2022). Plants response to light stress. *Journal of Genetics and Genomics, 49*(8), 735–747.

Smith, D. C., Richmond, M. H., and Smith, D. C. (1997). From extracellular to intracellular: The establishment of a symbiosis. *Proceedings of the Royal Society B, 204*(1155), 115–130.

Smith, S. E., and Smith, F. A. (2012). Fresh perspectives on the roles of arbuscular mycorrhizal fungi in plant nutrition and growth. *Mycologia, 104*(1), 1–13.

Stambler, N., and Dubinsky, Z. (2005). Corals as light collectors: An integrating sphere approach. *Coral Reefs, 24*(1), 1–9.

Stat, M., Morris, E., and Gates, R. D. (2008). Functional diversity in coral–dinoflagellate symbiosis. *Proceedings of the National Academy of Sciences of the United States of America, 105*(27), 9256–9261.

Stewart, A. D., Logsdon, J. M., and Kelley, S. E. (2005). An empirical study of the evolution of virulence under both horizontal and vertical transmission. *Evolution, 59*(4), 730–739.

Stouthamer, R., Breeuwer, J. A., and Hurst, G. D. (1999). *Wolbachia* pipientis: Microbial manipulator of arthropod reproduction. *Annual Review of Microbiology, 53,* 71–102.

Stouthamer, R., Russell, J. E., Vavre, F., and Nunney, L. (2010). Intragenomic conflict in populations infected by parthenogenesis inducing *Wolbachia* ends with irreversible loss of sexual reproduction. *BMC Evolutionary Biology, 10*(1), 229.

Sutton, D. C., and Hoegh-Guldberg, O. (1990). Host–zooxanthella interactions in four temperate marine invertebrate symbioses: Assessment of effect of host extracts on symbionts. *The Biological Bulletin, 178*(2), 175–186.

Swain, T. D., DuBois, E., Gomes, A., et al. (2016). Skeletal light-scattering accelerates bleaching response in reef-building corals. *BMC Ecology, 16*(1), 1–18.

Swain, T. D., Westneat, M. W., Backman, V., and Marcelino, L. A. (2018). Phylogenetic analysis of symbiont transmission mechanisms reveal evolutionary patterns in thermotolerance and host specificity that enhance bleaching resistance among vertically transmitted *Symbiodinium. European Journal of Phycology, 53*(4), 443–459.

Terán, E., Méndez, E., Enríquez, S., and Iglesias-Prieto, R. (2010). Multiple light scattering and absorption in reef-building corals. *Applied Optics, 49,* 5032–5042.

Thomas, L., and Palumbi, S. R. (2017). The genomics of recovery from coral bleaching. *Proceedings of the Royal Society B, 284*(1865), 20171790.

Thornhill, D. J., Fitt, W. K., and Schmidt, G. W. (2006). Highly stable symbioses among western Atlantic brooding corals. *Coral Reefs, 25*(4), 515–519.

Thornhill, D. J., Lewis, A. M., Wham, D. C., and LaJeunesse, T. C. (2014). Host-specialist lineages dominate the adaptive radiation of reef coral endosymbionts. *Evolution, 68*(2), 352–367.

Titlyanov, E. A., Titlyanova, T. V., Leletkin, V. A., Tsukahara, J., van Woesik, R., and Yamazato, K. (1996). Degradation of zooxanthellae and regulation of their density in hermatypic corals. *Marine Ecology Progress Series, 139,* 167–178.

Toft, C., and Andersson, S. G. E. (2010). Evolutionary microbial genomics: Insights into bacterial host adaptation. *Nature Reviews Genetics, 11*(7).

Tomascik, T., and Sander, F. (1987). Effects of eutrophication on reef-building corals. *Marine Biology, 94*(1), 77–94.

Tram, U., and Sullivan, W. (2002). Role of delayed nuclear envelope breakdown and mitosis in *Wolbachia*-induced cytoplasmic incompatibility. *Science, 296*(5570), 1124–1126.

Trautman, D. A., and Hinde, R. (2001). Sponge/algal symbioses: A diversity of associations. In *Symbiosis: Mechanisms and model systems,* 521–537. Springer.

Tremblay, P., Fine, M., Maguer, J. F., Grover, R., and Ferrier-Pagès, C. (2013). Photosynthate translocation increases in response to low seawater pH in a coral–dinoflagellate symbiosis. *Biogeosciences, 10*(6), 3997–4007.

van Rhijn, P., and Vanderleyden, J. (1995). The rhizobium-plant symbiosis. *Microbiological Reviews, 59*(1), 124–142.

van't Padje, A., Werner, G. D. A., and Kiers, E. T. (2021). Mycorrhizal fungi control phosphorus value in trade symbiosis with host roots when exposed to abrupt "crashes" and "booms" of resource availability. *New Phytologist, 229*(5), 2933–2944.

Venn, A. A., Loram, J. E., and Douglas, A. E. (2008). Photosynthetic symbioses in animals. *Journal of Experimental Botany, 59*(5), 1069–1080.

Verspoor, R. L., Price, T. A. R., and Wedell, N. (2020). Selfish genetic elements and male fertility. *Philosophical Transactions of the Royal Society B, 375*(1813), 20200067.

Vollmer, A. (2018). Rare parthenogenic reproduction in a common reef coral, *Porites astreoides. HCNSO Student Theses and Dissertations.* https://nsuworks.nova.edu/occ_stuetd/464.

Walker, N. S., Nestor, V., Golbuu, Y., and Palumbi, S. R. (2023). Coral bleaching resistance variation is linked to differential mortality and skeletal growth during recovery. *Evolutionary Applications, 16*(2), 504–517.

Wangpraseurt, D., Jacques, S. L., Petrie, T., and Kühl, M. (2016). Monte Carlo modeling of photon propagation reveals highly scattering coral tissue. *Frontiers in Plant Science, 7,* 1404.

Wangpraseurt, D., Larkum, A. W., Ralph, P. J., and Kühl, M. (2012). Light gradients and optical microniches in coral tissues. *Frontiers in Microbiology, 3,* 316.

Wedell, N. (2020). Selfish genes and sexual selection: The impact of genomic parasites on host reproduction. *Journal of Zoology, 311*(1), 1–12.

Werner, G. D. A., Strassmann, J. E., Ivens, A. B. F., et al. (2014). Evolution of microbial markets. *Proceedings of the National Academy of Sciences of the United States of America, 111*(4), 1237–1244.

Werren, J. H., Baldo, L., and Clark, M. E. (2008). *Wolbachia:* Master manipulators of invertebrate biology. *Nature Reviews Microbiology, 6*(10).

West, S. A., Griffin, A. S., and Gardner, A. (2007a). Evolutionary explanations for cooperation. *Current Biology, 17*(16), R661–R672.

West, S. A., Griffin, A. S., and Gardner, A. (2007b). Social semantics: Altruism, cooperation, mutualism, strong reciprocity and group selection. *Journal of Evolutionary Biology, 20*(2), 415–432.

West, S. A., Kiers, E. T., Pen, I., and Denison, R. F. (2002). Sanctions and mutualism stability: When should less beneficial mutualists be tolerated? *Journal of Evolutionary Biology, 15*(5), 830–837.

Wiedenmann, J., D'Angelo, C., Mardones, M. L., et al. (2023). Reef-building corals farm and feed on their photosynthetic symbionts. *Nature, 620*(7976), 1018–1024.

Wilkerson, F. P., Kobayashi, D., and Muscatine, L. (1988). Mitotic index and size of symbiotic algae in Caribbean reef corals. *Coral Reefs, 7*(1), 29–36.

Wilkinson, D. M., and Sherratt, T. N. (2001). Horizontally acquired mutualisms, an unsolved problem in ecology? *Oikos, 92*(2), 377–384.

Wooldridge, S. (2010). Is the coral-algae symbiosis really "mutually beneficial" for the partners? *BioEssays, 32*, 615–625.

Woolhouse, M. E. J., Webster, J. P., Domingo, E., Charlesworth, B., and Levin, B. R. (2002). Biological and biomedical implications of the co-evolution of pathogens and their hosts. *Nature Genetics, 32*(4), 569–577.

Xiang, T., Lehnert, E., Jinkerson, R. E., et al. (2020). Symbiont population control by host-symbiont metabolic interaction in *Symbiodiniaceae-cnidarian* associations. *Nature Communications, 11*(1).

Yakovleva, I., Baird, A., Yamamoto, H., Bhagooli, R., Nonaka, M., and Hidaka, M. (2009). Algal symbionts increase oxidative damage and death in coral larvae at high temperatures. *Marine Ecology Progress Series, 378*, 105–112.

Yamazaki, S. S., Nakamura, T., and Yamasaki, H. (2008). Photoprotective role of endolithic algae colonized in coral skeleton for the host photosynthesis. In *Photosynthesis: Energy from the Sun,* edited by J. F. Allen, E. Gantt, J. H. Golbeck, and B. Osmond, 1391–1395. Springer Netherlands.

Zanders, S. E., and Unckless, R. L. (2019). Fertility costs of meiotic drivers. *Current Biology, 29*(11), R512–R520.

Conclusion

The Eternal Disquiet Within

MANUS M. PATTEN AND J. ARVID ÅGREN

For this book, we asked the authors to take internal conflicts seriously. We wanted to know what the paradox of the organism meant for everyday organisms and for the very concept of organisms. Fully recognizing that organisms are the legacies of eons of conflict and cooperation may make you reconsider not only the way you look at biology, but also yourself.

In the mid-1960s, W. D. Hamilton went through just such an intellectual reevaluation. By then, he had published his major treatment on the evolution of social behavior, introducing the concept of inclusive fitness as its central organizing principle (Hamilton 1964). He next turned his attention to the problem of sex ratio (Hamilton 1967), and to his hero R. A. Fisher's argument for why populations tend to contain equal numbers of males and females (Fisher 1930/1999). In Fisher's classic explanation, one starts by considering a population where one sex is rarer than the other. Any parent that produces more of the rarer sex will birth offspring with higher fitness, and any allele associated with this adjustment will be favored. However, as the population approaches an even sex ratio the benefit will diminish. What Hamilton pointed out is that Fisher's argument does not hold equally for all genes. For example, a Y chromosome usually has no interest in females. The selective pressures for sex ratio will be different if the gene influencing it is located on an autosome, which spends an equal amount of time in males and females, versus on a chromosome that is inherited through only one sex, such as the Y chromosome (or the W in WZ systems).

To Hamilton, the implications of internal conflicts over the sex ratio—and of internal conflicts generally—were profound. As he would put it toward the end of his career:

Seemingly inescapable conflict within diploid organisms came to me as both a new agonizing challenge and at the same time as a release from a personal problem I had had all my life. In life, what was it I really wanted? My own conscious and seemingly indivisible self was turning out far from what I had imagined and I need not be so ashamed of my self-pity! I was an ambassador ordered abroad by some fragile coalition, a bearer of conflicting orders from the uneasy masters of a divided empire. Still baffled about the very nature of the policies I was supposed to support, I was being asked to act, and to act at once—to analyse, report on, influence the world about me. Given my realization of an eternal disquiet within, couldn't I feel better about my own inability to be consistent in what I was doing, about indecision in matters ranging from daily trivialities up to the very nature of right and wrong? . . . As I write these words, even as to be able to write them, I am pretending to a unity that, deep inside myself, I now know does not exist. I am fundamentally mixed, male with female, parent with offspring, warring segments of chromosomes. (Hamilton 1996, 134–135)

But what does this eternal disquiet within mean for evolutionary theory? Writing around the same time as Hamilton, John Maynard Smith reminded us that the central task of biology is to "explain adaptive complexity, i.e., to explain the same set of facts which the eighteenth-century theologian Paley used as evidence of a Creator" (Maynard Smith 1969, 2). The facts Maynard Smith had in mind were the extraordinary adaptations that characterize living organisms, only with an appeal to evolution by natural selection rather than divinity. Natural selection has traditionally been seen as a process that operates on organisms and leads to organismal adaptations. We now know, however, that there are several phenomena that Paley was unaware of—but that Hamilton and Maynard Smith had begun to appreciate—such as selfish genetic elements and selfish cells, which cannot be seen as representing organismal adaptations. Yet, these internal conflicts are as much the products of evolution by natural selection as any organismal adaptation.

One of the first attempts at reconciling internal conflicts with a Paleyan fascination with organismal design was Richard Dawkins's *The Extended Phenotype* (1982). The book represents the most ambitious articulation of the gene's-eye view of evolution and ends with a chapter titled

"Rediscovering the Organism." He borrowed this line from a critic of his approach, who was amused by Dawkins's seeming surprise that genes and adaptations are bundled up together into well-designed entities that we know as organisms. In light of the chapters in this volume, we are not convinced that this amusement at Dawkins's "discovery" was warranted; looking carefully at organisms reveals that the apparent cohesion of organisms is far from a given. Their existence—and, more importantly for us, persistence—needs to be explained.

That's where this book comes in. All chapters examine these problems in one way or another. In this concluding chapter, we take stock of what we have learned from the contributions and look to what is ahead in the study of internal conflicts.

Who Is the Organism?

Internal conflicts between the parts of an organism and the organism itself can arise in various ways. For parts like selfish genetic elements and selfish cell lineages, we can readily specify for them what would maximize their fitness—in other words, their "goal." As a concrete example, consider *Igf2*, a paternally expressed imprinted gene in mice. Given its pattern of relatedness to littermates and future siblings, it will have its greatest inclusive fitness when the placenta is, say, 110% the size of the optimum for *Igf2r*, a maternally expressed imprinted gene. For each of these two genes, we can specify some placenta size—though it differs for the two genes—that will maximize their inclusive fitness. The question, then, is this: What size of placenta is optimal *for the organism?*

Is the organismal optimum simply the average of these two? Or is the organismal optimum whatever would maximize the fitness of an unimprinted gene? If so, where should we find that unimprinted gene: on the autosomes, on the X chromosome, in the mitochondrial genome, in the nucleus of one of the microchimeric cells lurking in the mouse's body? It is possible that each of the genetic factions just named would see its fitness maximized for a unique size of placenta. Unless we are prepared to specify which faction speaks on behalf of the organism, we are hard-pressed to identify what the organism wants.

Several of the chapters in this volume confront this problem. In Scott and West's treatment (Chapter 1), the organism is analogized to a parliament of genes (*sensu* Leigh 1971). But, extending the analogy, what often passes for parliamentary will ends up being a cabal representing some privileged faction of the genome. And typically, this faction is the autosomal, unimprinted, nuclear, fairly transmitted (eumendelian) faction. One can imagine that the Y chromosome, a germline restricted chromosome, or a B chromosome would each look on at this move and wonder why they themselves are considered *extra*-organismal, despite being consistently present in the bodies of these organisms.

Clarke and Morgan (Chapter 4) use Dawkins's formulation of the paradox of the organism to identify what actually constitutes an organism. For them, any collection of genes that share a "stochastic expectation of the future" *is* the organism, at least in essence. Here again, one might be considering just that special faction of the genome—the well-behaved, autosomal, unimprinted, nuclear, eumendelian faction—and finding that they expect the same stochastic expectation of the future. But as before, other factions would look on and wonder why they have been excluded.

Okasha (Chapter 9) similarly grapples with the identification of who the organism really is. He has previously argued that it makes sense to speak of an organismal "agenda" when the interests of component parts are sufficiently aligned, and that internal conflicts from the sort of sub-agents detailed throughout this volume substantively threaten the notion that organisms can be said to have goals (Okasha 2018). Turning his attention to human psychology, he finds parallels between agency and coherence. Interestingly, he locates a similar struggle in psychology to find who the "true" self is. Is it the impulsive version of ourselves that might accept short-term rewards in exchange for long-term suffering? Or is it the deliberative version of ourselves that values long-term happiness over short-term rewards? We find it interesting that in psychology, the existence of internal conflict is met with a tendency to avoid deciding which self is the real one or at least a humility in attempting to do so. This is unlike the situation for internal evolutionary conflicts in organisms, where it seems biologists have by and large decided that they know who the real self is by fiat.

Huneman (Chapter 5) offers a way out of this jam by placing the organism on a spectrum from agent to ecosystem. While ecosystems may have functions (like nutrient cycling or carbon storage), we tend not think of them as having coherent goals or aims. Ecosystems comprise agential things but are not agents themselves. But when those agential things within cohere, one may speak of the organism as having aims and goals, of being an agent proper. And one may equate the aims and goals of an organism's parts with the organism's aims and goals in that case.

What Is Internal Conflict, and Is It an Existential Threat?

In Dawkins's original formulation of the paradox of the organism, he expressed a fear that internal conflicts might be capable of altogether subverting organisms. A number of chapters in this volume addressed the scope of conflict, aiming to clarify when we might reasonably expect such a fear to come true and, just as importantly, when not.

One reason to suspect the fear may not come true is that the number of selfish genetic elements that are actually capable of interfering with the general functioning of the organism is quite low. Many of the traits that are optimized by natural selection on organisms are untouched by selfish genetic elements. This is a theme in both Scott and West (Chapter 1) and Patten (Chapter 3). While selfish genetic elements may drag the fitness of their bearers down, they don't alter organismal phenotypes greatly. It's as though the actions of some selfish genetic elements are beneath the concern of the organism that hosts them. But then some genes *do* alter organismal phenotypes, and these, Patten argues, are the ones that Dawkins was right to worry about. However, Scott and West make the point that there is a vast genome beyond these selfish elements that would be aligned in overruling the efforts made by such rogue genes, and, as such, there is no need to worry about the status of the organism. Better still, the most consequential of these selfish elements are the ones most likely to spur the evolution of resistance from the genome beyond.

A further complication is raised by Bourrat (Chapter 2), who points out a certain slipperiness that accompanies how we talk about internal conflict. In fact, he argues, some of what is typically considered conflict between levels is not what it appears to be. Ordinarily, when we speak of internal conflict, we are describing a situation in which what a part (a gene, a cell) wants is different from what the organism wants. Bourrat points out that this construction unintentionally leaves the part out of the organism, as though there is some organism that exists outside of that part. It would be more correct to speak of some part being in conflict with other parts—that is, objects at the same level of selection.

Taking Bourrat's idea about conflict to heart might, yet again, leave us struggling to identify an organism. It would make the most sense to identify internal conflict as between one gene and another gene. That is, a selfish genetic element might want one thing, but a representative gene might want something else. This solves the measurement problem, as we are comparing two equivalent entities on the same timescale, but it again leaves us having to choose which gene serves as the representative to reflect the "true" organism.

Maybe all there is are genes and their interests, sometimes in conflict. One finds sympathy with the Dawkinsian temptation to dispense with the organism altogether.

When, Where, and How Much Do Conflicts Matter?

Setting aside these concerns over how to capture an organism or how to measure and identify conflict, we turn to what we have learned of the consequences of selfish genetic elements and selfish cell lineages. Several of the chapters highlight the necessity of confronting selfish elements if we want to explain some well-studied biological phenomena.

Schenkel, Ross, and Wedell (Chapter 7) present a gallery of selfish genetic elements that both owe their existence to, and subsequently exert their influence on, sexual reproduction. There is a certain irony to this, for sexual reproduction was identified by Maynard Smith and Szathmáry (1995) as one of the seminal cooperative acts in the history of life, as it makes reproduction a joint venture between partners. And while there has been an attempt at grounding the understanding of sexual reproduction in this cooperative light (Roughgarden et al. 2006), in the asymmetries that sexual reproduction imposes and the division of goods that accompanies meiosis, conflicts are inevitably stirred. The genetic systems underpinning all things related to sexual reproduction are regularly pulled into conflict. Although the origin of sexual reproduction may best be interpreted from a cooperative perspective, the great diversity that now characterizes sex-determination systems appears to have been driven by internal conflicts and are best interpreted accordingly.

Unlike the diversity of sex-determination systems, less attention has been paid to internal conflicts in the early stages of animal embryogenesis. Haig (Chapter 6) brings a paradox-of-the-organism perspective to this area of developmental biology. Again, there is an irony in how metazoan multicellularity, which should be viewed as a cooperative venture at its evolutionary onset, has since become the sine qua non for a variety of intragenomic conflicts that Haig details.

In addition to the concerns of fundamental biology, internal conflicts have much to offer those with more practical concerns. Boddy and Ågren (Chapter 8) highlight the clinical implications of internal conflicts, such as cancer and those that stem from live birth in mammals. Once more, this latter example represents something that begins as a cooperative venture evolutionarily, much like sexual reproduction and animal multi-cellularity, but as before we find that much of the evolutionary diversity that follows is generated by the pressure stemming from internal conflicts. Not only do Boddy and Ågren rely on internal conflicts to explain this diversity, but they also use the principles of internal conflict to point the way toward treatments. An understanding that cancerous tumors may themselves experience internal conflict inspires strategies to thwart their growth and maturation. Likewise, appreciating the conflicting interests of parents and offspring during pregnancy and beyond (e.g., through microchimerism) provides both an explanation for some of the maladies of parenthood and inspiration for how to best treat these.

Okasha (Chapter 9) ponders whether the kind of internal conflict that Hamilton voiced in the quote at the beginning of this chapter bears any formal resemblance to the kinds of conflicts that Dawkins spoke of with the paradox of the organism. Psychological internal conflicts are different from those featured in other chapters. In those conflicts, it is possible to point to some cooperative venture that the internal conflict subverts (e.g., sex is cooperative, multicellularity is cooperative, and pregnancy is co-operative—at least at their outset). But we cannot locate the cooperative venture in the case of the human psyche. Another notable difference be-tween psychological internal conflicts and the kind of evolutionary con-flicts we encounter throughout this volume is the timescale of their operation, as Okasha notes. For example, the tug-of-war that takes place between $Igf2$ and $Igf2r$ is not one that any mouse would consciously ex-perience. A developmental biologist following said mouse would find the placenta reaches some size, and supposing that developmental biologist was not clued in to imprinting conflicts, they might be completely un-aware of any conflict in the system. This is unlike psychological conflicts, in which we experience in developmental time a tug-of-war between com-peting impulses or emotions. However, there is something that unites evolutionary and psychological conflict, Okasha finds, something inspired by the paradox of the organism: It only makes sense to talk of a unity of purpose in organisms and a unity of self in humans when internal conflicts are negligible. When internal conflicts are severe in their effects, or when these conflicts aren't suppressed or tamped down, the unity of purpose necessary to think of organisms and people as unitary beings disappears.

Finally, McCoy and colleagues' (Chapter 10) treatment of internal conflict in coral-algal symbioses not only has something to say about the basic organismal biology of corals and long-term macroevolutionary trends but may be relevant to conservation practice on more immediate timescales. Corals are not equally extinction-prone, and knowing the kinds of internal conflicts corals face may help to prioritize conservation efforts.

Eternal Disquiet Within: Darwinian Paranoia or Darwinian Cynicism?

Despite the fact that organisms comprise what had formerly been free-living and competing replicating agents in the world, we often brush aside internal conflicts when thinking about the design of organisms and their status in evolutionary theory. As a consequence of this habit of thought, the adaptationist program has become a simple, almost reflexive, two-step process: First, find an evolutionary novelty; second, ask what benefits it confers *on the individual organism*. The possibility that organisms might be designed as they are to serve the interests of one of their internal parts is foreclosed in this formulation. All adaptations are treated as though they serve the interests of the organism, for what other interests could there be? The answer is the particular genic or cellular faction that underlies the novelty.

This way of thinking has echoes of selfish gene theory, in which all genes are conceptualized as acting in their own self-interest to promote their representation in future generations (Ågren 2021). The paradox of the organism grows out of this approach, but it also homes in on a narrower definition of what it means for genes and cells to act selfishly. For most genes and cells, most of the time, the best selfish strategy is to do what is best for the organism. As a consequence, there is often an equivalence between what the organism wants and what its component parts want. In many situations—likely for the vast majority of traits—this equivalence holds, and we are quite comfortable adopting a pluralistic position: One can view these cases as though genes are the lone beneficiaries of natural selection, or one can view these as though organisms are the real evolutionary agents.

In this book we have been focused entirely on those genes and cells that have interests separate and apart from the organism. However, to be clear, we are not advocating for the dismissal of the organism as a unit of selection—as in an *organizational* unit—or as a level of organization where adaptation manifests. We think it is fine to treat organisms as though they have goals and that they are the beneficiaries of adaptations that aid them in pursuit of these goals, provided care is taken in avoiding reflexively ascribing all adaptations to the organism. That said, as noted above, we recognize how challenging it can be to pin down exactly who or what it is that has these "organismal" goals.

Instead, we advocate against taking organismal unity as a given and against the reflexive and uncritical acceptance of organisms as the seat of adaptation. There is a very real possibility that the design we find in nature is not for the benefit of the organism but is, in many instances, for the benefit of some selfish element residing within the organism.

The habit of attributing agency and goals to genes has been labeled "Darwinian paranoia" (Francis 2004; Godfrey-Smith 2009), and many of our colleagues caution against it for fear that we will all too easily slide into causal attributions where they do not belong. On this view, it is paranoid (in a pejorative sense) to think, for example, that your high-functioning immune system is not working on your behalf but is instead doing the bidding of some hidden, shadowy gene. However, channeling Joseph Heller, just because you are paranoid doesn't mean they aren't after you. This is especially true when dealing with selfish elements, for they clearly have interests different from you (leaving aside the question of identifying the "you" in that statement). By taking internal conflicts seriously, paranoia gives way to a healthy dose of cynicism.

References

Ågren, J. A. (2021). *The Gene's-Eye View of Evolution*. Oxford University Press.

Dawkins, R. (1982). *The Extended Phenotype: The Gene as the Unit of Selection*. Oxford University Press.

Fisher, R. A. (1930/1999). *The Genetical Theory of Natural Selection. A Complete Variorum Edition*. Oxford University Press.

Francis, R. C. (2004). *Why Men Won't Ask for Directions: The Seductions of Sociobiology*. Princeton University Press.

Godfrey-Smith, P. (2009). *Darwinian Populations and Natural Selection*. Oxford University Press.

Hamilton, W. D. (1964). The genetical evolution of social behaviour I and II. *Journal of Theoretical Biology, 7*, 1–52.

Hamilton, W. D. (1967). Extraordinary sex ratios. *Science, 156*, 477–488.

Hamilton, W. D. (1996). *Narrow Roads of Gene Land*. Vol. 1, *Evolution of Social Behaviour*. Oxford University Press.

Leigh, E. G., Jr. (1971). *Adaptation and Diversity: Natural History and the Mathematics of Evolution*. Freeman, Cooper and Co.

Maynard Smith, J. (1969). The status of neo-Darwinism. In *Sketching Theoretical Biology*, edited by C. H. Waddington, 82–89. Edinburgh University Press.

Maynard Smith, J., and Szathmáry, E. (1995). *The Major Transitions in Evolution*. Oxford University Press.

Okasha, S. (2018). *Agents and Goals in Evolution*. Oxford University Press.

Roughgarden, J., Oishi, M., and Akçay, E. (2006). Reproductive social behavior: Cooperative games to replace sexual selection. *Science, 311*(5763), 965–969.

CONTRIBUTORS

EDITORS

J. ARVID ÅGREN, Cleveland Clinic Lerner College of Medicine.

MANUS M. PATTEN, Georgetown University.

CONTRIBUTORS

AMY M. BODDY, University of California, Santa Barbara.

PIERRICK BOURRAT, Macquarie University.

ELLEN CLARKE, University of Leeds.

BRENDAN CORNWELL, Stanford University.

JENNIFER A. DIONNE, Stanford University.

DAVID HAIG, Harvard University.

PHILIPPE HUNEMAN, CNRS/Université Paris I Panthéon-Sorbonne.

SÖNKE JOHNSEN, Duke University.

DAKOTA E. McCOY, University of Chicago.

WILL MORGAN, Perse School, Cambridge.

SAMIR OKASHA, University of Bristol.

LAURA ROSS, University of Edinburgh.

MARTIJN A. SCHENKEL, University of Groningen and Georgetown University.

THOMAS W. SCOTT, University of Oxford.

NINA WEDELL, University of Melbourne.

STUART A. WEST, University of Oxford.

INDEX

Acoelomorpha, 145, 300, 303, 308

Acropora, 285, 296, 302, 307

adaptation, 1–2, 10, 14, 22, 255, 258, 268–270; as adaptive compromise, 70, 235, 237, 242; beneficiary of, 3, 47, 201, 335–336; biological examples of, 133–135, 163, 243, 245–246, 285–286, 300–304; and common ancestry, 19; and design, 9–12, 33, 329; and fitness maximization, 24, 100, 109, 200. *See also* maladaptation

adaptationism: explanatory vs. methodological, 33; naive, 268

adaptive therapy, 239–241

Agaricia, 284, 307

agency, 99; and fitness maximization, 3, 60–63, 100–104; and internal conflicts, 63–67; Kantian, 113–114; organismal, 3, 58–63, 69, 97, 100–104, 172, 235; as paranoia, 356; of parts, 53, 112, 116–117, 129, 137, 235, 264; realism about, 104, 115–116; and the self, 264

aging (senescence), 115–117

Air, 18

altruism, 20; and levels of selection, 33, 42–43, 98, 110, 117; within bodies, 58, 149

Ambystoma, 308

anemones, 275, 290–291, 298, 306

aneuploidy, 157–159, 171

anisogamy, 205

anthropomorphism, 99, 268

arena germinalis, 126–128, 132, 141, 154–155, 159, 173–174

Aristotle, 1, 79–82, 84. *See* essentialism

Armadillidium, 215, 292

arms race, evolutionary, 111, 141; in corals, 299–300, 302–303; in parasites, 218, 277, 292–293

Arsenophonus, 21

ascidians, 308

asexuality, 36–38, 45, 145–151, 213–214, 275, 292; in corals, 281–282, 290–291. *See also* parthenogenesis

bacteriophages: and male-killing mechanisms, 217–219; and non-coral symbiosis, 287

B chromosome, 212, 331

bivalves, 277, 297, 300, 303, 308

body as machine, 1–3, 53–55, 172, 235–237, 241

Bombyx, 217

bottleneck: developmental, 86–90, 114, 259

Bradyrhizobium, 309

brooding, 274, 281–285, 291

cabals of the few, 12, 22, 26, 91–92, 130, 138–139, 160. *See also* parliament of genes

Caenorhabditis, 127, 140, 151–154, 157

cancer, 58, 69, 98, 103–105, 116–117, 235, 305, 334; and chimerism, 245–246; defense against, 48, 138; hallmarks of, 237–238; and multilevel selection, 33–35, 44–49; treatment of, 237–241, 248–249

Cassiopea, 287

CDKN1C, 18

centriole, 67, 68

centromeric drive, 209

Ceratodictyon, 308

cheating, 67, 90, 244; cellular, 235, 239, 244; in corals, 286–287, 305–309

chimerism, 244–245. *See also* microchimerism

chloroplast, 118, 212, 221; maternal inheritance of, 202, 205–206. *See also* cytoplasmic element

climate change, 277–278, 284, 300

Cliona, 305, 308

coevolution, 111, 202; host-symbiont, 217, 277, 282, 296–297, 302–303, 308; selfish genetic element and suppressors, 221–223. *See also* arms race, evolutionary

colonial organism, 139, 145, 281–282, 308

competition, 3, 98, 107–109; and cancer, 239–241, 248; difference from conflict, 35–38; between embryos, 136, 157–159; and endosymbionts, 216; between parts, 56–58, 62, 68, 76, 79, 105–111, 116–119, 176, 205–207; between ramets, 150; between sperm, 135, 239–241, 248, 293; within sexes, 21; and symbiosis, 286, 296, 299, 304, 309; and the veil of ignorance, 172

competitive release, 240–241

composition, 32, 40–41, 47–48, 81, 92, 100, 117, 201; special composition question, 76–77

conflict: defined, 35–36, 38, 43; as expectation, 2; quantified, 278–280, 295–296, 299, 300. *See also* cooperation; intragenomic conflict; selfish elements

conflict resolution. *See* suppression

conservation biology, 277

conventionalism, 110

convergent evolution, 283

Convolutriloba, 303

cooperation: bacterial, 15–17, 20, 26; and common ancestry, 19–20, 26, 152; and

conflict, 35–38, 76–77, 88, 126, 239, 247–248, 273, 328; and inclusive fitness, 10–11; and the major transitions, 62, 67, 236–237; and microchimerism, 247; among parts, 1, 14, 25, 53–54, 76–77, 130–132, 235, 239, 245, 255; in pregnancy, 241–242; quantified, 285, 298, 310; and sex, 333–334; and stress, 284–285; in symbiosis, 277, 298–299, 303, 306–308

coral-algal symbiosis: coral bleaching, 273, 276–279, 284–285, 300–302, 304, 306–307; housing, 305–308, 310; resource sharing, 294–300; and sex determination, 288–291, 293–294; and solar irradiation, 300–304. *See also* transmission

Corculum, 303, 305

coreplicon, 128

creator: arguments of evidence for, 9, 329. *See also* design

Ctenactis, 307

Cupriavidus, 309

cyanobacteria, 308

cytochrome oxidase II, 206

cytoplasmic bridges, 135, 139–140

cytoplasmic elements: cytoplasmic incompatibility, 213, 291–294; in eggs, 153–155; inheritance of, 202–207, 212, 221–222; selfish, 255; and sex determination, 212–216, 220; and symbionts, 209, 276; transfer to nucleus, 259; Weismann's theories about, 142, 144, 173–175. *See also* cytoplasmic male sterility

cytoplasmic male sterility, 59–60, 67, 207, 268–269

Danio. See zebrafish

Darwin, Charles, 9

Darwinism, 45, 97, 105, 125, 268–269; and cynicism vs. paranoia, 335–336

Daucus, 299

Dawkins, Richard: and the gene's-eye view, 79, 82, 84–88, 128, 241–242; on organisms, 1–2, 68–69, 75–77, 92–94,

255–260, 270, 329–330, 334–336.
See also gene's-eye view; paradox of the
organism
Democritus, 81–82
Dendrogyra, 307
Descartes, René, 235
design: biological, 9–12, 33, 235–237; and
the human eye, 236; and internal conflict, 12–13, 54, 88, 329–330, 335–336.
See also adaptation
developmental hourglass, 141
Dicrocoelium, 32
Dictyostelium, 112
dinoflagellates. *See* coral-algal symbiosis
Diploastrea, 290
Diptera, 217
dissociative identity disorder: parallels
with intragenomic conflict, 256,
265–267
division of labor: in multicellular organisms, 112, 117, 204, 239, 261
DNA elimination, 206–209, 221–223
DNA replication, 132–134, 155, 158
doublesex (dsx), 215, 217, 219
Drosophila: cytonuclear conflict in, 206,
219; development, 127, 139–140, 144,
153–158; *Segregation Distorter* in, 12,
57; sex chromosomes of, 21, 24, 212,
222, 269
Dryas, 212
DUX genes, 161–163

E. coli, 16
economics, 96, 98–104, 108–109, 112–114;
behavioral, 173, 264–265
ecosystems: and aging, 115–117; Clements-
ian view of, 104–106, 113; contrast to
agents, 96–99, 108–112, 119–120; coral,
273, 276–277, 286–287; Gleasonian
view of, 106, 113; and individuality,
105–109, 117–119, 300–301, 310; and
internal conflicts, 114–115; organicism,
104–105; realist and relativistic notions
of, 113–114; suborganismal, 204
embryogenesis, 143, 151, 153, 157–158,
163–166

endosymbiont. *See* chloroplast; coral-algal
symbiosis; cytoplasmic elements;
mitochondria
epigenetics, 18, 96, 133–134, 142–144,
175; and veil of ignorance, 169–172
essentialism, 79–86
eumendelian. *See* Mendelian inheritance
Eurema, 215, 292
Eusmilia, 307
evolutionary arms race, 111, 141; in
corals, 299–300, 302–303; in parasites,
218, 277, 292–293
Exaiptasia, 298
extended phenotype, 32, 35
extinction, 45, 55, 210, 284, 291, 293

fair meiosis. *See* Mendelian inheritance
feminization, 210–211, 213–216, 221, 278,
280, 288–294
fertility, 57, 209–210; infertility, 135, 160,
247; internal conflicts and, 64–66, 69,
206, 201, 268
fig wasps. See *Pegocapsus; Tetrapus*
Fisher, R. A., 9, 60, 97, 328
fitness, 35, 61–63, 88–89, 109, 116, 150,
156, 171, 204, 216, 246; alignment, 3,
103, 107–108, 112, 114, 118, 239,
247–248, 259; costs of internal conflict,
54–58, 64–70, 75, 91–92, 131–134, 169,
201, 206–209, 211, 235, 330; and definitions of conflict, 35–46; direct and
indirect, 9–11; genic vs. organismal,
128–129, 146; inclusive, 3, 9–11, 13–17,
19, 96, 200, 328; in the major transitions, 61–63, 117–119; maximization of,
3, 12–17, 19, 53–55, 59–61, 97, 100–103,
201; in mother vs. offspring, 17–19; and
the parliament of genes, 22–25; in the
Price Equation, 57; in symbiosis,
278–282, 286–287, 291–294, 299
Flavobacteria, 216
fluorescence, 276, 302, 303
foraminifera, 275, 303
Formal Darwinism, 53, 59–61, 64–66, 97,
100, 102, 113, 115
fundamental theorem of natural selection, 60

Gaia hypothesis, 105
gene's-eye view, 88, 128, 261, 329–330, 335–336. *See also* Dawkins, Richard
genetic conflict. *See* intragenomic conflict; selfish elements
genomic imprinting: and internal conflict, 17–19, 103, 168–169, 257–259, 330–331, 334; kinship theory for the evolution of, 55–56, 66; mechanisms of, 170; and relatedness, 14; as trait distorter, 90–91; in sex determination, 215, 292
Gerakladium, 308
germ granules, 143–144
germline: defined, 126, 173–176; developmental variation in, 141–155; germ crown and germ stem, 126–127, 142–143, 151, 153–154, 171, 176; selection, 131–132, 137, 166; sequestration from soma, 88, 125–127, 136, 140–142, 151, 158, 173–175, 204–205; specification of, 117, 142–144, 165–166; syncytia, 139–140. *See also* primordial germ cells; selfish elements; Weismann, August
germline-restricted chromosome, 208, 331
germ plasm, 142–143, 173–175
gestational drive, 159
global cooling. *See* climate change
global warming. *See* climate change
Glomus, 298, 299
glycine, 309
God, 9. *See also* creator
Gompertz curve, 116
gonochorism vs. hermaphroditism, 274, 283–284, 289–290
Gp-9, 14
Grafen, Alan, 53, 59–61, 64–66, 97, 100, 102, 113, 115. *See also* Formal Darwinism
Grb10, 18
greenbeards, 14–16, 19–20, 26, 135

Haig, David, 256–257, 270
Haliclona, 308
Hamilton, W. D., 328. *See also* fitness: inclusive
Hamilton's rule, 13

haplodiploidy, 212, 214–215, 293
Heliofungia, 307
helping behavior, 12–21, 25–26
Heraclitus, 77, 79, 81
heterochromatin, 133
holobionts, 111–112, 274, 285, 291
Holospora, 287
homing endonuclease, 55–57, 65, 69, 90
Homona, 218–219
Hordeum, 287
horizontal gene transfer, 14, 16–17, 19, 26, 82, 90
Hydra, 308–309
Hypolimnas, 210

Igf2 / Igf2r, 18, 330, 334
immune system, 89, 116, 159, 244, 246–248, 336
imprinted genes. *See* genomic imprinting
inclusive fitness. *See* fitness
Individual in the Animal Kingdom, The (Huxley), 2
individuality, 34, 97, 104, 256–258; of ecosystems, 105–108; evolutionary origins of, 2, 49, 61–63, 236–239; evolutionary transitions in, 39–40, 53–54, 67–70, 103, 117–119, 120, 236; of genets/ramets, 92, 147–150; of holobionts, 111; and the unit of selection, 106; weak individuality, 107–108, 113
innovation, 131–132, 149, 154–155
intentionality, 259, 263, 268–271
intragenomic conflict: in development, 126–128, 204–205, 223; and heredity, 203–209, 221–223; in individuality, 256–262; in mammals, 157–171; and psychological conflict, 259–269; in sex determination, 201–203, 209–221; significance of, 9, 11, 24–26, 220–224, 262. *See also* selfish elements

Keynes, John Maynard, 44
Kantianism, 113, 120
KCNQ1OT1, 18

lateral gene transfer. *See* horizontal gene transfer
Leigh, Egbert, 22, 26, 67, 87, 130. *See also* cabals of the few; parliament of genes
levels of selection, 66–69, 98, 110, 117, 150, 201; conflicts between, 32–35, 38–49; and the Price equation, 55–58
Lepidoptera, 217
lichen, 308
life cycle, 79, 116, 207, 259, 261, 281–282
light stress. *See* ultraviolet (UV) radiation
Lissoclinum, 308
Locke, John, 83

Madison, James, 130
Madracis, 307
major transitions. *See* individuality: major evolutionary transitions in
maladaptation, 32, 133, 160
Marginopora, 303
maximum tolerated dose, 238
Maynard Smith, John, 100, 329. *See also* individuality: evolutionary transitions in
Medawar, Peter, 244–245, 247–249
Medea, 136
Medicago, 298
meiosis: asymmetric, 209; as defense against selfish genetic elements, 103, 129, 259; fair, 19, 87–88, 103, 129; male vs. female, 134, 139–140, 202; and recombination, 128–129
meiotic drive, 12, 134–136, 173; of B chromosomes, 211–212; in females, 66, 209, 211; in males, 12, 60–61, 64–65, 140, 209, 269; of sex chromosomes, 21–24, 65, 209–211, 215–216, 220–223
Mendelian inheritance, 66, 127, 201–210, 220, 259, 331. *See also* meiosis: fair
metaphysics, 75–76, 104, 108, 120
microalgae. *See* coral-algal symbiosis
microbiome, 96, 105, 111, 244. *See also* holobionts
microchimerism, 235, 243–248; and the microchiome, 245–246. *See also* chimerism
Mimosa, 309

mitochondria, 85, 140, 330; inheritance of, 67, 89, 202, 205–206, 221, 259; and sex determination, 59–60, 207, 212–213, 268–269. *See also* cytoplasmic element
mobile elements, 16–17, 82–83, 86, 90, 102
Modern Synthesis, 61–62
Monostroma, 205
Montastraea, 298, 302, 307
mother's curse, 206
multicellularity, 40, 97, 125, 237; development of, 131, 137, 147, 204; evolution of, 70, 126, 239; as major transition, 2, 33, 62, 98, 103, 117–118, 236; vulnerability to internal conflicts and cancer, 1, 237–239, 333
multilevel selection. *See* levels of selection
Mus, 211
mutation: and cancer, 46, 131, 137, 238, 248; and development, 137–139, 149, 154, 165, 277; and the evolutionary process, 4, 126, 133; and identity, 85–87, 92; vs. innovation, 131, 150, 153
mutualism, 273, 276, 286–287, 295, 306; breakdown, 277, 287–288; as ecological process, 98, 105; empirical examples of, 109. *See also* coral-algal symbiosis; Symbiodiniaceae; symbiont
mycorrhizal fungi, 298–299
Myrionema, 309

Nasonia, 292
neoblasts, 145, 147–150, 174
niche construction, 96, 98, 102, 104
Nieuwentyt, Barnard, 235–236
nihilism: and understanding of the organism, 81–82
nitrogen fixation, 278, 298–299, 309
nutrients, 275–276, 278, 280, 288–289, 306, 309, 310

Obox, 161
Oenothera, 205
Oikopleura dioica, 140
Okasha, Samir: on agency, 53, 58–65, 97, 102; on levels of selection, 47; on rationality, 100–101

Orbicella, 296, 302

organism: as agent, 97, 99–104, 267–268; as biological individual, 256; concepts, 76–77, 97, 115–119, 244, 255–261, 328; as ecosystem, 97, 104–108; and organismality, 2, 127, 248, 262; perspectives on, 108–114, 172–173, 223, 235–237, 246–247, 270; as target of natural selection, 98, 128–129, 200–201, 269

Oscar, 217–218

Oskar, 144–145

Ostreobium, 302

Ostrinia, 217–218, 292

Oxypora, 307

Paley, William, 1, 9, 235–237, 329. *See also* design, biological

Pando, 92

paradox of change, 77–79, 83, 93

paradox of the organism: defined, 1, 11, 68–69, 75–76, 172–173, 255–256; as not actually a paradox, 47–49; perspective, 4, 200, 203, 219, 235–243, 247–249, 270; resolutions to, 9, 12, 24–26, 54, 69–70

paramecium, 35–36, 287

parasitism: analogies to internal conflicts, 44, 75, 86, 90–93, 130, 211; empirical examples, 32, 277–279

Parasitodiplogaster, 287

parental investment, 125, 155–159, 169–170, 241–243, 257, 281

parent-offspring conflict, 33, 66, 157, 242, 257

parliament of genes, 3, 9, 12, 22–26, 130–131, 159, 169, 172–173, 259, 330–331. *See also* cabals of the few

parthenogenesis, 213–215, 289–294. *See also* asexuality

paternal genome elimination, 207–208, 212, 293

pax somatica, 126, 133–134, 141, 168, 173–174

Pegocapsus, 287

Peto's paradox, 238

phagocytosis, 276

phenotypic plasticity, 96–97, 102, 104

philosophy of mind, 256

photosynthesis, 273, 278, 285, 294–301, 304, 306

phylogeny, 283, 285, 289, 291–292, 294

Physa, 207

piRNA, 135–136, 144

planarians, 126, 146–151, 159, 174

plasmids. *See* mobile elements

pluralism, 98, 108–110, 112–114, 119–120, 335

Pocillopora, 290, 297, 307

political theory, 172–173. *See also* parliament of genes; veil of ignorance

Porites, 284, 290, 301, 307

power, 19, 26, 305

preeclampsia, 91, 242–243, 249

preformation. *See* germline; germ plasm

pregnancy, 83–85, 243–248, 334; complications of, 160, 241–243, 249. *See also* preeclampsia

Price equation, 46, 56–58, 61–64, 67, 70

primoridal germ cells, 133, 138–139, 151–152, 154, 165–166. *See also* germline

public goods, 16, 309

rationality, 98, 100–102, 104, 108–109, 114, 263–265

Rawls, John. *See* veil of ignorance

recombination: homologous, 132–133, 158; meiotic, 129–130, 221–223, 259

repetitive DNA sequences, 134–135, 146, 151, 158, 161–162, 205

retroviruses, 166–168

Rhizobium, 298–299, 308–309

Rhizophagus, 299

ribonucleoprotein condensates, 143–144

Rickettsia, 216

RNA virus, 216, 219, 287

Roux, Wilhelm, 2, 58, 176

Schmidtea, 148

scleractinian coral, 282–283, 290, 297, 304, 306–307

Segregation Distorter (SD), 12, 57, 61

self, 255–256, 266–267; divided, 256–258, 270; dual, 264–265. *See also* unity of self

selfish cells, 56–58, 68, 127, 131–132, 138–139, 237–241, 247–249

selfish elements: and agency, 53, 59–60, 63–67, 260–262, 267–271; cytoplasmic, 205–207, 212–219, 289–294; and identity, 90–92; and individuality, 67–68, 260–262, 267–271; kinds of, 54–58, 70, 201–203; and levels of selection, 47, 56; postsegregational, meiotic, 125, 129, 134–136, 140, 155, 157, 159–161, 168, 171; postsegregational, mitotic, 125, 134–136; in relation to the paradox of the organism, 1–4, 9, 11–13, 54, 68–70, 75–76, 255–256, 329–336; and sex ratios, 21, 209–212; and sexual reproduction, 214–215; significance of, 220–223. *See also* germline; intragenomic conflict; parliament of genes; *specific genetic and cellular elements*

selfish gene theory. *See* gene's-eye view

senescence. *See* aging (senescence)

sex chromosomes: conflicts between, 21, 169, 220; evolution of, 221–223, 293; and meiotic drive, 24, 65; and sex determination, 202–203, 209–212, 216; and sex ratios, 21–22, 328

sex determination, 200–203, 209–212, 220–221, 291–293; conflict over, 67, 212–214, 289; in corals, 289–291, 293–294. *See also* feminization; sex chromosomes

sex reversal. *See* feminization

sexual reproduction, 125, 128, 281, 333–334; and its influence on internal conflict, 150–151, 201–202, 292–294. *See also* asexuality; parthenogenesis

Ship of Theseus, 92

sibling rivalry, 136, 155–159, 216, 330. *See also* parent-offspring conflict

Siderastrea, 284

soma. *See* germline

somatic evolution, 45, 125–128, 131–132, 150–154, 204–205

spawning, 140, 274, 281, 283, 285. *See also* brooding

spermatogenesis, 64–66, 134–136, 139–140, 150, 205–208, 222

Spiroplasma, 216, 218–219

split-brain phenomenon, 266–267

sponges, 145, 274–275, 277, 300, 305, 308

squid, 109, 111, 288

stem cells, 137–140, 145–154

Strombus, 304

Stylophora, 298

supergene, 15

superorganism, 34, 62, 285

supervenience, 41, 109

suppression: of cancer, 45, 235, 238; of cytonuclear conflict, 206–207, 221; of meiotic drive, 209, 220, 222; by the parliament of genes, 12, 22–24, 160; of selfish elements, 90–91, 130, 155, 237, 239, 258–262, 267, 270, 335

Symbiodiniaceae, 275–276, 281, 287, 295–296, 301, 304, 307–308

symbiosis. *See* mutualism

syncytia, 137, 139, 140, 150, 153

termites, 109

Tetrapus, 287

toxin-antitoxin system, 130, 135–136, 159, 162–163

TP53, 238

trA, 14

tragedy of the commons, 118

trait distortion: defined, 12, 54–58, 201; significance of, 12–13, 22–24, 25–26, 63–70, 90–92, 262. *See also* transmission distortion

transcription, 127, 132–137, 143, 151–155, 158–159, 167–168, 205. *See also* zygotic genome activation

transformer, 214

transmission: horizontal vs. vertical, 280–288, 304. *See also* coral-algal symbiosis

transmission distortion, 54–58, 63, 66–70, 90, 130, 262. *See also* trait distortion

transposable elements, 69, 84, 210, 223, 255; defenses against, 85, 135, 168, 205, 208; DNA transposons, 134; endogenous retroviruses, 166–168; retrotransposons, 132–133, 143–144, 165–168
Tribolium, 136
Trivers, Robert, 256–258, 270. *See also* parent-offspring conflict
twins, 84, 89, 244–245, 248

ultraviolet (UV) radiation, 280, 285, 300–304, 307
uniparental inheritance, 67, 202, 205, 207, 219, 221, 268
unit of selection, 98, 106–107, 110–112, 336
unity of purpose, 3, 53; and agency, 59, 63–69, 103–104, 115–119; defined, 263; and individuality, 62, 118, 264–271; and organismality, 91, 236–237. *See also* unity of self
unity of self, 263–264, 270

veil of ignorance, 129, 136–138, 141, 155, 169–173
Vibrio, 288

Weismann, August, 126, 142, 173–176
wmk, 217–218
Wolbachia, 215–219, 278, 291–292

X chromosome. *See* sex chromosomes
Xenopus, 127, 140, 153–155, 157–158, 170

Y chromosome. *See* sex chromosomes

zebrafish, 127, 133, 144, 153–158
zooxanthellae. *See* Symbiodiniaceae
Zyginidia, 215, 292
zygotic genome activation, 125, 154–155, 158, 160–163, 168, 171; delayed, 137, 145–146